Origins and development of adaptation

The Ciba Foundation is an international scientific and educational charity. It was established in 1947 by the Swiss chemical and pharmaceutical company of CIBA Limited—now CIBA–GEIGY Limited. The Foundation operates independently in London under English trust law.

The Ciba Foundation exists to promote international cooperation in biological, medical and chemical research. It organizes about eight international multidisciplinary symposia each year on topics that seem ready for discussion by a small group of research workers. The papers and discussions are published in the Ciba Foundation symposium series. The Foundation also holds many shorter meetings (not published), organized by the Foundation itself or by outside scientific organizations. The staff always welcome suggestions for future meetings.

The Foundation's house at 41 Portland Place, London W1N 4BN, provides facilities for all the meetings. Its library, open seven days a week to any graduate in science or medicine, also provides information on scientific meetings throughout the world and answers general enquiries on biomedical and chemical subjects. Scientists from any part of the world may stay in the house during working visits to London.

Origins and development of adaptation

Ciba Foundation symposium 102

1984

Pitman

London

ISBN 0 272 79749 9

Published in January 1984 by Pitman Publishing Ltd, 128 Long Acre, London WC2E 9AN, UK
Distributed in North America by CIBA Pharmaceutical Company (Medical Education Division), P.O. Box 12832, Newark, NJ 07101, USA

Suggested series entry for library catalogues:
Ciba Foundation symposia

Ciba Foundation symposium 102
viii + 273 pages, 36 figures, 23 tables

British Library Cataloguing in Publication Data
Symposium on Origins and Development of
Adaptation (*1983: Ciba Foundation, London*)
Origins and development of adaptation.
—(Ciba Foundation symposium; 102)
1. Adaptation (Biology)
I. Title II. Evered, David III. Collins, Geralyn M.
IV. Series
574.5′22 QH546

Printed in Great Britain at The Pitman Press, Bath

Contents

Symposium on Origins and development of adaptation held at the Ciba Foundation, London, 12-14 April 1983

Editors: David Evered (Organizer) and Geralyn M. Collins

B. C. CLARKE Chairman's introduction 1

A. D. BRADSHAW Adaptation of plants to soils containing toxic metals—a test for conceit 4
Discussion 14

C. D. FOY Adaptation of plants to mineral stress in problem soils 20
Discussion 32

E. A. BELL Plant–plant interactions 40
Discussion 48

T. C. HUTCHINSON Adaptation of plants to atmospheric pollutants 52
Discussion 67

J. GRESSEL Evolution of herbicide-resistant weeds 73
Discussion 87

GENERAL DISCUSSION Relative fitness and the stability of resistance 94

J. KUĆ Phytoalexins and disease-resistance mechanisms from a perspective of evolution and adaptation 100
Discussion 113

W. S. BOWERS Insect–plant interactions: endocrine defences 119
Discussion 132

T. R. E. SOUTHWOOD Insect–plant adaptations 138
Discussion 147

R. M. SAWICKI and I. DENHOLM Adaptation of insects to insecticides 152
Discussion 162

E. HODGSON and N. MOTOYAMA Biochemical mechanisms of resistance to insecticides 167
Discussion 183

S. G. GEORGOPOULOS Adaptation of fungi to fungitoxic compounds 190
Discussion 200

N. DATTA Bacterial resistance to antibiotics 204
Discussion 214

J. DAVIES and G. GRAY Evolutionary relationships among genes for antibiotic resistance 219
Discussion 228

D. L. HARTL, D. E. DYKHUIZEN and D. E. BERG Accessory DNAs in the bacterial gene pool: playground for coevolution 233
Discussion 243

FINAL GENERAL DISCUSSION 246

B. C. CLARKE Chairman's closing remarks 253

Index to contributors 259

Subject index 261

Participants

J. A. BAILEY Long Ashton Research Station, University of Bristol, Bristol, Avon BS18 9AF, UK

E. A. BELL Royal Botanic Gardens, Kew, Richmond, Surrey TW9 3AB, UK

W. S. BOWERS Department of Entomology, New York State Agricultural Experiment Station, Cornell University, Geneva, New York 14456, USA

A. D. BRADSHAW Department of Botany, University of Liverpool, P.O. Box 147, Liverpool L69 3BX, UK

B. C. CLARKE (*Chairman*) Department of Genetics, School of Biological Sciences, University of Nottingham, University Park, Nottingham NG7 2RD, UK

N. DATTA Department of Bacteriology, Royal Postgraduate Medical School, Hammersmith Hospital, Ducane Road, London W12 0HS, UK

J. E. DAVIES Biogen SA, 3 Route de Troinex, P.O. Box 1211, Geneva 24, Switzerland

V. DITTRICH CIBA–GEIGY AG, Building R1060, CH–4002 Basle, Switzerland

M. ELLIOTT Department of Insecticides and Fungicides, Rothamsted Experimental Station, Harpenden, Hertfordshire AL5 2JQ, UK

E. EPSTEIN Department of Land, Air and Water Resources, University of California, Hoagland Hall, Davis, California 95616, USA

Sir Leslie FOWDEN Rothamsted Experimental Station, Harpenden, Hertfordshire AL5 2JQ, UK

C. D. FOY Plant Stress Laboratory, Institute of Plant Physiology, Beltsville Agricultural Research Center, Beltsville, Maryland 20705, USA

G. P. GEORGHIOU Division of Toxicology and Physiology, Department of Entomology, University of California, Riverside, California 92521, USA

S. G. GEORGOPOULOS Laboratory of Plant Pathology, Athens College of Agricultural Sciences, Votanikos, Athens 301, Greece

I. J. GRAHAM-BRYCE East Malling Research Station, East Malling, Maidstone, Kent ME19 6BJ, UK

J. GRESSEL Department of Plant Genetics, The Weizmann Institute of Science, Rehovot 76100, Israel

J. B. HARBORNE Department of Botany, Plant Sciences Laboratories, University of Reading, Whiteknights, Reading, Berkshire RG6 2AS, UK

D. L. HARTL Department of Genetics, Washington University, School of Medicine, Box 8031, St Louis, Missouri 63110, USA

E. HODGSON Interdepartmental Toxicology Program, School of Agriculture and Life Sciences, North Carolina State University, Raleigh, North Carolina 27650, USA

T. C. HUTCHINSON Department of Botany, University of Toronto, Ontario M5S 1A1, Canada

J. KUĆ Department of Plant Pathology, S–305 Agricultural Sciences Building North, University of Kentucky, Lexington, Kentucky 40546, USA

R. M. SAWICKI Department of Insecticides and Fungicides, Rothamsted Experimental Station, Harpenden, Hertfordshire AL5 2JQ, UK

T. R. E. SOUTHWOOD Department of Zoology, University of Oxford, South Parks Road, Oxford OX1 3PS, UK

M. S. WOLFE Plant Breeding Institute, Maris Lane, Trumpington, Cambridge CB2 2LQ, UK

R. J. WOOD Department of Zoology, Victoria University of Manchester, Oxford Road, Manchester M13 9PL, UK

Chairman's introduction

B. C. CLARKE

Department of Genetics, School of Biological Sciences, University of Nottingham, University Park, Nottingham NG7 2RD, UK

As natural scientists, it is our duty to listen to what Nature says but, unfortunately, Nature does not always speak clearly, so that we have to do experiments in order to improve her diction. When I was an undergraduate we were given, for the good of our souls, an essay on adaptation by Peter Medawar. The burden of this essay was that adaptation is a term full of pitfalls, describing sometimes a process, sometimes a general state, and sometimes a particular phenotypic condition. In its general connotation, adaptation merely means appropriateness, or propriety in the old sense of that word. Because improper organisms do not survive, the word is liable, if it is used loosely, to conceal a tautology. We have to be careful not to describe something as an adaptation merely because it exists and because organisms that show it survive. Yet there are degrees of propriety, and we know that in some particular environments some genotypes survive or reproduce better than others. The causes of these differentials are proper subjects for scientific study.

The common thread that I see in this symposium is the study of differentials brought about by the presence in the environment of lethal substances, produced either by humans or by other organisms. The effects of these substances have given us some of the strongest evidence in favour of natural selection.

It still seems to me extraordinary that bacteria, plants and animals can make so rapid an evolutionary response to the poisons produced by the human race. The answer must surely be that every organism has a long history of attempts upon its life, and comes equipped with a battery of defensive weapons that can be modified for use against new threats. But none of these threats is wholly new; the human race, in creating its poisons, has most often

1984 Origins and development of adaptation. Pitman Books, London (Ciba Foundation symposium 102), p 1-3

copied nature. Herbicides, insecticides and antibiotics are often modifications of the offensive and defensive weapons that organisms have used against each other. Thus, it is likely that coevolution—the alternation of offence and defence—will be a recurring theme at this symposium, and rightly so.

Evolutionary geneticists, in attempting to explain change or stability, have a natural tendency to measure what are called 'environmental factors' or 'environmental variables', and these are usually physical characteristics of the environment such as temperature, humidity or shade. Yet the most powerful evolutionary variables are likely to be the effects of other organisms. Such effects have often been ignored because they are so difficult to measure. There is, however, a great deal of evidence for their importance. The remarkable examples of cryptic coloration among prey species testify to the effectiveness of predators as agents of natural selection. These protective resemblances not only involve colour and pattern but also morphology and behaviour. The great array of complex arrangements that organisms have evolved as defences against parasites, from hairy leaves to the production of antibodies, testifies to the importance of parasitism in affecting the course of evolution. Similarly, the phenomena of character displacement and ecological replacement testify to the importance of competition. With respect to my own parasitism have hardly yet been touched by the experimenter (or, in the protein and DNA—I believe that parasitism will turn out to be the dominant evolutionary force. Mathematical models suggest that the coevolutionary chase between host and parasite can very efficiently generate genetic diversity under an unusually wide range of conditions. But the evolutionary genetics of parasitism has hardly yet been touched by the experimenter (or, in the current vernacular, the 'experimentalist').

There has lately been a depressing tendency to label scientific points of view as if they were political persuasions, and sometimes to behave as if they were. We are blighted by many '-isms'. Thus, if you believe in natural selection you are a 'selectionist'; if you believe in random genetic drift you are a 'neutralist'. There does not seem to be a label for those who believe in both.

This symposium is about adaptation, and it is no surprise that people who study adaptation are called 'adaptationists'. 'Adaptationists' have been accused by Stephen Gould and Richard Lewontin (1979) of many misdemeanours: 'We fault the adaptationist programme for its failure to distinguish current utility from reasons of origin; for its unwillingness to consider alternatives to adaptive stories; for its reliance upon plausibility alone as a criterion for accepting speculative tales; and for its failure to consider adequately such competing themes as random fixation of alleles, production of non-adaptive structures by developmental correlation with selected features, the separability of adaptation and natural selection, multiple adaptive peaks, and current utility as an epiphenomenon of non-adaptive structures'.

While students of adaptation may, from time to time, have committed all these sins, it seems to me that Gould and Lewontin have missed the very great strength of the Darwinian approach to evolution—that it makes us formulate testable hypotheses. If you believe that the phenotype is only some kind of epiphenomenon, or a neutral character, you have no motivation or incentive to do an experiment. Gould and Lewontin have neglected even to mention the experimental evidence for the evolution of adaptations by natural selection. Nevertheless, their list of vices seems a useful way to start this meeting. We must resolve to avoid the delights of easy 'plausibility' (without, I hope, cultivating implausibility); we must eschew just-so stories, and what they call the 'Panglossian paradigm', where everything is for the best in the best of all possible worlds; and we must assiduously seek alternative explanations of our phenomena. But, above all, we must do experiments, observe the results, and listen to what Nature tells us.

REFERENCE

Gould SJ, Lewontin RC 1979 The spandrels of San Marco and the Panglossian paradigm: a critique of the adaptationist programme. Proc R Soc Lond B Biol Sci 205:581-598

Adaptation of plants to soils containing toxic metals—a test for conceit

A. D. BRADSHAW

Department of Botany, University of Liverpool, P.O. Box 147, Liverpool L69 3BX, UK

Abstract. Darwin, and many biologists afterwards, have seen few, if any, limits to the processes of adaptation by evolutionary change. Perhaps we have been conceited. A study of heavy-metal tolerance, and other conditions to which evolutionary adaptation has occurred, should overwhelm us with evidence for limits to the evolutionary process and limits to the adaptation it achieves. These limits clearly arise from restrictions in the supply of genetic variability. Nearly all species are in a condition of *genostasis*, in which there is a lack of appropriate variability for further evolutionary change. It is the molecular biologist who, by understanding the architecture of genes, will ultimately be able to explain what failures and limitations in genetic architecture at the molecular level cause the limits to adaptation itself.

1984 Origins and development of adaptation. Pitman Books, London (Ciba Foundation symposium 102), p 4-19

At any stage in the development of an idea, conceit can take over—not moral but intellectual conceit, coming not so much from people but from ideas. The idea of adaptation as a result of the action of natural selection or random variation is, because of its powerful simplicity, perhaps one of the best examples. It is so simple an idea, with such powerful implications, that even at the beginning Charles Darwin wrote, in The Origin of Species (1859), 'I can see no limit to this power, in slowly and beautifully adapting each form to the most complex relations of life'.

A hundred years later we continue with this view, stressing, with greater realization, the extraordinary range of adaptations that the simple process has produced. It is therefore well worthwhile taking stock of our position, to see whether we are guilty of conceit. The process might not be quite what we thought it was. Unfortunately, because evolutionary adaptation is such an immense topic, we cannot look at the whole of it. I will therefore take one single environmental condition which has profound effects on plants, and in which evolutionary adaptation clearly occurs, and will try to see what it can tell us about our understanding of the processes of adaptation.

Metal-contaminated environments

Environments have been contaminated by heavy metals such as lead, zinc and copper ever since the original magma of the earth solidified, but they have often been covered with innocuous superficial deposits. In recent years the occurrence of such environments has been enormously increased by mining, so that we now have areas all over the world, ranging from a few square metres to many hectares, where the soil is dominated by the presence of heavy metals. *Dominated* is a correct word, because heavy metals are extremely toxic to plants: 0.5 parts per million (p.p.m.) of copper, 12 p.p.m. of lead and 20 p.p.m. of zinc in solution will each completely stop plant root growth and cause plant death in a few weeks.

Heavy metals are immobile in soil, and the levels found in mining wastes are commonly of the order of 1% (i.e. 10 000 p.p.m.). Although only a small proportion of this is available to plants, these areas are extremely inhospitable and devoid of plants. If an ordinary plant species such as the common pasture grass, *Lolium perenne*, is established on these areas experimentally it may germinate and grow for a while, but it soon dies. We have, therefore, an environment which is not only extreme but also permanent. More importantly, because it is open and without plants, it is potentially available for colonization by plants. Since plants have effectively colonized every environment available to them, what happens?

Colonization as a result of evolution

It is an exaggeration to say that these metal-rich habitats are totally devoid of plants. In fact, a restricted number of species is to be found growing on them, species that are also found growing in normal environments uncontaminated by metal. It was Prat (1934) who first showed, for *Silene vulgaris* (red campion), that the populations growing on the metal-contaminated materials were different from populations of the same species growing on normal soils; they are tolerant to the metal toxicity, whereas normal populations are not (Fig. 1). We now know that this is true for all the species found growing on mine wastes which have been tested, such as *Agrostis tenuis* (bent grass), *Anthoxanthum odoratum* (sweet vernal grass), *Plantago lanceolata* (ribwort plantain) and *Rumex acetosella* (sheeps' sorrel). These species are only found in metal-contaminated sites because they have metal-tolerant populations;

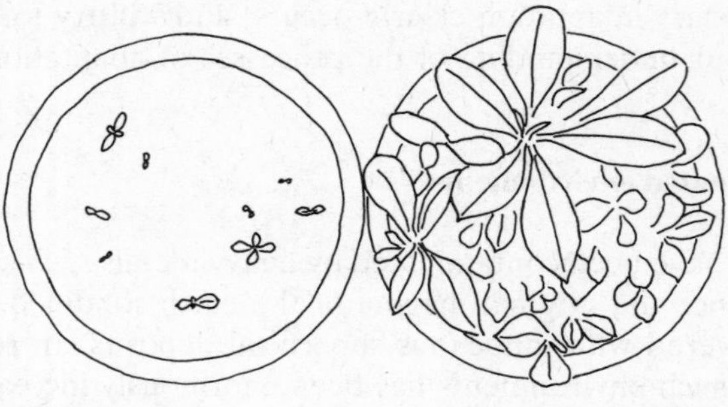

FIG. 1. The first evidence for the evolution of metal tolerance (or indeed for evolution in response to any pollutant): seedlings of normal (left) and copper-mine (right) populations of red campion (*Silene vulgaris*), growing on soil which has been treated with copper carbonate (drawn from Prat 1934).

their normal populations die rapidly when grown on metal-contaminated material. There is no clear evidence that any species is pre-adapted and thus has metal-tolerant normal populations.

The tolerance is largely specific to individual metals, and is almost certainly due to more than one mechanism, e.g. binding in the cell wall or isolation in the vacuole (Brookes et al 1981). It is generally determined by either a few or several nuclear genes, and has a high heritability (Bradshaw & McNeilly 1981).

The evolutionary process—rapid and local

Existence of tolerant populations is not by itself proof either of evolution or of a Darwinian process, although with our existing knowledge it is difficult to envisage any other origin. Although detailed testing has not been done, there is no evidence of a Lamarckian origin by induction—an individual plant cannot be trained to tolerate metals—and tolerance is not lost in culture in non-toxic conditions.

There is, instead, very good evidence that the character of tolerance can be rapidly developed in populations as a result of selection acting on heritable variation that occurs within populations which are not tolerant. If normal populations of a species such as *A. tenuis* are sown on a copper-contaminated soil, although nearly all the seedlings die, a few (about three in 1000) survive and grow successfully (Fig. 2). These can be shown to be copper-tolerant and to give rise to tolerant offspring when intercrossed (Walley et al 1974); this has now been demonstrated for at least six species.

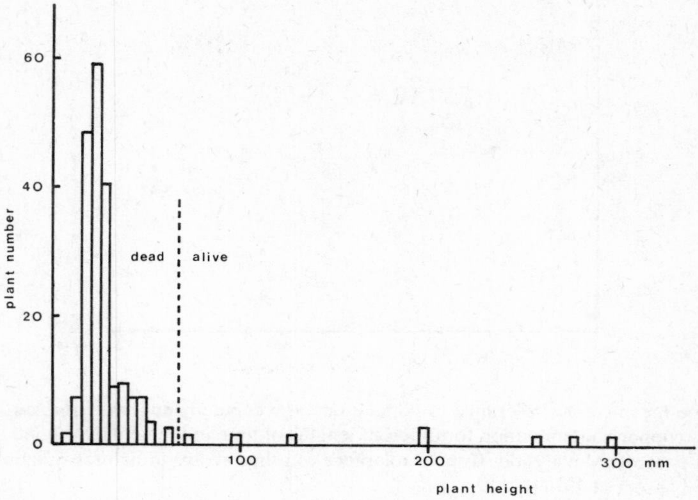

FIG. 2. The frequency distribution of growth of seedlings of a normal population of bent grass (*Agrostis tenuis*) on copper-contaminated soil, after 6 months, showing the survival of only a very few individuals. These seedlings are copper-tolerant and give rise to tolerant offspring, demonstrating that substantial copper tolerance can be selected in a single generation (from Walley et al 1974).

It is clear that exactly this process occurs in natural conditions. When a natural population is first subjected to metal contamination, enormous numbers of individuals die, but a few survive. There is selection for copper-tolerant individuals and a tolerant population is built up. This has been demonstrated in populations of *Agrostis stolonifera* adjacent to a copper refinery (Wu et al 1975a). The process can occur with great rapidity, in one or two generations, as in artificial selection experiments (Fig. 3). The rate-limiting factor appears to be the speed with which the tolerant survivors can develop vegetatively and sexually to give rise to a complete population (Bradshaw 1975). In less than 10 years, 50% of full tolerance can be achieved in *A. stolonifera*, despite its long generation time; this is quite sufficient for it to survive in levels of copper contamination that are lethal to normal populations of the grass.

At the same time the evolution can be extremely localized. Because strong selection for tolerance can counteract any immigration of genes from non-tolerant populations, contaminated areas as little as 20 m across can maintain tolerant populations. The transition from a tolerant to a non-tolerant population can occur over 5 m or less when there are sharp

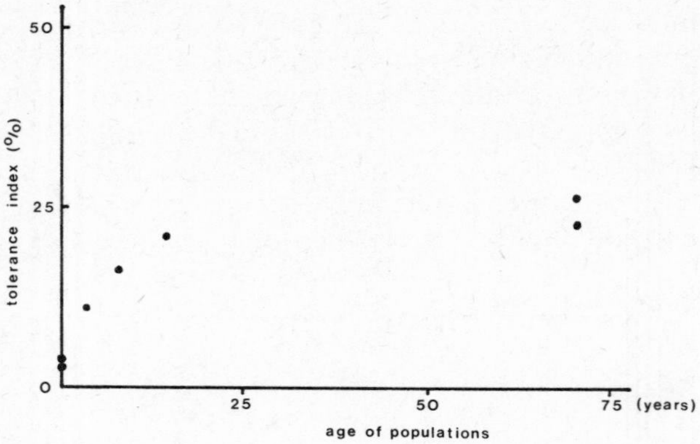

FIG. 3. The mean copper tolerance of populations of creeping bent grass (*Agrostis stolonifera*) exposed to copper contamination for different lengths of time in the neighbourhood of a copper refinery at Prescot, Merseyside. Copper tolerance can thus evolve in natural situations in a few years (from Wu et al 1975a).

boundaries between contaminated and normal soils (Antonovics & Bradshaw 1970). The adaptation can follow the pattern of the environment very closely.

The cost of tolerance

Tolerance does appear to have a cost in terms of fitness when plants are growing in the absence of heavy-metal contamination. In some cases, for instance zinc tolerance in *A. odoratum*, tolerant plants are distinctly slower growing than non-tolerant plants. Even in species where this does not happen, tolerant plants, when put into competition with non-tolerants on ordinary soils, perform significantly less well than non-tolerants (Cook et al 1972, Morishima & Oka 1977).

However, most of these differences come from observations on plants that derive from different populations. Their differences in growth rate or competitive ability may not be the direct effect of tolerance, but may be due to the evolution of differences in response to other characteristics of the environment; for example, slow growth is an adaptation to the low nutrient supply which is common in metal-contaminated habitats. For this reason it is interesting to discover that in populations on normal soils adjacent to, but outside, metal-contaminated areas, frequencies of metal-tolerant individuals are low (although they may be somewhat elevated from levels in normal populations). This implies a selection against tolerance and therefore against its accumulation due to gene flow. There is elegant evidence of such selection in *A. tenuis* outside a small copper mine in North Wales (McNeilly 1968) (Fig. 4). This fits in with the fact that tolerance has never been found in species on uncontaminated soils. There is clearly a cost to the tolerance that is at present available in plant species. For copper-tolerant individuals the explanation appears to be an enhanced requirement for copper under normal conditions, presumably brought about by the copper-tolerance mechanism.

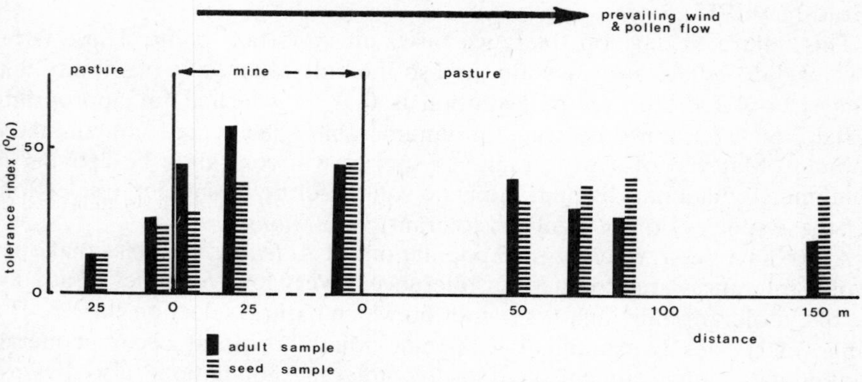

FIG. 4. Copper tolerance in adult and naturally produced seed populations of bent grass (*Agrostis tenuis*) at Drws-y-Coed copper mine: copper tolerance is more or less limited to the very small mine site, although the seed populations show that some genes for tolerance escape from the mine area and are selected against (from McNeilly 1968).

Limits to adaptation

The occurrence of tolerant plants in metal-contaminated environments might seem to imply that tolerance is all-powerful, but this is not so; it has distinct limits. This is manifest in the field by the fact that there are, commonly, areas of metal contamination that are not colonized, even by metal-tolerant plants. Sometimes this must be due to some other factor, such as extreme deficiency of a nutrient. But it can obviously be caused also by limits in the evolution of tolerance despite powerful selection pressures. Although comparisons of copper-tolerant and normal plants show that the tolerance mechanism has a remarkable ability to complex copper and to prevent its being translocated, there comes a point at which the tolerance mechanism is overcome; this can easily be demonstrated in solution culture (Wu et al 1975b). 'Super-tolerant' populations on areas of very high metal contamination have never been found.

The limits of variability

So far I have treated the adaptation of species for survival in metal-contaminated environments rather as if all species can evolve tolerance. But this is patently not true. Only certain species colonize these habitats (Antonovics et al 1971), and where a pre-existing flora has become subject to metal contamination, only a few species persist and evolve tolerance; although the rest have had the opportunity to do so, they appear to have failed to achieve it (Bradshaw 1975).

This failure to develop tolerance raises an important point. These latter species did not lack for selection, and so the only alternative possibility that will account for the lack of evolution is that they lacked for appropriate variability. This must be true for innumerable species and contaminated habitats. Material of a wide range of species will constantly be arriving in contaminated habitats by migration and will therefore be subject to selection; yet these species do not establish tolerant populations.

As we have seen, when normal populations of *A. tenuis*, a species that does evolve tolerance, are screened for tolerance, a very low frequency of at least partially tolerant individuals is found, on which natural selection can act. If a range of species is examined, it is found that those that occur in metal-contaminated environments and evolve tolerant populations also possess variability for tolerance in their normal populations (Gartside & McNeilly 1974). We have now looked at this in more detail, using a more sensitive screening technique (Table 1) involving the ability of individual seedlings to root in copper solution, in water culture, when supported on a raft of polyethylene beads. For the 15 species tested there is no case of a species that evolves tolerance but which does not possess variability for tolerance in its normal populations—although, in some species that possess variability for tolerance, tolerant populations have not yet been found. Evolution cannot proceed if the appropriate variability is not present, and for tolerance this variability is apparently not found in all species.

Relevance to other adaptive situations

Because metal contamination provides a rather extreme habitat, it is tempting to dismiss what we can learn from the evolution of tolerance as irrelevant to the processes of adaptation to other, more normal, environments. But it must be remembered that most habitats are extreme in one characteristic or another; there is severe competition in fertile habitats, severe nutrient deficiency in calcareous habitats, lack of light in woodlands, and high salinity in salt marshes. Indeed, the fact that only a limited number of characteristic

TABLE 1 The percentage of copper-tolerant individuals found in normal populations of various grass species, in relation to the presence of the species on copper-polluted waste and whether the plants collected were tolerant of copper

Species	*% occurrence of tolerant individuals*	*Presence of species on mines:*		*Tolerance of collected adult plants*
		On waste	*At margins*	
Holcus lanatus	0.16	+	+	+
Agrostis tenuis	0.13	+	+	+
Festuca ovina	0.07	−	+	−
Dactylis glomerata	0.05	+	+	+
Deschampsia flexuosa	0.03	+	+	+
Anthoxanthum odoratum	0.02	−	+	−
Festuca rubra	0.01	+	+	+
Lolium perenne	0.005	−	+	−
Poa pratensis	0.0	−	+	−
Poa trivialis	0.0	−	+	−
Phleum pratense	0.0	−	+	−
Cynosurus cristatus	0.0	−	+	−
Alopecurus pratensis	0.0	−	+	−
Bromus spp.	0.0	−	+	−
Arrhenatherum elatius	0.0	−	+	−

+, presence; −, absence. Only those species which possess copper-tolerant individuals in their normal populations evolve tolerant populations which colonize mine waste. (Data of C. Ingram.)

species occurs in every habitat is proof that the habitats are severe for other species, which would otherwise be present.

All the characteristic features of the evolutionary process that allows plants to grow in metal-contaminated environments can be matched by examples from other environments. Table 2 gives some examples, although it cannot do full justice to unpublished information that is within people's experience, particularly plant breeders. Indeed, the whole experience of plant breeding is that once the genes for the development of a new character, or the enhancement of an old one, are available, rapid progress is possible. But normally the great difficulty lies in finding the desirable genes.

Discussion—the limits to evolution

If we put all this information and the deductions from it together, there are certain unavoidable conclusions which bear on the nature of adaptation due to evolutionary processes.

We must realize that all habitats, without exception, have the potential to generate selection pressures. This is obvious when we consider all the species

TABLE 2 Some examples of adaptation to environments other than those contaminated by metals, which illustrate the same principles of adaptation

Colonization as a result of evolution	
High altitudes	*Achillea borealis*
Low-fertility habitats	*Trifolium repens*
Serpentine soils	*Gilia capitata*
Rapid and local evolution	
Park-grass-fertilizer experiment	*Anthoxanthum odoratum*
Vernal pools	*Veronica peregrina*
Grazing	*Dactylis glomerata,* *Trifolium repens*
Evolutionary cost	
Life-cycles	*Poa annua*
Soil calcium	*Festuca ovina*
Latitude	*Oxyria digyna*
Limits to adaptation	
Yield	chickens
Disease resistance	*Triticum, Solanum, Hordeum*
Alpine climate	*Melandrium rubrum, Ranunculus acris*
Limits to variability	
Herbicide resistance	*Stellaria media, Bellis perennis*
Disease resistance	*Triticum, Solanum, Oryza*
Flower colour	*Delphinium*

that do not occur in a given habitat; those which continue to immigrate but fail to establish demonstrate the existence of coefficients of selection against them equal to 1.0.

Such species cannot have failed to adapt, and therefore to survive, because of a lack of selection. Their failure must be due to a lack of appropriate variability, that is, of such variability as would allow them to survive in that environment. The species that do occur in a particular environment are those that possess the appropriate variability. This variability is thus not universal to all species; only certain species possess it.

If the appropriate variability is present, then it is selected rapidly. Because of high coefficients of selection, evolutionary adaptation will occur within a few generations. If it does not occur rapidly, because the variability allows only very small improvements in fitness or because there is some cost due to pleiotropy or linkage, for instance, then it is, again, because the *appropriate* variability is not present. Inappropriate variability includes genes with weak effects, genes with disadvantageous pleiotropic effects, or genes linked too closely to other disadvantageous genes.

It follows that, for every character, any population has effectively reached an evolutionary plateau, determined not by selection but by lack of appropriate variation—a condition of *genostasis* (Bradshaw 1984). This genostasis must be almost universal, and the incidence and degree of adaptation, whether found on old mine workings or generally in nature, must be the outcome of what genetic variability is available. The genostatic condition applies, of course, only to the particular character on which selection has been acting: there can be plenty of variability in other parts of the genotype.

Lack of progress in achieving a particular goal in an artificial selection programme, or lack of further improvements in fitness characters in natural populations, often appears to be due to overdominance or to other forms of non-additive gene action. Fitness characters often show striking genetic variability but low heritability (Falconer 1981). While this may be the immediate cause of lack of adaptation, it is not the ultimate cause, which is lack of appropriate, in this case additive, variation. Similarly, where stabilizing selection appears to be the cause of a lack of change in a character, it is perfectly possible that there is really a lack of variability that would allow directional selection to have effects. A better turtle, after all, would surely be possible!

Of course it must follow that where a population or species has not already been subject to particular selection pressures it is likely to possess potential variability for selection—that is, for the characters in question. It is perhaps significant that most of the best examples of evolution in action are where new selection pressures are operating: for instance, in industrial melanism in moths, in resistance to pesticides in insects and in heavy-metal tolerance (Bishop & Cook 1981).

Conclusion

I would conclude that we should not allow the original Darwinian concept of evolution to lead us to conceit. Evolution may be able to achieve remarkable results, but what should bother us is how much it does *not* achieve, despite all the pressures of natural selection.

This suggests that to understand adaptation in the future our attention should turn even more from what is outside to what is inside organisms. The key to understanding evolution lies in understanding the architecture of characters or, more particularly, the construction of characters. It is the molecular biologists who, by understanding the ways in which genes that produce new functions (and therefore new adaptations) are built up by random events within a pre-existing structure, may be able to explain how some things are possible and others are not.

It is unlikely that we shall ever be able to predict future evolution because of the stochastic elements of the mutation process. But surely the key to understanding how adaptations do, and do not, occur must lie within the architecture of genes.

REFERENCES

Antonovics J, Bradshaw AD 1970 Evolution in closely adjacent plant populations. VIII: Clinal patterns in *Anthoxanthum odoratum* across a mine boundary. Heredity 25:349-362

Antonovics J, Bradshaw AD, Turner RG 1971 Heavy metal tolerance in plants. Adv Ecol Res 7:1-85

Bishop JA, Cook LM 1981 Genetic consequences of man made change. Academic Press, London

Bradshaw AD 1975 The evolution of heavy metal tolerance and its significance for vegetation establishment on metal contaminated sites. In: Hutchinson TC (ed) Heavy metals in the environment. Toronto Univ Press, Toronto, p 599-622

Bradshaw AD 1984 The importance of evolutionary ideas in ecology—and vice versa. In: Sharrocks B (ed) Evolutionary ecology. Blackwell, Oxford, p 1-25

Bradshaw AD, McNeilly T 1981 Evolution and pollution. Arnold, London

Brookes A, Collins JC, Thurman DA 1981 The mechanism of zinc tolerance in grasses. J Plant Nutr 3:695-705

Cook SA, Lefebvre C, McNeilly T 1972 Competition between metal tolerant and normal plant populations on normal soil. Evolution 26:366-372

Falconer DS 1981 Introduction to quantitative genetics (2nd edn). Longman, London

Gartside DW, McNeilly T 1974 The potential for evolution of heavy metal tolerance in plants. II: Copper tolerance in normal populations of different plant species. Heredity 32:335-348

McNeilly T 1968 Evolution in closely adjacent plant populations. III: *Agrostis tenuis* on a small copper mine. Heredity 23:99-108

Morishima H, Oka HI 1977 The impact of copper pollution on barnyard grass populations. Jpn J Genet 52:357-372

Prat S 1934 Die erblichkeit der resistenz gegen kupfer. Ber Dtsch Bot Ges 102:65-67

Walley KA, Khan MSI, Bradshaw AD 1974 The potential for evolution of heavy metal tolerance in plants. I: Copper and zinc tolerance in *Agrostis tenuis*. Heredity 32:309-319

Wu L, Bradshaw AD, Thurman DA 1975a The potential for evolution of heavy metal tolerance in plants. III: The rapid evolution of copper tolerance in *Agrostis stolonifera*. Heredity 34:165-187

Wu L, Thurman DA, Bradshaw AD 1975b The uptake of copper and its effect on respiratory processes of roots of copper tolerant and non-tolerant clones of *Agrostis stolonifera*. New Phytol 75:225-9

DISCUSSION

Clarke: A.J. Cain (1964), in his essay on 'The perfection of animals', said that evolution could more or less do what was necessary, when required. Perhaps you would disagree about this. Among the grasses, is there any ecological or taxonomic sense in your list (Table 1) of species that have or do not have available variation in tolerance?

Bradshaw: We have been unable to make any sense of the occurrence of variability. When we reviewed what was known about tolerance to different metals in all species (including all the species found on metal-contaminated sites whether or not they were known to be tolerant) previous ideas that we had had about links with particular families of flowering plants evaporated. There are no legumes that are metal-tolerant in Europe, but in other parts of the world they can grow at very high metal levels, e.g. in Zimbabwe, where legumes are some of the most common species on metal-contaminated sites. Yet certain families contain no examples of metal-tolerant individuals. The numbers involved are, however, rather small so the results could just be a matter of chance.

Hartl: If genes for heavy-metal tolerance are detrimental in uncontaminated soil, then why do certain species maintain the genetic variation?

Bradshaw: This is an annoyingly difficult problem to answer; it is a contradiction. We don't know enough about the processes that produce variability. There are two possibilities. There is some evidence (Urquhart 1971) that metal tolerance is variable in its dominance and that initially it may be recessive but may vary in different populations (there is evidence, almost, of an evolution of dominance). So in this case the variability would be 'sheltered' in those populations where the genes are recessive. The other possibility is that mutation rate is important: mutation would 'feed in' genes against selection, the effects of which would be quite low when the genes are rare. The difficulty in this case is to measure the mutation rate in outbreeding species, which we have been working with until now. We would like to study this further by looking at mutation in tissue culture, where the tolerance is also expressed.

Gressel: Some genera, such as *Lolium* (the ryegrasses) seem to have picked up resistance (or tolerance) to many different things, and may therefore possess more adaptivity than the 'average' species. *L. perenne* has adapted to the herbicides paraquat and dalapon (Faulkner 1982) and to sulphur dioxide (Horsman et al 1978). *Lolium rigidum* has adapted to diclofop methyl (S. Knight 1983, personal communication). *Poa annua* has also evolved tolerance to a few herbicides. Is there something special about these?

Bradshaw: Because *L. perenne* is an important cultivated crop plant, people have simply studied it more. Any species is able to evolve many different things (which, in a sense, argues against what I said in my paper) but, on the other hand, not all species can do everything—of that there is clear evidence.

Davies: In the experiments where you found that one group gave rise to tolerant derivatives and another group did not (Table 1), were you testing all species on a soil with a constant concentration of the copper ion?

Bradshaw: Yes. The concentration of copper was $0.2\,\mu g\,cm^{-3}$.

Davies: Did the group that do not give rise to tolerant derivatives have a much lower basal level of resistance (in terms of parts per million) to the copper ion?

Bradshaw: Gartside & McNeilly (1974) found that this was true to a certain extent, but the results did not go far enough to be conclusive. They tested *L. perenne* for copper tolerance and found, at what appeared to be low copper levels, certain signs of tolerant individuals. But, interestingly, that variability turned out not to be heritable.

Davies: It's really a question of the level of metal susceptibility of the plant. One cannot expect to find a species becoming tolerant to a concentration that is simply much too high for it.

Bradshaw: I agree. But in our work we found that if we started screening at lower levels of contamination, the individuals that we picked out as being tolerant possessed a variability that was not heritable. It is only at the higher levels of contamination that one sees variation that is distinctly heritable. We have never found, as I mentioned, any species that clearly evolve tolerance in nature and which do not have variability that is well expressed within their normal populations.

Davies: What is the difference in susceptibility to copper, lead or zinc for those two groups of plants?

Bradshaw: In their normal populations, the two groups have roughly similar susceptibilities, but one group of species has individuals in their population that are tolerant and the other group of species does not.

Davies: But that is only true at the concentration of copper that you tested; tolerance depends on concentration.

Graham-Bryce: You emphasized that the process of colonization is rapid, but it is important to establish the nature of the process to assess the significance of this rapidity. In your illustration (Fig. 1) of red campion (*Silene vulgaris*) growth you ask whether a local tolerant population evolved or whether it was there in the first place. Is this not merely a semantic point, because that work presumably demonstrates the selection of highly tolerant individuals from a heterogeneous population as a result of an extreme selection pressure? One would surely expect such a process to happen as soon as the selection pressure is applied. What seems more interesting is that examination of subsequent performance under heavy-metal stress indicated what might be termed a 'drift in susceptibility'—for example, the increased root growth in the population from a mine site, compared with that in the initially selected individuals. Is that a result of some subsequent adaptation?

Bradshaw: As far as our work can show (and we have never studied many cycles of selection), in one cycle of selection one can build up a population that has 50–70% of the tolerance of a mine population, but it depends on the species. This perhaps relates to the genetic basis of the variability that is available in the population. Fig. 3 (p 8) shows that, after the first cycle of selection, further selection gives rise to a predominantly tolerant population.

Georgopoulos: You were speaking about flowering plants, but fungi that are pathogenic to plants have been controlled with copper for about the last hundred years. This has never caused any problems. Do you believe that the appropriate variability is not present in the fungi?

Bradshaw: I would like to know that myself! It provides a very good example of what I'm arguing for. Thankfully, these important pathogens do not seem able to evolve copper tolerance. *Scytalidium*, on the other hand, is well known as a species that *can* evolve copper tolerance, although in the somewhat different environment of liquid culture (Starkey 1973). Yeasts can also do this.

Elliott: What is known about the mechanism of the tolerance? Are natural chelating agents present to dispose of the metal ions, rendering them inactive?

Bradshaw: There are two lines of evidence, as I mentioned. One is that tolerance is connected with substances such as proteins in the cell wall, and that the metal is complexed by them in the cell wall (Turner & Marshall 1972). Another line of evidence suggests that the metal enters the cell and is isolated in the vacuole. Other, general tolerance mechanisms have been suggested to depend on malate (Ernst 1975). All the suggestions made about mechanisms have, naturally but unfortunately, been suggested for specific cases. The final answer is not clear. But whatever happens, metal accumulation in the plant is restricted to the root. In tolerant plants there is normally more metal in the root than in non-tolerant plants.

Hutchinson: Not only does metal accumulate in the root but the actual physiological basis of tolerance resides there, in several investigated cases. Some interesting reciprocal grafting experiments have been done by J.C. Brown in the USA. Sensitive and tolerant clones to iron-deficiency were grafted. Irrespective of which graft was made, the tolerance was really determined by the root stock, i.e. sensitive root stocks gave sensitive tops even if the top was from a resistant plant.

What we have been discussing reminds me of the development, in rats, of resistance to warfarin, in that it has originated independently in several separate geographic locations. The same plant species, in both North America and Europe, have been shown to develop metal-tolerant populations. There is clearly very widespread genetic potential in some of these genera, such as the grass *Agrostis*, which has produced resistance to copper, lead and zinc in the USA, Canada, the UK and Germany quite independently.

I have another point about tolerance specificity, with which Professor Bradshaw may not agree. Quite regularly we are finding, i.e. with two grass and two unicellular algal species, that the specificity of metal tolerance breaks down considerably. Co-tolerances seem to develop; that is, one can select for tolerance to one particular metal, e.g. zinc, and coincidentally one will also find tolerances to lead and cadmium associated. Co-tolerances occur for which the metals were not elevated in the environment and could not, therefore, have been selected for in the usual way (see Hutchinson, this volume). So, rather than having to have an energy-expensive physiological mechanism with absolute specificity for each different metal, the plant can have some general biochemical mechanisms that can handle groups of chemically similar elements. One such example is when copper confers an enhanced tolerance to silver (Cox & Hutchinson 1980, Stokes et al 1973).

Bradshaw: I don't disagree with you. We have been looking at populations by crude analysis and have found that tolerance to individual heavy metals appears to be specific. But more refined techniques reveal that a copper-tolerant population that has never had any experience of lead or zinc has a low, but significant, level of tolerance to lead and zinc also. It seems that there is perhaps both a generalized tolerance operating at a low level of contamination and specific tolerance able to operate at higher levels. This rather suggests that there may be two different mechanisms operating.

Kuć: I find it difficult to come to grips with the terms tolerant, intolerant, and lack of tolerance. This is because many of these metal ions are actually *nutrients*, in trace quantities, and are absolutely essential to the metabolism of the plant. At one extreme one could say that all plants have not only a tolerance but a need for the ions. In defining tolerance one should indicate the amount of the contaminant concerned. Thus, plants can be considered tolerant if they grow and develop normally (whatever 'normally' means) at a certain level of the contaminant; if they do not, then they can be considered intolerant. The difficulty is in deciding what this level should be. Furthermore, tolerance to a given contaminant will also depend on environmental conditions other than the quantity of the contaminating substance in the environment of the plant.

Bradshaw: Obviously one is guilty of simplification when one starts to talk about a *tolerant* individual. That same individual could be found *not* to be tolerant by dying when exposed to a little more copper. But nevertheless, relative to normal plants, the individual can be considered to have a level of tolerance that is distinguishable. This problem also applies to any consideration of tolerance to antibiotics or herbicides. It also goes back to my consideration of the limits of tolerance. In areas such as the middle of the copper mine at Parys Mountain, copper-tolerant *Agrostis* fails to grow in the most toxic areas. One might have expected to find plants with super-tolerance to copper, but they do not occur. Plants from that habitat don't seem to be any better than plants that come from less extreme habitats. There seems to be one level of tolerance that the mechanism can achieve, and no more, implying a limit to the adaptation that can be achieved.

REFERENCES

Cain AJ 1964 The perfection of animals. Viewpoints Biol 3:36-63

Cox RM, Hutchinson TC 1980 Multiple metal tolerances in the grass *Deschampsia cespitosa* (L.) from the Sudbury smelting area. New Phytol 84:631-647

Ernst W 1975 Schwermetallvegetation der Erde. Fischer, Stuttgart

Faulkner J 1982 Breeding herbicide-tolerant crop cultivars by conventional means. In: LeBaron HM, Gressel J (eds) Herbicide resistance in plants. Wiley-Interscience, New York, p 235-256

Gartside DW, McNeilly T 1974 The potential for evolution of heavy metal tolerance in plants. II. Copper tolerance in normal populations of different plant species. Heredity 32:335-348

Horsman DA, Robert TM, Bradshaw AD 1978 Evolution of sulphur dioxide tolerance in perennial ryegrass. Nature (Lond) 276:493-494

Starkey RL 1973 Effect of pH on toxicity of copper to *Scytalidium* sp., a copper-tolerant fungus, and some other fungi. J Gen Microbiol 78:217-225

Stokes PM, Hutchinson TC, Krauter K 1973 Heavy-metal tolerance in algae isolated from contaminated lakes near Sudbury, Ontario. Can J Bot 51:2155-2168

Turner RG, Marshall C 1972 The accumulation of zinc by subcellular fractions of roots of *Agrostis tenuis* Sibth in relation to zinc tolerance. New Phytol 71:671-676

Urquhart C 1971 Genetics of lead tolerance in *Festuca ovina*. Heredity 26:19-33

Adaptation of plants to mineral stress in problem soils

C. D. FOY

US Department of Agriculture, Agricultural Research Service, Plant Physiology Institute, Plant Stress Laboratory, Beltsville, MD 20705, USA

Abstract. Crop production depends on interactions among genetic potentials of plants and various stress factors in the environment. Mineral stresses in the form of toxicity or deficiency are not always economically correctable with current technology. But plant species, and genotypes within species, differ widely in tolerance to different mineral stresses, and some of these differences are known to be genetically controlled. Hence, a viable alternative or supplemental approach is to adapt plants more precisely to soils. Soil fertility programmes of the past have emphasized 'changing the soil to fit the plant'. As a result, many crop cultivars have been developed under nearly ideal conditions of pH and soil fertility and hence cannot tolerate the stresses encountered in marginal soils. There is, therefore, a great need for stress-tolerant plant genotypes for use in a 'minimum input' system of agriculture for problem soils. The merits of tailoring the plant to fit the soil have only recently been recognized, even by the scientific community. However, within the last decade multidisciplinary teams of scientists have been formed in several countries, and considerable progress is being made in selecting and breeding plants for greater nutrient efficiency and for tolerance to such factors as aluminium toxicity and salinity. In view of problems in world food and energy production and distribution, this approach should receive the highest priority.

1984 Origins and development of adaptation. Pitman Books, London (Ciba Foundation symposium 102), p 20-39

Crop production depends on interactions between the genetic potential of plants and various environmental factors. Until recently, relationships between plant genotype and mineral stress have been largely ignored. In the past, our approach to soil fertility problems has emphasized 'changing the soil to fit the plant' and, as a result, many crop cultivars have been developed for nearly ideal conditions of soil fertility and pH. Such cultivars are like babies who thrive inside the incubator but cannot tolerate the stresses of the outside world. These plants frequently show symptoms of mineral toxicity or deficiency when grown on soils that are only slightly different from those on which they were developed.

For example, the famous 'green revolution' wheat (*Triticum aestivum* L.) cultivar 'Sonora 63', developed in Mexico, will not tolerate the strongly acidic, aluminium-toxic soils of Brazil. In the United States, the wheat cultivar 'Gaines', which holds the world yield record, is also extremely sensitive to aluminium ions and performs poorly when moved from the Palouse soils of Washington State, on which it was developed, to adjacent, slightly more acidic soils. Wheat cultivars developed by the Ohio Agricultural Experiment Station at Wooster are generally more tolerant to acidic, aluminium-toxic soils than those developed in the adjacent state of Indiana. This difference has been attributed to the fact that the soils in eastern Ohio, particularly subsoils, are more strongly acidic than those in Indiana (Campbell & Lafever 1976).

The 'green revolution' rice (*Oryza sativa* L.) cultivar 'IR-8' is more sensitive to Fe^{2+} toxicity than native cultivars of South-east Asia. 'Wayne' (*Glycine max* L.) soybean, developed for high yield in the midwestern United States, develops iron-deficiency chlorosis on certain calcareous soils of that region. Some inbreds of corn (*Zea mays* L.) and tomato (*Lycopersicon esculentum* Mill.) cannot absorb or use iron efficiently, even in soils of pH 5.0, or below. One inbred line of corn could not obtain adequate molybdenum from a strongly acidic, usually aluminium-toxic soil and, hence, was limited more by molybdenum deficiency than by aluminium toxicity. (For original references see reviews by Foy et al 1978, Foy & Fleming 1978, Clark 1981, 1982.)

The practice of liming and fertilizing soils to 'optimum' levels was profitable in the moderately acidic soils of the USA when lime, fertilizer and energy were cheap. However, in many 'developing' countries this approach has never been practical, and even in the 'developed' countries, current costs of lime, fertilizers and energy, and fear of environmental pollution are causing re-examination of such practices. In all countries, some soil conditions are not economically correctable with current technology, and there is a growing feeling that we should seek greater accommodation with nature. For example, scientists in some regions are developing techniques of crop production based on 'minimum inputs' rather than 'maximum outputs' for marginal soils (Dambroth & El Bassam 1982, Sanchez & Salinas 1981). 'Minimum input' means using modest amounts of fertilizers and lime to produce a profitable but not necessarily maximum yield. The 'maximum output' system uses large amounts of soil amendments to achieve maximal or near-maximal yield, also at a profit. In marginal land, tailoring the plant to fit the soil may be more economical than changing the soil to fit the most demanding plant. Individuals who farm marginal soils must often live with a 'minimum-input' system and, in these situations, 'high-input' technology is often not only unprofitable but harmful.

Soil problems amenable to a plant-adaptation approach

Soil situations in which a plant-adaptation approach seems promising are as follows: acidic, aluminium-toxic subsoils that are difficult to lime; strongly acidic mine spoils or other disturbed sites where rapid plant cover and soil stabilization are desired at minimal cost; steep pasture lands that are strongly acidic, infertile and difficult to lime even in the surface layer; vast areas of strongly acidic, phosphorus-fixing surface soils and subsoils of the tropics (Campo Cerrado Region of Brazil and the Llanos of Colombia); saline soils; soils polluted with heavy metals; calcareous or other alkaline soils with unavailability of iron or other micronutrients; wet soils; dry soils; and even hard soils.

Plant selection or breeding can also be used to improve the effectiveness of fertilizer nutrients, particularly phosphorus and nitrogen, on good soils. For example, nitrogen-efficient cultivars would make more efficient use of energy and fertilizer and reduce ground-water pollution. Phosphorus-efficient plants are needed, along with cheaper phosphorus fertilizers that are less subject to fixation in soils. Plant breeding can regulate the mineral composition of crops and thereby improve the quality of food and feed. For example, forages that absorb magnesium efficiently could reduce the incidence of grass tetany in grazing animals, and plants that exclude cadmium would reduce the accumulation of this element in humans.

Steps in adapting plants to problem soils

The first step is to choose a stress-tolerant species when possible e.g. weeping lovegrass, *Eragrostis curvula* (Schrad.) Nees., can be used instead of alfalfa, *Medicago sativa* L., on acidic mine spoils. Second, one could select the most tolerant cultivar within a species—e.g. cultivars of bluegrass, *Poa pratensis* L., differ by 25-fold in tolerance to an acidic aluminium-toxic soil. Next, one could select and propagate the most tolerant individual plants which may appear within established ‘uniform’ cultivars when stress is applied. Finally, when necessary, one can breed specifically for various traits, but many plant adaptational problems can be solved before this stage is reached.

Benefits of adapting plants to problem soils

The proposed plant-adaptation approach is ecologically ‘clean’, energy-conserving and often cheaper than amending the soil to fit the most exacting plants. It is thus compatible with national and international goals of economical food production, conservation of fertilizers and energy and control of pollution. The approach is particularly appropriate for countries having little foreign exchange. More specifically, the possible benefits are as follows: (1) increased yields of a given crop species on the present acreage—e.g. introduction of a cultivar more specifically adapted to prevailing stresses, such as the development of an iron-efficient soybean cultivar for calcareous soils (Fehr 1982); (2) expansion of crop acreage of a species to marginal soils currently unsuited to the crop—e.g. extension of wheat acreage into the Campo Cerrado of Brazil by the development of aluminium-tolerant cultivars (da Silva 1976); (3) introduction of new and more profitable species for specific needs—e.g. introduction of an aluminium-tolerant and cold-tolerant strain of limpograss, *Hemarthria altissima* (Poir.) Stapf & C. E. Hubb., for possible use on strongly acidic, high-altitude mine spoils or on acidic sites in more northern climates (A. J. Oakes & C. D. Foy, 1980, Am Soc Agron Abstr p 103–104; and (4) crop improvement through the gradual introduction of crops having more exacting growth requirements and greater value—e.g. on an acidic mine spoil one could begin with aluminium-tolerant weeping lovegrass and end with aluminium-sensitive barley, *Hordeum vulgare* L., or with alfalfa, because the growth of plants will generally improve both the physical and chemical properties of such soils.

Objections to adapting plants to problem soils

It has been said that stress-tolerant plants may yield less in the absence of stress. Although this is valid for many native plants that colonize hostile sites, it is not necessarily true for crop plants. For example, a high degree of aluminium tolerance has been found in certain cultivars of snapbean (*Phaseolus vulgaris* L.), cotton (*Gossypium hirsutum* L.), tomato, wheat and barley that also produce high yields in the absence of aluminium stress. But, even if some yield potential is sacrificed in acquiring stress tolerance, the result may still be profitable because lower inputs of lime and fertilizer may be needed to produce a given yield of a given quality crop. Yet, the adaptation of plants to soils need not discourage the use of lime and fertilizer because stress-tolerant crops will actually increase the use of marginal land that received no treatment previously. Stress-tolerant genotypes could make such land economically productive with low to moderate inputs of lime and fertilizers.

Another objection is that aluminium-tolerant wheat cultivars are tall-strawed, and hence, may 'lodge' (i.e. fall over, and hence be missed by the combine) under high nitrogen fertilization (which also tends to encourage tall plants). But Camargo et al (1980) have now identified a source of tolerance to aluminium in the short-strawed, high-yielding wheat cultivars 'Tordo' and 'Siete Cerros' and suggested ways of breeding for this trait; so aluminium-tolerant cultivars need not be tall-strawed.

A third objection is that stress-tolerant plants may be lower in mineral content, and hence lower in quality for food or animal feed, but this is not necessarily true. For example, the aluminium-tolerant 'BH 1146' wheat accumulates higher concentrations of magnesium in its foliage than does the aluminium-sensitive 'Sonora 63' (C. D. Foy, personal observation); hence, the former would provide greater protection against grass tetany than would the latter when both were used for grazing. There is also evidence that aluminium-tolerant plants are more efficient in absorbing calcium and phosphorus at low concentrations of these elements in the growth medium.

A further objection is that breeding nutrient-efficient cultivars could impoverish the soil. The correct approach, however, would not bleed soil fertility to zero levels but would promote the effective use of fertilizers already fixed in the soil and those used as amendments. The goal is to produce profitable (not necessarily maximal) yields of acceptable quality with lower inputs per unit of output.

Critics have said that breeding for tolerance to one stress factor may increase plant vulnerability to others, and this is a valid concern. Certain trade-offs may be necessary, and the approach must be applied within a framework of economic and other constraints. Ultimately, it could be argued that an endless number of plant genotypes would be needed for specific soil conditions, and there is obviously a practical limit. However, we can still develop better soil–plant management packages than we have at present. For example, to grow wheat successfully in the Campo Cerrado of Brazil, one needs an aluminium-tolerant cultivar which requires less lime and phosphorus fertilizer than an aluminium-sensitive cultivar to produce a profitable yield. The fact that breeders have successfully selected for plant tolerance to multiple stresses means that fewer cultivars would be needed (Ponnamperuma 1982). When higher soil amendments could be afforded, the grower could switch to a cultivar that was less tolerant to stress but perhaps higher yielding and more profitable.

Objectives of research on mineral stress

Plant species and genotypes within species differ widely in their tolerances to the stresses of mineral toxicity or mineral deficiency. (For details see reviews by Foy & Fleming 1978, Foy et al 1978, Clark 1981, 1982, and Foy 1983a.) Our research has been concerned primarily with aluminium and manganese toxicities in acidic soils and with iron unavailability in calcareous soils. Our aims are: (1) to identify current and potential mineral stress factors in problem soils; (2) to screen plant germplasm for stress tolerance; (3) to collaborate with plant breeders in developing superior plant genotypes for specific problem soils; (4) to determine the physiological and biochemical mechanisms of differential plant adaptation to mineral stress; (5) to use plant physiological traits to refine screening procedures and improve soil–plant management practices; (6) to use plant genotypes as indicators of potential mineral stress problems; and (7) to characterize interactions among mineral stress and other environmental factors such as water, light, air pollutants, temperature, pathogens, rhizobia and mycorrhizae.

Genetic control of tolerance to mineral stress in plants

Aluminium tolerance

In barley, tolerance to aluminium is controlled by one major, dominant gene (Reid 1971), whereas in wheat two and possibly three major, dominant genes and several modifiers appear to be involved (Campbell & Lafever 1979, Lafever & Campbell 1978). In maize, aluminium tolerance is inherited via one locus with multiple alleles (Rhue 1979). Lafever & Campbell (1978) found no genetic difference in reaction to aluminium among the sensitive cultivars 'Gaines' from Washington State and 'Redcoat' and 'Arthur' from Indiana, nor among tolerant cultivars 'Atlas 66' from North Carolina and 'Seneca' and 'Thorne' from Ohio. A. M. Prestes et al (1975, Am Soc Agron Abstr, p 60) concluded that in wheat the gene for aluminium tolerance is at a locus on chromosome 5D.

Lopez-Benitez (1977) suggested that the D genome from hexaploid wheat and the R genome from rye (*Secale cereale* L.) contain the loci that condition aluminium tolerance for triticale (× *tritosecale*). Sapra et al (1979) found that 'Beagle' triticale, which had a full complement of rye chromosomes, showed high aluminium tolerance, but 'Armadillo' and 'Rosner', which lack chromosome 2R, and 'Cimmaron', which lacks 2R and 4-7R, also showed considerable tolerance. Apparently, the genes on rye chromosomes that control or modify aluminium tolerance interact with the D genome of wheat to express the trait in substituted hexaploid triticales. Hanson & Kamprath (1979) found no clear-cut genetic control of aluminium tolerance in soybeans. Differential aluminium tolerance has been reported in callus cell cultures of tomato (Meredith 1978a, 1978b). Callus formation from 'VNT Cherry' tomato was less inhibited by aluminium than that from 'Marglobe'. The aluminium-resistant variants probably resulted from mutation, but they may have an epigenetic basis.

Manganese tolerance

In alfalfa, tolerance to manganese has been attributed to additive genes having little or no dominance (Dessureaux 1959), whereas in lettuce manganese tolerance is controlled by one to four genes, depending on the species (Eenink & Garretsen 1977). In soybeans, also, manganese tolerance is under polygenic control with evidence for cytoplasmic inheritance (Brown & Devine 1980).

Efficiency in use of iron

Resistance to iron-deficiency chlorosis (and hence, iron efficiency) in oats is controlled by one major, dominant gene with modifiers (McDaniel & Brown 1982); in dry beans by major genes with complete dominance at two loci (Coyne et al 1982); and in tomato by one major, dominant gene (Brown & Wann 1982). In soybean, Weiss (1943) found one major gene involved in iron efficiency, but Fehr (1982) concluded that this trait was quantitatively inherited.

Efficiencies in use of other nutrients

Genetic control of nitrogen, phosphorus and calcium efficiencies in plants has been demonstrated by Gabelman & Gerloff (1982), and Clark (1981, 1982) has reviewed the literature on plant genetic variation in the use of both macro- and micronutrients.

Development and release of stress-tolerant plant germplasm

Aluminium tolerance

Selection for high tolerance to aluminium in Brazilian wheat cultivars has been conducted on strongly acidic, aluminium-toxic soils of Brazil for the past 50 years (da Silva 1976). 'Titan' wheat, released for use on acid soils in eastern Ohio, is not as tolerant to aluminium as is 'BH 1146' from Brazil but is much more tolerant than Indiana cultivars like 'Abe', 'Arthur' and 'Redcoat'. Two aluminium-tolerant and drought-tolerant sorghum (*Sorghum bicolor* L. Moench) cultivars have been released in Brazil (R. Schaffert, personal communication 1981), and Duncan (1981) developed an acid-tolerant (and presumably aluminium- and manganese-tolerant) sorghum germplasm population 'GPIR' in Georgia, USA. Reid et al (1980) have released the aluminium-tolerant barley population (composite cross XXXIV) for experimental purposes.

In alfalfa, recurrent selection increased tolerance to acidic, aluminium-toxic Bladen and Tatum soils (Devine 1981). J. H. Elgin (personal communication, 1982) subjected 'Arc' alfalfa to four cycles of recurrent selection (two each in acidic soil and nutrient solutions containing aluminium), and found that the resulting population was significantly more tolerant to aluminium than was Arc. However, this difference did not persist in field studies on acidic soils. Individual clones that survived in field plots at pH 4.2 are being tested further. J. H. Bouton & M. E. Sumner (1981, Am Soc Agron Abstr p 173) found that an alfalfa population developed by recurrent selection on an acidic (pH 4.4) aluminium-toxic soil produced significantly higher yields on a soil with pH 4.8 than did a population selected on a limed soil with pH 6.5 or the cultivar 'Apollo' used as a check. For subsequent results from these continuing studies see also Brooks et al (1982) and Bouton et al (1982). J. J. Murray (personal communication, 1982) used recurrent selection to increase the acid-soil (aluminium) tolerance of tall fescue (*Festuca arundinacea* Schreb.) by 40% in the first cycle and by an additional 8% in the second cycle; a third cycle is in progress.

Efficiency in use of iron

Fehr & Cianzio (1980) released a soybean population (AP9(S1)C_2) that had a superior resistance to iron-related chlorosis (and thus a high iron efficiency). In addition, soybean cultivars (for example 'Weber') that have both high yields and improved resistance to iron-deficiency chlorosis have been developed for calcareous soils in northern Iowa (Bahrenfus & Fehr 1980). Fehr (1982) has stated that 'No genetic limitations to future progress have been identified in developing new cultivars with high yield and high resistance to iron-deficiency chlorosis.' More recently, Iowa State University has made plans to release 'A_7' soybean which has the highest known resistance to iron-deficiency chlorosis (W. R. Fehr, personal communication, 1983). This is an F_4 plant selection from the population AP9 Fe (S1)C_4 which has been used in recurrent selection for resistance to iron-deficiency chlorosis. Samples of 50 seeds are available to institutions. Iron-efficient strains of weeping lovegrass have also been developed for calcareous soils (Voigt et al 1982).

The physiology of differential tolerance to mineral stress

Knowledge about the physiology and biochemistry of differential tolerance to stress would be useful in modifying both plant and soil to achieve a desired result. However, the identification of stress-tolerance mechanisms has lagged considerably behind plant-breeding progress.

Aluminium tolerance

Aluminium tolerance has been variously associated with the following factors: plant-induced pH increases in root zones (nutrient solutions and soils); trapping of aluminium ions in non-metabolic sites within plants; higher internal concentrations of organic acids which may chelate and detoxify aluminium; greater stability of the root-cell plasmalemma in the presence of aluminium; lower cation-exchange capacity in the roots; lower concentrations of aluminium in roots but not in tops; greater efficiency in the uptake, transport and use of calcium, phosphorus, magnesium and iron; higher root phosphatase activity; higher nitrate reductase activity; higher internal concentrations of silicon; preferential uptake of NO_3^- over NH_4^+; and greater resistance to drought. Current views on the physiology of differential plant tolerance to aluminium and other metallic cations are covered in several reviews (Foy et al 1978, Foy 1983b).

Manganese tolerance

Manganese tolerance has been associated with the following factors: higher oxidizing powers of plant roots (or associated microorganisms); lower manganese absorption and translocation rates; trapping of excess manganese in non-metabolic centres; higher internal tolerance to excess manganese (possibly through a chelation–detoxification mechanism); and more efficient uptake and distribution of iron and silicon. Current evidence from our laboratory suggests that in genotypes of cotton, wheat, soybean and snapbean, tolerance to high manganese concentrations within the plant is more important than reduced manganese uptake in determining manganese tolerance (Foy 1973, 1983c, Foy et al 1978).

Efficiency in use of iron

Superior iron-efficiency in the 'FQ 22' strain of weeping lovegrass (compared with 'FQ 71') has been associated with the following factors: the ability to maintain a low pH in its root zone (which prevents iron precipitation); a greater affinity for NH_4^+ than NO_3^- (NH_4^+ lowers pH by release of H^+); more effective transport of iron from roots to tops; and restricted transport of manganese and calcium (Foy et al 1981). In soybean and tomato, iron-efficiency has been associated with a lowering of pH in the root zone and the release of a 'reductant' that converts Fe^{3+} to Fe^{2+}, the form most available to plants. In tomato this reductant has been identified as caffeic acid or one of its derivatives (Olson et al 1981).

Conclusions

Mineral stress in plants on problem soils cannot always be economically corrected with current technology. An alternative or supplemental approach is to fit plants more precisely to soils. This approach does not propose the elimination of lime and fertilizer or the bleeding of soil fertility to zero levels. Instead, it advocates a more effective use of plant genetic variability in solving difficult problems of soil fertility. The result will be combinations of plant genotype, soil, lime and fertilizer which may be even better matched than drug doses and people in medical prescriptions.

The most productive soils of the world are not always in areas where food needs are greatest. Furthermore, world-wide attempts to move food from regions of plenty to those in need have not been successful. The only workable crop-production strategy is to produce more food in localities where it is needed. This often entails using marginal land and low-input technology, at least initially. Hence, breeding crop cultivars with greater tolerance to soil-induced stress is especially valuable to developing countries.

The plant genetic approach proposed here involves the identification of present and potential growth-limiting factors in soils, the screening of available germplasm for stress tolerance, the elucidation of genetic, physiological and biochemical mechanisms of stress tolerance, and the selection or breeding of superior cultivars for specific purposes. Such multidisciplinary research, which is already in progress at several research centres, has great potential for alleviating present and anticipated food shortages throughout the world (Christiansen 1979, Vose 1982).

REFERENCES

Bahrenfus JB, Fehr WR 1980 Registration of Weber soybean (Reg no 137). Crop Sci 20:415-416

Bouton JH, Hammel JE, Sumner ME 1982 Alfalfa, *Medicago sativa* L., in highly weathered acid soils. IV. Root growth into acid subsoil of plants selected for acid tolerance. Plant Soil 65:187-192

Brooks CO, Bouton JH, Sumner ME 1982 Alfalfa, *Medicago sativa* L., in highly weathered acid soils. III. The effects of seedling selection in an acid soil on alfalfa growth at varying levels of phosphorus and lime. Plant Soil 65:27-33

Brown JC, Devine TE 1980 Inheritance of tolerance or resistance to manganese toxicity in soybean. Agron J 72:898-904

Brown JC, Wann EV 1982 Breeding for Fe efficiency: use of indicator plants. J Plant Nutr 5:623-635

Camargo EEO, Kronstad WE, Metzger RJ 1980 Parent–progeny regression estimates and association of height level with aluminum toxicity and grain yield in wheat. Crop Sci 20:355-358

Campbell LG, Lafever HN 1976 Proper wheat variety selection for production on acid soils. (Ohio Agric Res Dev Cent, Wooster, Ohio) Ohio Report 61:91-92

Campbell LG, Lafever HN 1979 Heritability and gene effects for aluminum tolerance in wheat. In: Ramanujam S (ed) Proc 5th Int Wheat Genet Symp. Indian Society of Genetics and Plant Breeding, Indian Agricultural Research Institute, New Delhi 110012, India, vol 2, p 963-977

Christiansen MN 1979 Organization and conduct of plant stress research to increase agricultural productivity. In: Mussell H, Staples RC (ed) Stress physiology in crop plants. John Wiley, New York, p 1-14

Clark RB 1981 Plant response to mineral element toxicity and deficiency. In: Christiansen MN, Lewis CF (eds) Breeding plants for less favorable environments. John Wiley, New York, p 71-142

Clark RB 1982 Plant genotype difference to uptake, translocation, accumulation and use of mineral elements. In: Saric MR (ed) Genetic specificity of mineral nutrition of plants. Serb Acad Sci Art, Scientific Assemblies, Dept Nat Math Sci, Vol XIII No 3, Belgrade, Yugoslavia, p 41-55

Coyne DP, Corban SS, Knudsen D, Clark RB 1982 Inheritance of iron-deficiency in crosses of dry bean (*Phaseolus vulgaris* L.) cultivars. J Plant Nutr 5:575-585

Dambroth M, El Bassam N 1982 Low input varieties: definition, ecological requirements and selection. In: Saric MR (ed) Genetic specificity of mineral nutrition of plants. Serb Acad Sci Art, Scientific Assemblies, Dept Nat Math Sci, Vol XIII No 3, Belgrade, Yugoslavia, p 325-336

da Silva AR 1976 Application of the plant genetic approach to wheat culture in Brazil. In: Wright MJ, Ferrari SA (eds) Plant adaptation to mineral stress in problem soils. Spec Publ Cornell Univ Agr Exp Stn, Ithaca, New York, p 223-231

Dessureaux L 1959 Heritability of tolerance to manganese toxicity in lucerne. Euphytica 8:260-265

Devine TE 1981 Genetic fitting of crops to problem soils. In: Christiansen MN, Lewis CF (eds) Breeding plants for less favorable environments. John Wiley, New York, p 143-173

Duncan RG 1981 Registration of GPIR acid soil tolerant sorghum population (Reg No Gp 73). Crop Sci 21:637

Eenink AH, Garretsen G 1977 Inheritance of insensitivity of lettuce to a surplus of exchangeable manganese in steam sterilized soils. Euphytica 26:47-53

Fehr WR 1982 Control of iron deficiency chlorosis in soybeans by plant breeding. J Plant Nutr 5:611–621

Fehr WR, Cianzio SR 1980 Registration of AP9 (Sl)C_2 soybean germplasm (Reg No Gp 33). Crop Sci 20:677

Foy CD 1973 Manganese and plants. In: Lieben J (ed) Manganese. Natl Acad Sci, Wash DC, p 51-76

Foy CD 1983a Plant adaptation to mineral stress in problem soils. Iowa State J Res 57:339-354

Foy CD 1983b The physiology of plant adaptation to mineral stress. Iowa State J Res 57:355-391

Foy CD 1983c Physiological effects of hydrogen, aluminum and manganese toxicities in acid soils.

In: Adams F (ed) Soil acidity and liming (2nd edn). Am Soc Agron, Madison, WI (Agronomy Monographs 12), in press
Foy CD, Chaney RL, White MC 1978 The physiology of metal toxicity in plants. Annu Rev Plant Physiol 29:511-566
Foy CD, Fleming AL 1978 The physiology of plant tolerance to excess available aluminum and manganese in acid soils. In: Jung GA (ed) Crop tolerance to suboptimal land conditions. Am Soc Agron Spec Publ 32, p 301-328
Foy CD, Fleming AL, Schwartz JW 1981 Differential resistance of weeping lovegrass genotypes to iron-related chlorosis. J Plant Nutr 3:537-550
Gabelman WH, Gerloff GC 1982 The search for and interpretation of genetic controls that enhance plant growth under deficiency levels of a macronutrient. In: Saric MR (ed) Genetic specificity of mineral nutrition of plants. Serb Acad Sci Art, Scientific Assemblies, Dept Nat Math Sci, Vol XIII No 3, Belgrade, Yugoslavia, p 301-312
Hanson WD, Kamprath EJ 1979 Selection for aluminum tolerance in soybeans based on seedling root growth. Agron J 71:581-586
Lafever HN, Campbell LG 1978 Inheritance of aluminum tolerance in wheat. Can J Genet Cytol 20:355-364
Lopez-Benitez A 1977 Influence of aluminum toxicity in intergeneric crosses of wheat and rye. PhD thesis, Oregon State Univ (Diss Abstr No 77-20153)
McDaniel ME, Brown JC 1982 Differential iron-chlorosis of oat cultivars. A review. J Plant Nutr 5:545-552
Meredith CP 1978a Response of cultural tomato cells to aluminum. Plant Sci Lett 12:17-24
Meredith CP 1978b Selection and characterization of aluminum resistant varieties from tomato cell cultures. Plant Sci Lett 12:25-34
Olson RA, Bennett JH, Blume D, Brown JC 1981 Chemical aspects of the iron stress response mechanism in tomatoes. J Plant Nutr 3:905-921
Ponnamperuma FN 1982 Genotypic adaptability as a substitute for amendments in toxic and nutrient deficient soils. In: Scaife A (ed) Proc Ninth Int Plant Nutr Colloq (Warwick Univ). UK Commonwealth Agricultural Bureaux, Vol 2, p 467-473
Reid DA 1971 Genetic control of reaction to aluminum in winter barley. In: Nilan RA (ed) Barley genetics II. Proc Int Barley Genet Symp 1969, Washington State Univ Press, Pullman, Washington, p 409-413
Reid DA, Slootmaker LAJ, Craddock JC 1980 Registration of barley composite cross XXXIV. Crop Sci 20:416-417
Rhue RD 1979 Differential aluminum tolerance in crop plants. In: Mussell H, Staple RC (eds) Stress physiology in crop plants. John Wiley, New York, p 61-80
Sanchez PA, Salinas JG 1981 Low input technology for managing Oxisols and Ultisols in tropical America. Adv Agron 34:280-406
Sapra VT, Mugwira LM, Choudhry MA, Hughes JL 1979 Screening triticale, wheat and rye germplasm for aluminum tolerance in soil and nutrient solution. In: Ramanujam S (ed) Proc 5th Int Wheat Genet Symp. Indian Society of Genetics and Plant Breeding, Indian Agricultural Research Institute, New Delhi 110012, India, vol 2, p 1241-1253
Vose PB 1982 Rationale of selection for specific nutritional characters in crop improvement with *Phaseolus vulgaris* L. as a case study. In: Saric MR (ed) Genetic specificity of mineral nutrition of plants. Serb Acad Sci Art, Scientific Assemblies, Dept Nat Math Sci Vol XIII No 3, Belgrade, Yugoslavia, p 313-323
Weiss MG 1943 Inheritance and physiology of efficiency in iron-utilization in soybeans. Genetics 28:253-268
Voigt PW, DeWald CL, Matocha JB, Foy CD 1982 Adaptation of iron efficient and inefficient lovegrass strains to calcareous soils. Crop Sci 22:672-676

DISCUSSION

Clarke: The soils of interest here are ones that have not been physically standardized. How much variation in the soils occurs over short distances? For example if a soil has patches with low as well as high aluminium concentrations, would one need more than one variety of a crop to produce a profitable maximum growth in such a field?

Foy: There can be tremendous variation within any one field. One of the things that causes variation in soil fertility, as well as the soil's physical properties, is erosion. Where erosion occurs, a more acid subsoil would be exposed. I understand that in Holland a mixture of rye, wheat, barley and oats is sometimes sown, and that each crop is allowed to find its own niche. The order of tolerance to an acid site is generally rye > oats > wheat > barley. A mixture of varieties would probably produce better overall vegetative growth than a single variety in such conditions. Whether this practice is profitable would, of course, depend on many other factors. For example, varieties having different maturity dates might be suitable for growing in mixtures for forage use but not for grain production. Many types of soil infertility or toxicity are not related to human activity but simply occur naturally. For example, in high-rainfall regions the soils are leached of their basic cations (Ca^{2+}, Mg^{2+}, K^{+}, Na^{+}) and thus become acid. But soil clay minerals, which are aluminosilicates, are unstable when strongly acid (H^{+}-saturated), and they therefore decompose, giving rise to soluble and exchangeable aluminium which may be toxic to plants grown and fertilized in the normal way.

Fowden: In the United Kingdom the soils are probably more variable than those found in many other parts of the world. So we are indoctrinated, in some ways, to the sort of question that Professor Clarke has asked.

Clarke: I would have supposed that in developing countries the soils would be more locally heterogeneous than the ones we have here in the UK, which tend to be homogenized by mechanized agriculture.

Fowden: It is true that UK soils have been mixed much more by intensive working and cultivation methods in the last 50 years or so. But the nature of the minerals that have led to UK soils over aeons is perhaps more variable than in many other parts of the world, because of our complicated geology.

Bradshaw: The need that Dr Foy mentioned to develop a range of cultivars, all specifically adapted to different soil conditions, could arise in relation to what we have just been discussing. But it could also arise where areas of soil differ on a large scale, over hundreds of kilometres. Whether or not such a range is required will depend on whether the tolerant cultivars are lower in yield when grown on normal non-toxic soils. The question is whether there is an inescapable cost to tolerance.

Foy: At present there is probably some cost. For example, the wheat (*Triticum aestivum*) cultivar 'Sonora 63' from Mexico, which is aluminium-sensitive, would probably out-yield the aluminium-tolerant wheat 'BH (Belo Horizonte) 1146' from Brazil in the absence of aluminium stress. But BH 1146 is often more profitable in Brazil because it will produce a good yield at a soil pH of 4.7, whereas Sonora 63 will not even produce seed.

Wolfe: BH 1146 is a Brazilian wheat that was released in 1946; hence its name. In current breeding programmes, varieties are being selected to eliminate most of the deficiencies that varieties like BH 1146 have, such as stem-rust susceptibility, tall-straw, and so on. From the current breeding programme it is not at all evident that there is much of a cost to including aluminium-tolerance, other than the delay caused by selecting another character.

Foy: That is especially true if cost is defined as profitability under the present system. Minimum input is the key factor.

Epstein: We should not be unduly inhibited by possible sacrifices in yield in less than optimal conditions. A good barley farmer in the San Joaquin Valley in California can get 6000 kg/ha, and we have grown barley selections at full sea-water salinity and have obtained 1000–1500 kg/ha, which is very little and in California would not be an economically feasible yield. But on many farms in India, 1000–1500 kg/ha would be quite acceptable. So it all depends on the particular socioeconomic conditions. If one can grow something where otherwise nothing would grow, then that is progress.

Gressel: You mentioned, in considering the genetic basis for these metal tolerances, that only one or two alleles may be involved. I believe that the problems of selecting for tolerance to salt are quite different genetically; they are probably polygenic with low heritability.

Epstein: You are right. Tolerance to salt is more complex. The heavy-metal tolerance can be handled perhaps more simply by the evolution of one or two appropriate chelators, as Professor Bradshaw already mentioned. The genetics of this can be fairly simple. We do not, however, know of any ligands that would complex sodium or chloride, so the mechanism is to sequester the salt in the vacuole, or to keep it out of the plant top almost entirely, or to recirculate it via the phloem, perhaps even causing it to be excreted into the soil. So, genetically, salt tolerance is probably more complex than heavy-metal tolerance and is governed by more genes, involving more physiological processes.

Foy: Many of these metal toxicities are manifested as iron disorders. That is, iron deficiency or imbalance can be induced by many pathways—by excess phosphate, by excess heavy metals or by low iron solubility *per se*. One can contrast the phenomenon of 'avoidance' with that of 'tolerance' (D.T. Clarkson, personal communication); a plant can modify its root environment in such a way that it doesn't make contact with the toxic element. For example, aluminium-tolerant varieties of wheat and barley raise the pH of their root zones and precipitate aluminium in the medium. The plant is therefore protected from the aluminium toxicity. On the other hand, an iron-efficient weeping love-grass strain lowers the pH of its root zone and makes iron available, which is helpful in calcareous soils. Two different plant traits are required in the two different environments (acid vs. neutral or alkaline).

Clarke: Perhaps you are describing threshold phenomena. When these phenomena occur, very often a genetic analysis will initially suggest a simple inheritance, but more detailed analysis will reveal that a lot of genes and a lot of processes are involved. This is certainly true of many inherited medical disorders. The genetic distinction that Dr Epstein made between salt tolerance and heavy-metal tolerance may therefore turn out, eventually, not to be real.

Hutchinson: In natural ecosystems extensive areas of serpentine soils are found, and these often have a high content of nickel and chromium, which are strongly phytotoxic, yet the soils are entirely vegetated. Enormous areas of salt marshes are also vegetated, despite their high salinity. It may be semantic and somewhat pointless to consider whether one type of tolerance is more difficult for plants to handle than another. Clearly the genetic variability to handle many tolerances is available.

Foy: The International Rice Research Institute in the Philippines is doing an effective job of selecting rice genotypes for tolerances to the multiple stresses of salinity, metal toxicities and low levels of available essential nutrients in soils (Ponnamperuma 1982). Investigators are finding that some varieties already have a 'built-in package' of multiple stress tolerances. In Brazil aluminium-tolerant varieties of wheat and beans are proving to be phosphorus-efficient also (Salinas & Sanchez 1976). Other work indicates that aluminium-tolerant lines of wheat, tomato and corn are more tolerant to low levels of phosphorus than are aluminium-sensitive lines, both in the presence and absence of excess aluminium (Foy et al 1978). This can result in significant savings in phosphorus-based fertilizers. Other factors that are being associated with aluminium tolerance are calcium and magnesium efficiencies and tolerance to drought (Foy 1983). In Atlas 66 wheat, aluminium tolerance is genetically linked to leaf-rust resistance which is attributable to a Brazilian parent.

Hutchinson: Have you done any experiments on the pH change around roots by using buffering systems to maintain the environmental pH? What would happen then to the tolerant varieties?

Foy: Wheat varieties differing in aluminium tolerance tend to change the pH differentially, even when pH is adjusted twice daily. For example, the net amount of acid or base required to maintain an equal pH for Atlas 66 and monon varieties of wheat can differ by seven-fold, and we have never been able to find a way to use buffers successfully to prevent such differential pH changes. Buffers may be appropriate for operating within half a pH unit, but a decrease of 0.1 pH units will double the solubility of aluminium, and this is uncontrollable in practice.

Hutchinson: I suppose that aluminium hydroxides are some of the best buffers themselves.

Hartl: I was a little surprised to hear your emphasis on selecting for desirable traits within cultivars rather than on crossing between cultivars, followed by directional selection. Why did you choose this perspective?

Foy: It is simply a matter of doing first what is easiest. The easiest step in obtaining appropriate plants for a given purpose is to select a tolerant species. Another possibility is to select a cultivar that is already superior to others in tolerance to a given stress factor. For example, blue grass varieties differ by 25-fold in tolerance to an acid, aluminium-toxic soil. A third approach is to select superior individual plants from within a cultivar that shows segregation under stress. Standard varieties of sorghum and tall fescue, which appear to grow uniformly in a limed soil, show differential growth in a strongly acid soil. By selecting superior individuals in such populations one can make rapid progress in acquiring stress-tolerant germplasm for a given purpose. A fourth resort is to combine desirable plant traits by breeding. For example, a desirable combination would be aluminium tolerance and short straw in the same wheat variety.

Bradshaw: In choosing the right approach, a lot would depend on the nature of the cultivar. If it was a highly selected cultivar from an inbred species one wouldn't find the variability. Presumably the sorghum that you mentioned, which was inbred, was not so highly selected that it had lost all variability.

Gressel: That would surely depend also on the number of genes involved. If the property of aluminium tolerance is polygenic, one would need interbreeding; and if it is monogenic one would stay within the multiple alleles.

Bradshaw: Are there any circumstances in which aluminium tolerance cannot be found, for example in barley?

Foy: The range of aluminium tolerance is wide in barley and wheat but much narrower in soya bean. As a species, soya bean is much more tolerant to aluminium than is alfalfa (lucerne). Species of the genus *Stylosanthes* are generally quite tolerant to acid soils and aluminium (L.J.C.B. Carvalho and P.R. Mosquim, personal communication), and several ecotypes have been found growing wild and well nodulated at soil pH values as low as 3.8 in Brazil. Some differences in aluminium tolerance have been found among *Stylosanthes* ecotypes but these are smaller than those in wheat.

Bradshaw: How tolerant is alfalfa?

Foy: Alfalfa is extremely sensitive to aluminium toxicity. In our recurrent selection work on acid soils and nutrient solutions containing aluminium, only 2% of the plants were placed in the first category of tolerance, after several screening cycles. Clonal testing of alfalfa for acid soil and aluminium tolerance is now being done by Dr James Elgin of the Crops Research Laboratory at Beltsville.

Epstein: On the question, implied by Professor Bradshaw, of whether there is enough variability, the answer is that we haven't looked for it enough. In the world collection of barley there are 22 000 entries, and no-one has screened the entire collection for any particular property, including aluminium tolerance. In the world collection of wheat there are 33 000 entries, and these also have not been totally screened. For rice the number is 56 000 entries. So for no single species has it been possible to make a total survey for the tolerance of any one factor.

Foy: Reid et al (1980) have released an aluminium-tolerant composite cross of barley for experimental use. This population was obtained by screening a mixture of 1295 varieties on an acid, aluminium-toxic soil and by selecting and crossing superior individual plants.

Wolfe: This makes me wonder what Professor Bradshaw meant, in his paper earlier, by a 'normal' population. The population that surrounds a particular mine dump has itself been selected for certain characters, so that it may not be representative of the range of variation that is available for tolerance of a particular heavy metal.

Bradshaw: I agree. Dr L. Simeonidis in our lab has recently shown that a species like *Agrostis tenuis* which has copper-tolerant individuals in its 'normal' populations does not have them in all populations. This is most odd and we don't understand why. But in my paper I was using the evidence that even with selection occurring over a long period of time evolution of tolerance has not occurred in some species and that we can't find the variability for it in these species. In Britain, barley is a plant of calcareous soils that has never been forced to grow in acid soils, and I had assumed that it did not possess the necessary variability to grow on acid soils. But, from what you say, Dr Foy, it clearly has.

Foy: Dayton barley from Ohio is much more tolerant to aluminium than Kearney barley from Kansas. We have used these two varieties successfully as indicator plants to show that aluminium toxicity limits root penetration in certain acid subsoils (Long & Foy 1970).

In Ohio the development of aluminium-tolerant wheat varieties can be explained by the occurrence of strongly acidic (<pH 5.0) subsoils in the eastern part of the state. However, we also find aluminium tolerance in unexpected places. Examples are tomato varieties developed in alkaline soils of Idaho and a bean variety developed on marl at pH 8.1 in Florida. Hence, many plants may have tolerances that are not needed in their present environ-

ments and that are just awaiting the application of the proper stress for expression.

Hutchinson: The last example you gave of tolerance to a pH of 8.1 is intriguing. The U-shaped solubility curve for aluminium falls off rather steeply below pH 4.0 and then increases rapidly again just above pH 8.0. So the selection for aluminium tolerance at that pH (8.1) might have been the accidental consequence of exposure to available aluminium.

Foy: We checked the possibility that this Dade snapbean had been selected for tolerance to aluminate toxicity at pH 8.0 or above. But Kearney barley, which is extremely sensitive to aluminium, grows well on the marl (pH 8.1) on which Dade snapbean was developed. Furthermore, levels of soluble aluminium in the marl were so low as to be only faintly detectable. Hence, we concluded that Dade snapbean has not been subjected to aluminium stress recently but still carries the aluminium tolerance trait. Dade is known for its ability to make reasonable growth on both acid and marl soils in Florida.

Clarke: I have a general question about the extent to which possible genetic manipulations can 'go one better' than nature. Crops in developing countries are often re-sown from seed, and adaptations to the local environments must be evolving *in situ*. Can we do any better by creating such varieties somewhere else and then bringing them into the area?

Foy: I know of totally foreign species or strains of plants that have been brought in and filled a particular niche. For example, one strain of limpograss (*Hemarthria altissima*), collected in Africa at an altitude of 7000 feet, was introduced to the United States and found to have good forage qualities plus exceptional cold tolerances (Oakes 1979). It has survived temperatures as low as −15 °C at Beltsville. It occurred to me that if this grass also had metal tolerance, it might be used to vegetate strongly acid, high-altitude (or northern latitude) mine spoils. Many plants that tolerate the metal toxicities associated with strongly acid soils (bermuda grass, weeping lovegrass) are not sufficiently cold-tolerant for this purpose). This strain of limpograss turned out to be exceptionally tolerant to excess aluminium in both acid soils and nutrient solutions and is being field-tested on acid mine spoils in West Virginia and Pennsylvania. A wide range of germplasm from the genus (*Stylosanthes*) (Townsville lucerne), originally from Australia, has been successfully established on the strongly acid, aluminium-toxic and phosphorus-deficient soils of Brazil. Hence, human intervention can certainly accelerate the plant adaptation process and make better use of the plant genetic variability that is available.

Sawicki: But isn't it true that when such a plant has adapted to a particular environment its 'normal' population could not be considered 'normal' in another environment?

Clarke: I have seen populations of snails that are very strongly locally adapted, and if they are moved from one environment to another only a few hundred yards away they all die. One might expect that not all plants would survive well when moved from one part of the world to another.

Wood: In the eighteenth century in England there were very many varieties of sheep, each apparently adapted to a local area. Then Robert Bakewell and others began to practise a systematic selection, which produced a tremendously rapid change in the appearance of these animals and in their ability to grow rapidly and fulfil various other economic functions (see Pawson 1957, Wood 1973). Robert Bakewell's particular interest was selection for rapid growth rate and, associated with that, there was a deterioration in the wool. But later, breeders were able to overcome that problem. Breeds like the Corriedale, developed in New Zealand, were able to grow very rapidly and also to produce excellent wool. So animal breeders have shown that one can, in fact, surpass nature very considerably.

Clarke: Such developments have, of course, coincided with our ability to adapt the environment. Dr Foy's point is that we should be changing the organism rather than changing the environment. It is perhaps only after we have standardized and improved our environment that we can attempt to select, say, for a highly productive sheep, which develops fast and produces lots of wool. The sheep otherwise might not survive in a less suitable environment.

Wood: This is true, and new foodstuffs were also developed at the same time. The higher-plane nutritional environment formed the basis for this kind of genetic selection.

Bell: I believe that there are differences in haemoglobin type between the lowland sheep of the Pennine hills in the UK and sheep that live on the hill tops. Evans et al (1958) studied different breeds of sheep and showed differences between the environments in which the sheep had evolved.

Wood: Yes. The basis of local adaptation is not understood well, but there are indications for differences between highland and lowland varieties. We have hear-say evidence from the eighteenth century that many lowland and many highland varieties existed and that they could be very different indeed. Different varieties were said to exist even on either bank of the same river, for instance (White 1789).

Gressel: Breeders and farmers may disagree about the usefulness of moving new varieties of crops around. Throughout the Middle East, where wheat shows the greatest variability, being native to the area, each farmer or Bedouin tribe had its own indigenous and highly adapted variety. With the introduction of new, high-yielding varieties much of the adaptability has been lost. The farmers are happy but the breeders have lost genetic material for the future. I believe this is also true for maize growth in South America.

REFERENCES

Evans JV, Harris H, Warren FL 1958 Distribution of haemoglobins and blood potassium types in British breeds of sheep. Proc R Soc Lond B Biol Sci 149:249-262

Foy CD 1983 Physiological effects of hydrogen, aluminum and manganese toxicities in acid soils. In: Adams F (ed) Soil acidity and liming, 2nd edn. Am Soc Agron, Madison, WI (Agronomy Monograph 12), in press

Foy CD, Chaney RL, White MC 1978 The physiology of metal toxicity in plants. Annu Rev Plant Physiol 29:511-566

Long FL, Foy CD 1970 Plant varieties as indicators of Al toxicity in the A_2 horizon of a Norfolk soil. Agron J 62:679-681

Oakes AJ 1979 Winter hardiness in limpograss, *Hemarthria altissima*. Soil Crop Sci Soc Fla Proc 39:86-88

Pawson HC 1957 Robert Bakewell. Crosby, London

Ponnamperuma FN 1982 Genotypic adaptability as a substrate for amendments in toxic and nutrient deficient soils. In: Scaife A (ed) Proc Ninth Int Plant Nutr Colloq (Warwick University). UK Commonwealth Agricultural Bureaux, Slough, vol 2:467-473

Reid DA, Slootmaker LAJ, Craddock JC 1980 Registration of barley composite cross XXXIV. Crop Sci 20:416-417

Salinas JG, Sanchez PA 1976 Soil plant relationships affecting varietal and species differences in tolerances to low available soil phosphorus. Cienc Cult (Sao Paulo) 28:156-168

White G 1789 The natural history and antiquities of Selborne in the county of Southampton. Swan Sonnenschein, London (1875 edn, p 187)

Wood RJ 1973 Robert Bakewell (1725-1795) pioneer animal breeder and his influence on Charles Darwin. Folia Mendeliana 8:231-243

Plant–plant interactions

E. A. BELL

Royal Botanic Gardens, Kew, Richmond, Surrey TW9 3AB, UK

Abstract. Higher plants show three types of biochemical adaptation which enable them to combat pathogenic organisms in the form of lower plants. Firstly they may synthesize antibacterial or antifungal compounds in concentrations that prevent the invasion of the higher plant by the bacteria or fungi. Secondly they may synthesize such compounds in less than adequate amounts for defence in healthy tissues but respond to invasion by increasing the synthesis; and, thirdly, they may respond to invasion by synthesizing antibacterial or antifungal compounds *de novo*. Higher plants also show biochemical adaptations that enable them to compete with individuals of the same or different higher plant species. These include the synthesis of volatile and water-soluble phytotoxins which suppress the germination, growth, or both, of competitors.

1984 Origins and development of adaptation. Pitman Books, London (Ciba Foundation symposium 102), p 40-51

Interactions between plants take many forms. Some interactions are to the apparent benefit of the interacting species; others clearly benefit one individual or species at the expense of another. When two higher plants are in competition with one another, factors such as the ability of one competitor to absorb nutrients more efficiently or to scatter its seeds more widely than the other may ensure its success. When a higher plant is exposed to possible invasion by a fungal pathogen an anatomical feature such as the thickness of leaf cuticle may be critical to the survival of one species or the other. In this paper I shall consider primarily the development of our understanding of interactions between plants at the chemical and biochemical, rather than at the anatomical or physiological, level.

Higher plant–lower plant interactions

In reviewing the biochemical basis of disease resistance in higher plants Ingham (1973) divided compounds that confer resistance into two classes: the first class comprises compounds that are present in the plant before infection (pre-infectional compounds); the second class comprises compounds that are detectable only after infection (post-infectional compounds).

Of the pre-infectional compounds, some (the prohibitins) are able, at their pre-infectional concentrations in the higher plant, to halt the *in vivo* multiplication of microorganisms. Others (the inhibitins) must increase in concentration after infection before they can exercise effective control over the invading pathogen.

Of the post-infectional compounds, some (the post-inhibitins) are formed by the modification of pre-existing non-toxic precursors while the phytoalexins (dealt with elsewhere; see Kuć, this volume) are synthesized *de novo* after infection.

Prohibitins

Plant phenolic compounds probably constitute one of the most important groups of prohibitins. Although the ability of a compound to inhibit spore germination or growth of a potential pathogen *in vitro* does not prove that the compound has a defensive role *in vivo*, much circumstantial evidence suggests that phenolics do have this role in some plant species. The dead outer scales of the varieties of onion that are resistant to the fungus *Colletotrichum circinans* have been shown to contain catechol and protocatechuic acid. These compounds dissolve in water and the antifungal solution protects the living parts of the plant (Walker & Stahmann 1955). The resistance of *Phaseolus lunatus* (lima bean) varieties to *Phytophthera phaseoli* (downy mildew) is related to the bean's content of a resorcinol-type compound (Valenta & Sisler 1962) while the resistance of potato varieties to *Phytophthera infestans* is determined by levels of *o*-diphenol. In *Lupinus albus* (the white lupin), the fungitoxins luteone and 2′-deoxyluteone have been identified both in the living leaf and on the leaf surface (Harborne et al 1976), suggesting that prohibitin synthesis no less than phytoalexin synthesis may be a significant factor in protecting some plant species.

Tannins, which are themselves condensation products of phenolic compounds of lower molecular mass (M_r), are known to be toxic to a range of pathogenic fungi at the concentrations in which they occur in bark, cork and heartwood. The ability of some varieties of chestnut to resist *Endothia parasitica* (the chestnut blight) is also known to be related to the nature and solubility of the tannins that they contain (Nienstaedt 1953). However, it should be emphasized that secondary compounds may have more than one defensive role in a plant, and tannins considered here as potential antifungal agents have also been shown to discourage herbivory by animals. The alkaloids of the Solanaceae, also toxic to animals, have been shown to act as prohibitins with respect to *Alternaria solani*, a fungus which attacks *Solanum* and *Lysopersicum* species (Sinden et al 1973), while the terpene gossypol from cotton has been identified as an antifungal factor (Margalith 1967).

For further examples of prohibitin synthesis and accumulation the reader is referred to Levin (1976), Swain (1977) and Harborne (1982).

Inhibitins

In some plants, fungicidal compounds can be detected in only low concentrations in the healthy tissue but are found in much higher concentrations after infection. The antifungal compounds scopolin and chlorogenic acid provide examples; both increase in concentration markedly in potatoes after infection with *Phytophthera infestans* (Harborne 1982).

Post-inhibitins

The post-inhibitins are toxic compounds produced from non-toxic precursors in response to infection.

Glucosinolates, which are found widely distributed in the Cruciferae, can give rise to fungitoxic isothiocyanates by a two-step process involving the removal of the sugar moiety and the subsequent rearrangement of the aglycone (Fig. 1). Tissue disruption is, however, a prerequisite for the formation of the isothiocyanate as the enzymes and substrates do not come into contact in the undamaged tissues.

In a similar way cyanogenic glycosides can give rise to hydrogen cyanide which is toxic to many species of fungus. One of the most convincing pieces of evidence that cyanogenic glycosides do act as post-inhibitins is to be found, paradoxically, in the failure of cyanogenic forms of *Lotus corniculatus* (bird's foot trefoil) to resist the fungal attack of *Stemphylium loti* (Millar & Higgins 1970). This fungus synthesizes the enzyme formamide hydrolyase (cyanide

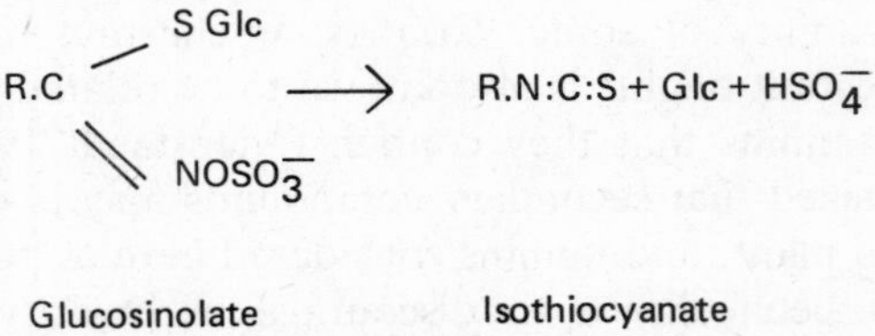

FIG. 1. The formation of an isothiocyanate from a glucosinolate.

hydratase, EC 4.2.1.66) which detoxifies hydrogen cyanide. Clearly, the existence of this specific adaptation in the successful fungus adds weight to the belief that hydrogen cyanide is an influential factor in the inability of non-adapted species of fungus to invade the cyanogenic plant. Biochemical adaptations identifiable in successful predators (plant or animal) may often provide the most convincing evidence of a protective role of a secondary compound, for if no such role existed then selectionary pressure would not have favoured a species adapted to circumvent a purely hypothetical defence.

Although space does not permit a discussion of symbiotic relationships between higher and lower plants it is appropriate to note that not all invasions of higher plants by lower plants are detrimental to the higher plants. The associations between nitrogen-fixing bacteria and legumes, and between many higher plant species and the mycorrhizal fungi, provide us with examples of mutually beneficial associations.

Higher plant–higher plant interactions

It is common knowledge that some species of higher plant grow well together and others do not. The failure of one species to thrive in close proximity with another may be due to a number of different factors. One plant may deprive its neighbour of sunlight; another may compete more successfully than the neighbour for water or mineral nutrients in the soil; and a third may liberate into the environment chemical compounds that adversely influence the germination, growth, or both, of competing individuals or species (*allelopathy*).

One of the earliest investigations of this phenomenon of allelopathy was concerned with the adverse effects produced by *Juglans nigra* (the black walnut tree) on surrounding vegetation. Massey (1925) reported the effects of the toxin from walnut on tomatoes and alfalfa. The toxin was subsequently shown to be juglone, a quinone that is formed by the hydrolysis and oxidation of the 4-glucoside of 1,4,5-trihydroxynaphthalene washed from the walnut leaves by rain or dew. Interestingly, *Poa pratensis* (Kentucky blue grass) is unaffected by juglone and grows well with *J. nigra*, thus providing an example of a specific adaptation to a general phytotoxin.

In 1966 Müller published an account of the role of allelopathy on the vegetational composition of chaparral and adjacent areas of natural uncultivated grassland in Southern California. He described how the shrubs *Salvia leucophylla* (sage brush) and *Artemisia californica* completely inhibit the growth of grasses in their immediate vicinity and partially inhibit the growth of grasses at greater distances. On sloping ground the successive zones of total and partial inhibition were equally pronounced on the uphill and downhill sides of *Salvia* thickets, suggesting the existence of a volatile airborne inhibitor rather than a water-soluble toxin which would have been carried downhill. Subsequent experiments confirmed that leaves of *S. leucophylla* contain the terpenes α-pinene, β-pinene, camphene, 1,8-cineole and camphor. Müller demonstrated release of these terpenes by the intact leaves and suppression of the growth of herb seedlings exposed to an atmosphere containing them. The terpenes were also shown to be absorbed onto dry particles of the soil adjacent to *S. leucophylla* plants and to remain there during the summer drought of Southern California, thus providing a reservoir of phytotoxins able to inhibit the herb-seedling development, which would otherwise occur at the onset of the winter rains in this area of Mediterranean climate. Of additional interest was the finding that 1,8-cineole and camphor were common ingredients of the volatile mixtures released by both *S. leucophylla* and *A. californica*.

In contrast to the volatile phytotoxins released by *S. leucophylla* and *A. californica*, two non-aromatic shrubs of the chaparral, *Adenostoma fasciculatum* and *Arctostaphylos glandulosa*, were found to exercise an inhibitory effect on herb growth by releasing water-soluble phenolic compounds that are washed from leaf to ground. However, the production of toxins in the two species followed different patterns; root extracts and leaf litter of *A. fasciculatum* did not inhibit herb-seedling growth, whilst extracts of roots, living foliage, fallen fruit and, particularly, freshly fallen leaves of *A. glandulosa* were all highly inhibitory to seedling growth (Müller & Chou 1972).

Where phytotoxins are present in the root of a plant, as in *A. glandulosa*, they may be liberated into the soil *via* root exudates and, indeed, the suggestion has been made (Harborne 1982) that the walnut tree may release the 4-glucoside of 1,4,5-trihydroxynaphthalene (the precursor of juglone) through the root as well as through the leaf.

The autotoxic effect of *trans*-cinnamic acid, which occurs in the root exudate of *Parthenium argentatum* (the guayule rubber plant), is well known (Bonner & Galston 1944), and it is thought that the compound may confer a selective advantage on the species by ensuring that individual plants are sufficiently well spaced to take maximum advantage of whatever rainfall is available in their semi-arid habitat. The toxin may, of course, reduce competition from other species simultaneously, but this possibility has not been fully explored.

The synthesis of compounds that are themselves toxic to the plants that synthesize them is probably the exception rather than the rule. Peterson & Fowden (1965) have shown that azetidine-2-carboxylic acid (the lower homologue of proline), which occurs in high concentrations in the leaves of

Convallaria majalis (lily of the valley), in the rhizomes of *Polygonatum multiflorum* (Solomon's seal) and in the seeds of species of the tropical legume genus *Bussea*, is not incorporated into the protein of *C. majalis* or *P. multiflorum*. Azetidine-2-carboxylic acid is, however, incorporated into the protein of *Phaseolus aureus* and *Triticum aestivum*: the prolyl-tRNA synthetases (EC 6.1.1.15) in the azetidine-2-carboxylic acid-producing species can discriminate against the lower homologue, whereas the prolyl-tRNA synthetases of *P. aureus* and *T. aestivum* cannot. A second example of toxicity avoidance by high specificity of enzymes is found in *Caesalpinia tinctoria*. This legume synthesizes 3-(3-hydroxymethylphenyl) alanine, a close analogue of tyrosine. It is notable, however, that the tyrosyl-tRNA synthetase (EC 6.1.1.1) isolated from *P. aureus* (a non-producer species) activates 3-(hydroxymethylphenyl) alanine five times faster than does the tyrosyl-tRNA synthetase from *C. tinctoria* (Fowden et al 1979).

The accumulation of high concentrations of seleno compounds with low M_r, such as seleno-methyl-selenocysteine, is typical of the so-called selenium accumulator or indicator plants which grow on selenium-rich soils. These plants are not only highly toxic to grazing livestock (the 'loco' weeds of the American West are selenium-accumulating *Astragalus* species) but may play an indirect role in the toxicity of non-accumulator species, including crop plants used for human food. This may come about because the accumulator species mobilizes inorganic selenium from the soil and incorporates it into small organic molecules. When these are returned to the soil (possibly via root exudates or when leaf litter decomposes), they are taken up by the crop plants which then incorporate the selenium into protein in the form of selenocysteine or selenomethione. The incorporation of these residues is damaging to the crop species but they can nevertheless tolerate levels of selenium that render them toxic to grazing animals and humans.

Non-protein amino acids and related compounds in allelopathy

The wide distribution of non-protein amino acids and the high concentrations in which they are found in the seeds of some plant species (Bell 1980), together with the knowledge that non-protein amino acids such as canavanine and azetidine-2-carboxylic acid can be toxic to plant species other than those that synthesize them (Fowden et al 1979), led us to investigate the effects of these, and 24 other related compounds that were isolated from plants, on the seed germination and the subsequent seedling growth of *Lactuca sativa* (lettuce). These experiments showed (Wilson & Bell 1978, 1979) that, at 1 mM concentration, six of the compounds tested produced minor (<20%)

decreases in germination rate compared with controls. At the same concentration, however, all the following compounds—mimosine, homoarginine, 3,4-dihydroxyphenylalanine, α,γ-diaminobutyric acid, α,β-diaminopropionic acid, canavanine, azetidine-2-carboxylic acid, α-amino-β-oxalylamino-propionic acid, α-amino-β-methylaminopropionic acid, 4-hydroxyarginine, enduracididine, 4-hydroxyproline, tetrahydrolathyrine, 5-hydroxynorleucine and baikiain—suppressed either hypocotyl growth, or radicle growth, or both, by more than 50% as compared with controls.

In a parallel series of experiments we showed that the effect of canavanine on the seedling growth of a canavanine-synthesizing species was minimal, as was the effect of homoarginine on a homoarginine-synthesizing species and that of β-aminopropionitrile on a β-aminopropionitrile-synthesizing species (Wilson & Bell 1977).

To establish whether any of the potentially phytotoxic compounds were liberated in the root exudates (Rovira 1969) of plants that synthesized these compounds, seedlings were germinated in a liquid flow system which allowed the content and phytotoxicity of the exudate to be determined (Wilson & Bell 1978). Seeds of *Neonotonia* (*Glycine*) *wightii* (Leguminosae), which contain canavanine and 3-carboxytyrosine, were germinated and the ninhydrin-reacting compounds liberated from the seed and seedlings were monitored. During inhibition, a number of different amino acids 'leaked' from the seeds, including major concentrations of 3-carboxytyrosine and glutamic acid and minor concentrations of canavanine. Immediately before germination, very little ninhydrin-reacting material was detected. After germination, major concentrations of canavanine were detected in the exudate, accompanied by lower concentrations of protein amino acids and 3-carboxytyrosine, which steadily declined. The canavanine concentration of the eluate continued high throughout the experiments (310 hours) and the eluate produced marked inhibition of hypocotyl and radicle growth in seedling lettuce.

Subsequent experiments, in which we used a modified experimental system, showed that β-aminopropionitrile (present in the seed as the γ-glutamyl derivative) was present in the root exudates of *Lathyrus odoratus*, and that mimosine was present in the root exudates of *Leucaena leucophylla*. β-Aminopropionitrile, although without effect on lettuce seedlings, was found to be highly toxic to seedlings of *Vicia benghalensis* and lethal to seedlings of *Lathyrus aphaca*. Kuo et al (1982) have recently reported that the root exudate of *L. odoratus* contains γ-glutamyl-β-aminopropionic acid itself and no fewer than six isoxazolin-5-ones, and that the root exudate of *Pisum sativum* contains isowillardine [β-(uracil-3-yl)alanine] together with four isoxazoline-5-ones. These findings suggest that low M_r nitrogen compounds may be widespread in root exudates but the allelopathic roles, if any, of most of them remain to be investigated.

REFERENCES

Bell EA 1980 Non-protein amino acids in plants. In: Bell EA, Charlwood BV (eds) Secondary plant products. Springer-Verlag, Berlin (Encyclopedia of plant physiology, New Series), vol 8:403-432

Bonner J, Galston AW 1944 Toxic substances from culture media of guayule which may inhibit growth. Bot Gazz 106:185-198

Fowden L, Lea PJ, Bell EA 1979 The Non-protein amino acids in plants. Adv Enzymol Relat Areas Mol Biol 50:117-175

Harborne JB 1982 Introduction to ecological biochemistry, 2nd edn. Academic Press, London

Harborne JB, Ingham JL, King L, Payne M 1976 The isopentenyl isoflavone luteone as a pre-infectional anti-fungal agent in the genus *Lupinus*. Phytochemistry (Oxf) 15:1485-1487

Ingham JL 1973 Disease resistance in plants: the concept of pre-infectional and post-infectional resistance. Phytopathol Z 78:314-335

Kuo Y-H, Lambein F, Ikegami F, Van Parijs R 1982 Isoxazolin-5-ones and amino acids in root exudates of pea and sweet pea seedlings. Plant Physiol (Bethesda) 70:1283-1289

Levin DA 1976 The chemical defenses of plants to pathogens and herbivores. Annu Rev Ecol Syst 7:121-159

Margalith P 1967 Inhibitory effect of gossypol on micro-organisms. Appl Microbiol 15:952-953

Massey AB 1925 Antagonism of the walnuts (*Juglans nigra* L. *J. Cinerea* L.) in certain plant associations. Phytopathology 15:773-784

Millar RL, Higgins VJ 1970 Association of cyanide with infection of bird's foot trefoil by *Stemphylium loti*. Phytopathology 60:104-110

Müller CH 1966 The role of chemical inhibition (allelopathy) in vegetational composition. Bull Torrey Bot Club 93:332-351

Müller CH, Chou C 1972 Phytotoxins: an ecological phase of phytochemistry. In: Harborne JB (ed) Phytochemical ecology. Academic Press, London, p 201-216

Nienstaedt H 1953 Tannin as a factor in the resistance of chestnut, *Castanea* spp to the chestnut blight fungus *Endothia parasitica* (Mun) A. and A. Phytopathology 43:32-38

Peterson PJ, Fowden L 1965 Purification, properties and comparative specificities of the enzyme prolyl-transfer ribonucleic acid synthetase from *Phaseolus aureus* and *Polygonatum multiflorum*. Biochem J 97:112-124

Rovira A 1969 Plant root exudates. Bot Rev 35:35-57

Sinden SL, Goth RW, O'Brien MJ 1973 Effect of potato alkaloids on the growth of *Alternaria solani* and their possible role as resistance factors in potatoes. Phytopathology 63:303-307

Swain T 1977 Secondary compounds as protective agents. Annu Rev Plant Physiol 28:479-501

Valenta JR, Sisler HD 1962 Evidence for a chemical basis of resistance of lima bean plants to downy mildew. Phytopathology 52:1020-1037

Walker JC, Stahmann MA 1955 Chemical nature of disease resistance in plants. Annu Rev Plant Physiol 6:351-366

Wilson MF, Bell EA 1977 Amino acids and β-aminopropionitrile as inhibitors of seed germination and growth. Phytochemistry (Oxf) 17:403-406

Wilson MF, Bell EA 1978 The determination of the changes in the free amino acid content of eluate from the germinating seeds of *Glycine wightii* (L.) and its effect on the growth of lettuce fruits. J Exp Bot 29:1243-1247

Wilson MF, Bell EA 1979 Amino acids and related compounds as inhibitors of lettuce growth. Phytochemistry (Oxf) 18:1883-1884

DISCUSSION

Clarke: To what extent can the susceptibility of plants be related to their ecological situation? Are races or species of plants that grow next to *Salvia* more susceptible to *Salvia* than those that occur somewhere else?

Bell: We have not yet done this with *Salvia* species. We did try it with certain *Lathyrus* spp. that can grow together and which are closely related morphologically but have chemical differences.

Clarke: One could predict either outcome (i.e. greater or less susceptibility); it depends on who is winning the evolutionary race!

Bowers: Pathogens can elicit a response in attacked plants, causing them to produce protectants such as phytoalexins (see Kuć, this volume). Does this occur also in interactions between plants? In other words does a plant neither defend itself nor secrete a protectant until it receives an elicitor from a competitive plant?

Bell: No; I don't think there is any evidence of this.

Foy: In adaptation to the unavailability or to a low level of iron, some plants produce a reducing compound that has been identified as caffeic acid (or a relative of it) which is produced only under iron-deficiency stress. This compound reduces Fe^{3+} to Fe^{2+}, which the plant can then use. Some iron-efficient plants also lower the pH of their root zones, and thereby increase the solubility and availability of iron.

Bell: One also sees an increase in proline under salt stress in some species, but that is a primary compound rather than a secondary one.

Georgopoulos: Some higher plants parasitize other higher plants, and certain substances are presumably involved in this. Many of these higher plant parasites have a wide host range. Do you know what toxic materials may be involved here?

Bell: No. We haven't looked at those ourselves, so I chose not to include them. The chemistry can be quite different in the parasite and in the host. In *Striga* spp., for instance, there are higher concentrations of carbohydrates.

Kuć: We should be cautious in interpreting the role in disease resistance of plant compounds that may be fungistatic, fungicidal, bacteriostatic or bacteriocidal. The tobacco plant, for example, is notorious for containing alkaloids, phenolics, coumarins and isocoumarins that have a broad spectrum of activity against many organisms. Yet the tobacco plant is susceptible to many diseases. We would certainly not eat tobacco plants but we encourage our children to eat lettuce. Yet in the overall scheme of things lettuce is every bit as resistant to most organisms in its environment as is tobacco. I shall discuss mechanisms for disease resistance further in my paper (p 100-118).

Bell: You are quite right, and one must be cautious. In my paper I have referred to *Stemphylium loti* which detoxicates hydrogen cyanide enzymically. The tobacco horn worm does the same with the alkaloids. During evolution a plant might have, by chance, synthesized a particular compound which may, for a length of time, give it protection against a whole range of potential predators until one or more of these in turn becomes adapted in a manner that enables it to circumvent the plant's defence. In a given environment a compound may protect the plant against 99% of potential predators but it can never protect against all of them indefinitely because of the changes that can take place in the predators. In looking for an ecological argument, one must look at the reactions and the biochemistry of those organisms that have successfully overcome what one supposes to be a chemical defence.

Bailey: Professor Bell quoted that most fungi are toxic, but this is misleading. The reason that many microorganisms are pathogenic is because they are not toxic. Many microorganisms colonize plants by getting into cells that remain alive and establish biotrophic relationships. This point pertains also to some earlier points about plant–plant interactions. Parasitic plants also operate without killing cells and set up harmonious relationships. This is an important adaptive process.

Southwood: The best way to test the effect of plant compounds on herbivorous insects is to do a bioassay on a wide range of polyphagous insects. Lettuce is, on the whole, very palatable to an enormous range of insects, for these purposes. In my view lettuce represents that class of plants that are ruderals, which escape in space and time rather than staying in a particular place. They do not have many toxins, but rabbits are susceptible to various effects if they eat large quantities of lettuce.

Bowers: The tobacco horn worm successfully exploits the tobacco plant, but at some cost. If the horn worm is grown on a tobacco plant, it requires 21–25 days under normal temperatures to complete its development. However, if the worm is placed on a defined diet containing no alkaloids such as nicotine, it will complete its development in 12–13 days. Nicotine, anabasine, nornicotine and many alkaloids are produced in the roots of the tobacco plant. If the tobacco plant is grafted onto a potato it does not accumulate nicotine and it can be exploited by many insects. We have something like 12 000 species of plant in storage that we extract and assay for insecticidal, insectistatic and insect growth-regulatory compounds. We also search amongst them for compounds that have fungicidal activities. We find such compounds with considerable frequency, and some of them have a rather broad spectrum of activity. We once found, in a wild horseradish plant, a potent fungicidal component which we isolated. On purification we obtained some beautiful yellow crystals that were highly fungicidal. Infrared and nuclear magnetic resonance spectra were unrevealing, but the mass spectrometer told us that we had isolated elemental sulphur from this plant—one of the oldest known fungicides! We found that wound damage to the plant caused secretion of sulphur into the wound, and this seems to be a striking response to injury (W.S. Bowers & P.H. Evans, unpublished results). Could you provide any alternative explanation for this compound, Professor Kuć, in the light of what you have just been saying?

Kuć: It would be very difficult to demonstrate that the secretion has a role in disease resistance to pathogens. The sulphur could be part of a general defence mechanism or wound response which protects the plant against some microorganisms. The role of the sulphur in disease resistance to pathogens would be strengthened if a plant that did not produce sulphur was found and if it was susceptible to diseases to which the sulphur-producing plant was resistant.

Bell: It has been found that lupins that contain abnormally low levels of alkaloids are more susceptible to insect attack than lupins containing normal levels. Would you accept this as evidence of one possible role for alkaloids, Professor Kuć?

Kuć: Yes; perhaps I should confine my statements to fungi, bacteria and viruses. However, in breeding for disease resistance, there have been notable failures in looking for increased preformed phenolics, alkaloids and other antibiotic compounds and in relating them to disease resistance in the field. I grant you that even in animal systems the passive, preformed and non-specific mechanisms for resistance probably account for better than 95% of the resistance to most infectious agents in the environment. The very specific response mechanisms are probably used as a last resort, to make a difference only when organisms have breached all the other defences.

Harborne: In trying to draw a parallel between lettuce and tobacco we are not really comparing like with like. Lettuce has been deliberately selected to breed out all the toxins (e.g. the bitter-tasting sesquiterpene lactones) so that we can eat it, whereas in tobacco we have deliberately retained the alkaloids (e.g. nicotine) because people enjoy them as stimulants or sedatives. Perhaps one should compare tobacco with chicory—and how many insects will feed on chicory?

Kuć: If we compare lettuce with tobacco in terms of their pathogens, and the diseases to which they are each susceptible or resistant, we find that each is resistant to the great majority of microorganisms in their environment.

Bell: This, perhaps, means that these plants contain substances that cannot be metabolized by the microorganisms. The lack of nutrients for the microorganisms may be as effective as the presence of toxins in protecting the plant.

Kuć: Of the many different plants that I have worked with and analysed, there is not one in which I have failed to find a substance that is inhibitory to microorganisms. Often the compounds are active at very low concentrations.

Sawicki: There has been a large programme of breeding for plant resistance. Is anything known about the biochemical reasons for resistance to fungi, for example, in wheat, rice or any other crop? Have any chemicals been isolated from those resistant varieties?

Bell: Yes; there is evidence, referred to in my paper, that the resistance of some varieties of lima bean to downy mildew is related to the levels of phenolics in the beans.

Bowers: In the United States, cabbages have been successfully bred for resistance to the cabbage looper. Such resistant varieties have high levels of thiocyanate and so they are, of course, unsuitable for eating and hence resistant to us, as well!

Sawicki: But with crops this is quite different because there is first a resistance to the pathogen, and then the pathogen is able to break down this resistance again, through mutations. Is anything known about the biochemistry of this?

Clarke: Resistance to *Helminthosporium sacchari* in sugar cane has been reported to be associated with a change in a membrane protein, and it is believed that resistant forms differ from susceptibles in four amino acids (Strobel 1973).

Kuć: I will be dealing with some of these problems in my paper later (p 100-118).

REFERENCE

Strobel GA 1973 Biochemical basis of the resistance of sugarcane to eyespot disease. Proc Natl Acad Sci USA 70:1693-1696

Adaptation of plants to atmospheric pollutants

THOMAS C. HUTCHINSON

Department of Botany, University of Toronto, Toronto, Ontario M5S 1A1, Canada

Abstract. Man-made air pollutants are a recent phenomenon in the evolutionary experience of plants and animals although natural air pollutants from volcanic eruptions, forest fires and dust storms have accompanied evolution for geological eras. Plants have responded to increasing concentrations of such pollutants as sulphur dioxide, fluorides, photochemical oxidants and acid rain at the community, species, population and individual levels. The lichens and bryophytes have shown particularly dramatic changes in urban and industrial areas. Many species have had their distribution severely limited. Tolerances to sulphur dioxide have evolved in populations of a number of grasses and herbs, and some sulphur dioxide-tolerant lichens have invaded inner city areas. Sensitivity to pollutants is partly a function of substrate chemistry. Synergistic interactions occur between various pollutants and also between pollutants and pathogens. A good deal of genetic variation occurs within crops, and this allows for selection of pollution-tolerant varieties. The nature of specific adaptations is not generally well known although, for sulphur dioxide, recent studies in poplar and spinach strongly suggest that increased production of the enzyme superoxide dismutase may be a key factor. In other adaptations, morphological and anatomical features play a part.

1984 Origins and development of adaptation. Pitman Books, London (Ciba Foundation symposium 102), p 52-72

Plants have been exposed to air pollutants in urban environments for many centuries, both as a result of coal and wood burning and from local industry. Since the industrial revolution, the intensity of pollution of towns and surrounding areas has greatly increased, as has the area affected. The choking smogs have continued until quite recently in Western Europe and North America, and are still increasing in other parts of the world. Reducing smogs were a regular feature in such British cities as Manchester, Birmingham, Sheffield, Leeds, London and Glasgow, and the consequences for human health were severe. During such smog episodes, the acidic atmosphere contained high concentrations of soot-laden particulate matter, together with sulphur dioxide, sulphuric and hydrochloric acid aerosol, and other gases.

These smogs developed when still air was trapped in an inversion, into which smoke from domestic coal burning and from heavy industry was discharged. Very few plant species were able to survive such phytotoxic smogs. Even in closed greenhouses, the polluted air penetrated and caused lesions on the leaves of plants such as lettuce and coleus. It was, however, generally recognized that species differed sharply in their sensitivity to such pollution episodes and to city rain. It was really because air pollution acted as a potent selective force that aspidistras were almost the sole plant grown indoors in many inner city areas, and also that privet hedges were the norm for industrial cities in the north of England. The amount of soot and dirt that such plants could accumulate on their leaves, and yet survive, was remarkable. The extensive planting of plane trees, a New World species, in London was also largely because of their tolerance to air pollution.

Although air pollution has been significant in cities for several hundred years, it can nevertheless be viewed as a very recent phenomenon in the evolutionary experience of plants and animals. Parallel experiences from natural sources of pollution are infinitely more ancient. These include particulate emissions, and sulphur dioxide, hydrogen sulphide and hydrochloric acid, from volcanic eruptions and from thermal sources such as hot springs. Forest fires produce very large amounts of particulates, which can reduce the sunlight and be deposited hundreds of miles away, while dust storms can have similar effects. Around the ancient hot springs at Yellowstone, in the United States, clumps of monkey flower (*Mimulus guttatus*) and tufted hair grass (*Deschampsia cespitosa*) grow, both species being well described in polluted sites elsewhere in North America. These species do not distinguish between natural and anthropogenic air pollution.

Clearly, air pollutants interact with plants, often harmfully, and selective and adaptational responses can be expected. Some species or populations will be selected against, whilst either others may be pre-adapted to such conditions or tolerant populations may evolve, or tolerant species may invade. A number of cases have been described in which air pollutants have altered the composition and stability of plant communities. This is true of: (a) forests around major smelters, where emissions of sulphur dioxide and heavy metals occur, such as the nickel-smelting site in Canada at Sudbury; (b) lichen and bryophyte floras in and near urban conurbations; and (c) plant communities near limestone quarries or cement works.

Sinclair (1969) refers to polluted air as 'a new highly potent selective force acting on forest ecosystems'. It should not be assumed that the plant response is static. Bennett et al (1974) suggested that 'it is unreasonable to assume that plants cannot adapt to air pollution stress, since they adapt to other environmental stresses'. The marked differences in response between individuals of the same population have prompted foresters to establish clones of pollution-resistant trees, for example of white pine (*Pinus strobus*), tolerant to SO_2, and of ponderosa pine (*Pinus ponderosa*), tolerant to oxidant pollutants. It is the responses of communities and populations, however, that I shall consider in this article.

Responses of communities and taxa to air pollutants

In extreme cases, air pollution can entirely eliminate a flora, and the accumulation of persistent toxic material can perpetuate this condition. This has happened in the inner zone around the Sudbury smelters in Canada, where the original forests have been devastated by acidic, sulphur dioxide emissions over a 70-year period as well as by the accumulation of toxic heavy metals, such as nickel and copper, in the soils and litter (Hutchinson & Whitby 1974, Freedman & Hutchinson 1980). Yet the SO_2 has acted selectively, with a decrease in the number of species growing, as the smelter-source is approached. Some 'weedy' species, such as the herb *Polygonum cilinode* and the moss *Pohlia nutans*, have shown strong tolerances to the strongly acidic conditions—both species occur in highly polluted areas—while many other species, including most other lichens and mosses, recover in significant numbers only at distances in excess of 50 km from the smelter. The white pine (*Pinus strobus*) is highly sensitive to elevated SO_2 levels, and has been eliminated within 15 km of the smelters. Similar but smaller devastated areas occur around many other smelter locations, and the denudation often persists long after the pollution levels are reduced or the smelter closes e.g. in Ducktown, Tennessee and in Butte, Montana.

Limestone dust accumulations, although chemically quite different from smelter emissions, have general effects in common with them. The effect of quarry dust exposure, since 1967, on a southern forest in Virginia, USA was described by Brandt & Rhoades in 1972. Large amounts of dust have accumulated nearby, and foliage is constantly coated with it. At the particular site studied, dust deposition was 426 metric tons km^{-2} $month^{-1}$, and suspended matter was 824 $\mu g\,m^{-3}$ 24 h compared with values of 21 and 75, respectively, for a rural control. Even though large-scale pollution had occurred for only five years, profound changes were already evident. Trees, shrubs and saplings were dying in the dusty area. The control forest had 11 305 stems per hectare, while the polluted site was down to 7595 stems/ha. Selective changes were taking place, and some examples of these are shown in Table 1. The dominant trees in the control site were two species of oak (*Quercus prinus* and *Quercus rubra*) and one of red maple (*Acer rubrum*). When the control and quarry sites were compared, however, the number of seedlings of sugar maple (*Acer saccharum*), hickory (*Carya* spp), tulip tree

TABLE 1 Relative responses of saplings and shrubs to dust deposition from a large limestone quarry in Virginia, USA

Species	*Control site*[a] *(Number/hectare)*	*Dusty site*[b] *(Number/hectare)*
Increased		
Acer saccharum (sugar maple)	10	845
Carya spp. (hickory)	140	235
Liriodendron tulipifera (tulip tree)	25	125
Ostrya virginiana (ironwood)	320	1075
Cercis canadensis (redroot)	75	915
Decreased		
Acer pennsylvanicum (striped maple)	660	30
Acer rubrum (red maple)	740	160
Betula alleghaniensis (birch)	25	0
Hamamelis virginiana (wychhazel)	1315	195
Oxydendrum arboreum	335	90
Pinus strobus (white pine)	45	0
Quercus prinus (chestnut oak)	735	30
Quercus rubra (red oak)	520	80
Rhododendron maxima (rhododendron)	470	25
Sassafras albidum (sassafras)	855	175
Tsuga canadensis (hemlock)	220	0
Viburnum acerifolium (maple-leafed viburnum)	345	0

Quarry operated from 1947 with major expansion in production in 1967. Data collected 1972 (modified from Brandt & Rhoades 1972). [a,b] Density at the sites, counting only those seedlings and shrubs less than 5 cm diameter.

(*Liriodendron tulipifera*), red-root (*Cercis canadensis*) and iron-wood (*Ostrya virginiana*) had greatly increased in the quarry site, while the numbers of dominant species at this dusty site had declined, as had many other tree and ground-cover species, including the three coniferous species in the forest (*Pinus strobus*, *Tsuga canadensis* and *Pinus longifolia*).

The reasons for these changes may relate to selective responses to dust accumulation on leaves, to the consequent reduction in light penetration, and to changes in the surface chemistry of the leaves. They may also be due to the effects on the soil, which has become more alkaline, and to possible effects on pollination and fertilization, as a result of chemical changes affecting pH at the time of pollen deposition onto the stigma. R. M. Cox (personal communication) has shown that many boreal forest conifers require acidic conditions at the stigma surface.

Effects of air pollution on lichens and mosses have been known for at least 100 years and have been studied extensively in the past 20 or 30 years. Both groups have acquired the reputation of being especially sensitive to a wide range of air pollutants and being useful as pollutant indicators. Since they lack a cuticle and are directly exposed to gases, particulates and rain, they act as ion exchangers and are very sensitive to atmospheric chemistry. Many of their nutrient requirements are obtained from the atmosphere, and their ion-exchange properties have been used as biological monitors of the environment. Lichens are especially useful, growing throughout the winter and in many habitats. Since many species exist—e.g. 18 000 lichen species belonging to 500 genera, of which 1368 species are known in Britain alone (Hawksworth & Rose 1976)—and since these show a wide range of sensitivities to different air pollutants, the patterns of distribution of lichen and bryophyte communities reflect rather accurately the local pollutant loads. Around many cities the number of species is very low. The term 'lichen deserts' has been applied. Barkman (1969) reviewed some historical trends, and suggested that such 'deserts' were due to high SO_2 levels, acid rains and urban dryness. Amongst epiphytes in Amsterdam, from 1900 to 1968, 55 species became extinct. The Dutch flora in the last 100 years has lost 3.8% of its flowering plants, 15% of its terrestrial bryophytes, 13% of its epiphytic bryophytes and 27% of its epiphytic lichens. Even within a genus, species differ in their responses, however. The lichen *Caloplaca cerina* disappeared from 14 locations where it had been common, while *Caloplaca phlogina* showed no such effect.

As early as 1891, lichens growing on wood cores taken from the trunks of trees were transplanted experimentally into or out of city areas in Munich. F. Arnold observed which species died, and how rapidly. Similar experiments were done in New York by Brodo (1961), and at Sudbury by Leblanc & Rao (1966), who noted that the death of species moved to polluted areas was related directly to SO_2 levels, and that breakdown of chlorophyll *a* to phaeophytin accompanied their demise.

Gilbert (1969) studied the lichen and moss flora in the Newcastle-upon-Tyne conurbation in north-east England. Aside from the community responses described above, he also noted that some species penetrated the city centres more deeply when they could grow on alkaline substrata, such as asbestos roofs, limestone walls and on cement and mortar. Skye (1968), in a similar Swedish study, noted that lichen species with a high capacity to buffer strong acids penetrated more deeply into Stockholm. Many lichens occur on tree trunks, and often they are specific to one or a few tree species. Coker (1967) noted that the buffer capacity of tree bark and its ion-exchange capacity are lowered by exposure to SO_2; and such bark becomes less satisfactory as a lichen substrate. In Sweden, Barkman (1969) noted in the 1960s that in Vlaardingen the rain varied in pH from 2.4 to 4.4. In Leiden, the pH of tree bark of elderberry was lowered from 7.0 to 5.2 and, for black oak, from 4.0 to 2.9. In Stockholm, Skye (1968) found pH as much as 3 units lower than normal in the bark of elm and ash, and as low as pH 2.4 in oak. Such major substrate changes must have influenced the bark epiphytes.

Not all species have declined on exposure to pollutants. Hawksworth & Rose (1976) point out that the increase in SO_2-tolerant species growing on rocks has been spectacular but, in addition, some of the species confined to trees have also increased: '*Lecanora conizaeoides*, a species with an efficient SO_2 avoidance mechanism (and water repellant qualities) was not collected anywhere in the world prior to the middle of the last century. By 1900 it was becoming widespread and it is now extremely common on trees and wood in most of Britain and oceanic Europe, where mean winter SO_2 levels are in the range 55–150 $\mu g/m^3$; at SO_2 levels above this it becomes rare, and at levels below 30 $\mu g/m^3$ it is very rare.' Other species are simply very tolerant of pollution. The lichen *Lecanora dispersa* was one of the few survivors in inner London, and also in New York.

Ferguson & Lee (1980) have suggested that the *Sphagnum* mosses, which formerly covered much of the wet, poorly drained hills of the Pennines in northern England, and which comprise much of the blanket peat in this region, have been largely eliminated by acidic pollutants from the major industrial centres nearby. This loss of *Sphagnum* species on such a wide scale is not repeated elsewhere in Britain, and seems to have occurred within the past 200 years. Experimental additions of bisulphite and sulphate to bogs in Wales showed that *Sphagnum* was sensitive at deposition rates that have occurred in the Pennines.

It is not only the mosses and lichens that are affected in Britain. Much recent work suggests that the growth of most species in areas near cities is suppressed to some extent by the air pollution loads experienced. Experiments on a variety of grasses, collected from both rural and urban environments and then exposed in environmental growth chambers to carbon-filtered air, showed that they grew much better in filtered than in ambient air (Bleasdale 1973, Crittenden & Read 1979).

In North America, the natural selection exercised by air pollutants in the western part of the USA is in the form of photochemical oxidants from the Los Angeles Basin. Their effects on the forest ecosystem of the San Bernardino Mountains have been studied intensively for many years. Ponderosa pine, one of the five major species of that mixed conifer forest, has been severely affected and mortality has been extensive, amounting to 8–10% during 1968–1972. Interestingly, oxidant-weakened trees usually succumb to bark-beetle infestations. In the Blue Ridge Mountains of Virginia there appears to be a shift from oxidant-sensitive species, such as white pine (*Pinus strobus*), to more tolerant species. Skelly (1980) noted that, of 315 white pine trees tagged in 1977, 3%, 80% and 17% were considered to be sensitive, intermediate or tolerant, respectively, to oxidants. Roots of dying trees were found to be infected with the destructive fungus *Verticicladiella procera.* During the period 1955–1978, radial growth for sensitive trees was significantly lower than for tolerant trees, and there was a general decline in growth for all classes.

Species and population responses

Species differences to SO_2 and to aerial oxidants have already been noted. The variability in response available for selection is often quite large. Ryder (1973) quoted a large number of crops in which tolerant varieties had been identified. Many crops have not yet been examined in detail nor have they been exposed systematically to a wide range of pollutants. Studies on SO_2 and photochemical oxidant effects have tended to predominate. Nevertheless, the summary on Table 2 (derived from Ryder's article) indicates that, should the

TABLE 2 Crops and trees in which varieties tolerant to air pollution have been described (after Ryder 1973)

Tolerance to: *Ozone*	*Fluoride*	*Sulphur dioxide*	*Oxidant*
Alfalfa	Citrus	Turfgrass	Spinach
Cucumber	Grain sorghum	Petunia	Sweet corn
Green beans	Gladiolus	Douglas fir	White bean
Lettuce	Norway spruce	Eastern white pine	Petunia
Oats	Ponderosa pine	Larch	Ponderosa pine
Onion	Scots pine	Lodgepole pine	
Potato		Norway spruce	
Radish		Scots pine	
Red Clover			
Tobacco			
Tomato			
Turfgrass			
Coleus			
Petunia			
Eastern white pine			

need arise to breed additional pollutant tolerance into crops, the genetic-based variability will not be found wanting. It can also be seen that genetically based tolerances occur to a range of pollutants, i.e. ozone, fluorides, sulphur dioxide and oxidants.

This genetic variability has been utilized already in some breeding programmes. Cultivars of shade tobacco have been developed, in Connecticut and Ontario, which are tolerant to so-called weather fleck, which is caused by ozone exposure and which, in sensitive varieties, causes silver streaking on the upper surface of the leaves, reducing their marketable value. Aycock (1972) showed that ozone tolerance in tobacco is controlled by several genes, each with additive effects.

Relatively little work has been done so far on the evolutionary responses of natural populations to air pollutants. The exceptions are the series of current studies on populations of perennial ryegrass (*Lolium perenne*) growing in polluted and unpolluted habitats in Britain and New York.

In early work, Dunn (1959) studied colonies of the leguminous herb *Pupinus bicolor*, collected from the Los Angeles area of California, as well as collections of the perennial and annual lupine species. He found that populations taken from unpolluted coastal areas of California were more sensitive to photochemical smog than were populations from the Los Angeles area. He also found that at least two gaseous compounds present in smog affect survival and reproduction in *Lupinus*; one was water-soluble while the other was virtually insoluble in water.

Briggs (1972) found that, when grown from spores, the liverwort *Marchantia polymorpha*, collected up to 13 km from the centre of Glasgow, UK, was significantly more tolerant to lead than was a population of the same species collected in the unpolluted vicinity of Chatsworth in Derbyshire. The lead was an air pollutant in Glasgow and had accumulated in the soil and plants *in situ.* Studies with the common plantain (*Plantago lanceolata*) have also indicated lead-tolerance in roadside populations in North Carolina compared with populations collected at a distance from roadsides.

Populations of the dock (*Rumex obtusifolia*), originating from areas of different mean annual SO_2 exposures, were compared under conditions of exposure to filtered air or to SO_2 fumigations, at 520 μg/m^3 for 11 days. Adaptations in SO_2 tolerances related to the area of origin of the plants. Differences in responses to SO_2 concentrations, which relate to the exposures experienced by different populations, have also been recorded for a number of grasses, including Italian ryegrass, cocksfoot and meadow grass.

A detailed study of the response of heritability in the annual Carolina cranesbill, *Geranium carolinianum*, has been made by Taylor & Murdy (1975), Taylor (1978) and Murdy & Ragsdale (1980). They selected populations growing close to a small power station in Georgia, USA, which discharges SO_2 from a low smokestack (Bradshaw & McNeilly 1981), and compared them with a population from a rural Georgian habitat. Seed populations were studied. They found (a) that the seed populations from near to the power station were significantly more tolerant to SO_2 than those from the rural controls (Fig. 1); (b) that in both populations a good deal of variability of responses existed; and (c) that by raising seed from parents whose response was known and then testing them in SO_2 environments, the

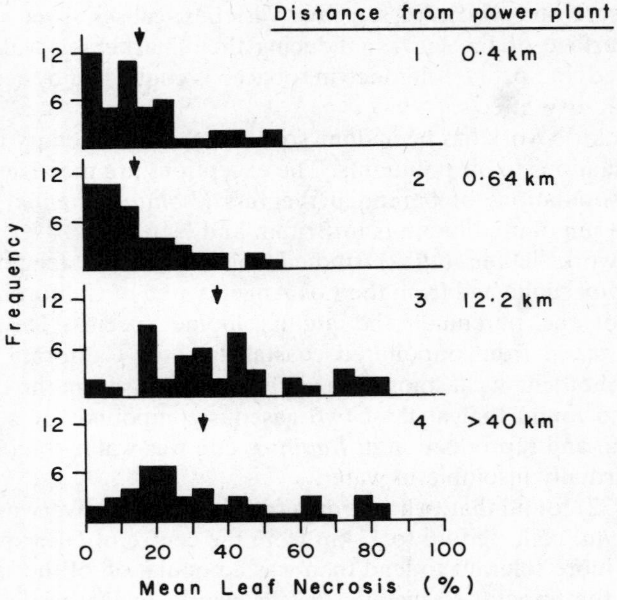

FIG. 1. The mean leaf-necrosis response of 1972 populations of *Geranium carolinianum* from sites 1, 2, 3 and 4 to sulphur dioxide exposure. Solid black arrow represents mean of respective population. Frequency is defined as the number of individuals. (From Taylor & Murdy 1975.)

heritability of SO_2 tolerance was found to be about 50% (Figs. 2 & 3). This was based on the regression of the mean SO_2-susceptibility of the offspring on the mean of their parents (Fig. 3). The control populations did have a very few tolerant individuals, which are presumably the basis on which selection can act.

The studies on SO_2 tolerance in urban populations of *Lolium perenne* have already been mentioned. Bell & Clough (1973) found that a widely grown commercial perennial ryegrass variety, S23, when exposed to 0.12 p.p.m. SO_2 for 9 weeks reduced its yield by 48%, compared with populations given filtered air, or compared with a population collected from Helmshore (a site a few km north of Manchester, UK) which had been exposed for many years to elevated SO_2 levels. Similar observations have been made on populations from the Liverpool, UK, area and, indeed, on the ryegrass cultivar 'Manhattan', collected from Central Park, New York City.

Interestingly, a consensus seems to be developing that the response of a genotype to acute SO_2 levels may not give any indication of its response to

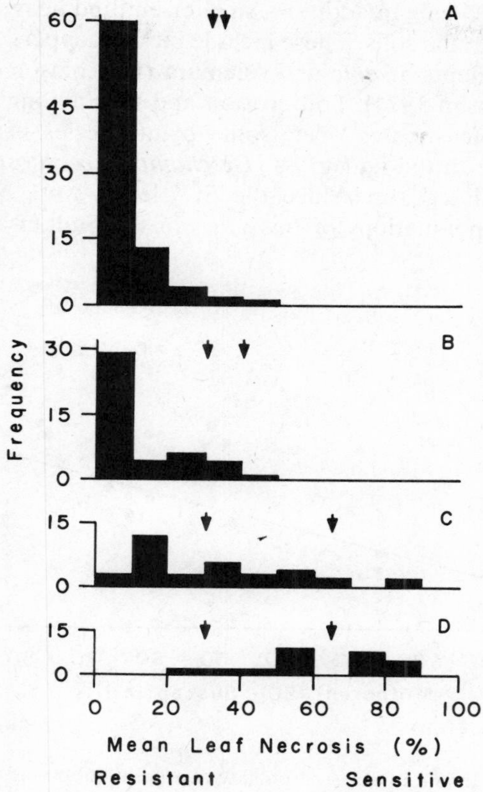

FIG. 2. The distribution of mean leaf-necrosis in progeny resulting from paired matings between a resistant plant and individuals of increasing susceptibility to sulphur dioxide. Solid black arrow indicates SO_2 susceptibility of parental types. Frequency is defined as the number of individuals. (From Taylor 1978.)

low levels of the pollutant. This appears to be true for several coniferous tree species, as well as for several grasses.

In the Sudbury area of Ontario, Canada, mentioned already, three major nickel smelters have operated for 70 years. Enormous quantities of SO_2 have been emitted, reaching 3 million short tons per year in the 1960s and early 1970s. The vegetation has been eliminated by the phytotoxic gases over hundreds of square kilometres. Regeneration of the forests has been further handicapped by the severe soil acidification, which has been caused by SO_2 and acidic aerosols. This acidification also releases toxic quantities of aluminium from the soil. In addition, smelter-emitted heavy metals accumulate to toxic levels in the soils. These include nickel, copper, cobalt and silver, as well as the elements arsenic and selenium (Hutchinson & Whitby 1974, Whitby & Hutchinson 1974). Soil erosion and fires amongst the dead trees add to the multiple stresses. Yet, some plant species have survived and others, such as the tufted hair grass (*Deschampsia cespitosa*), have invaded the innermost polluted zone since the SO_2 levels were reduced in 1972. Interestingly, the populations of this grass in the Sudbury area now cover

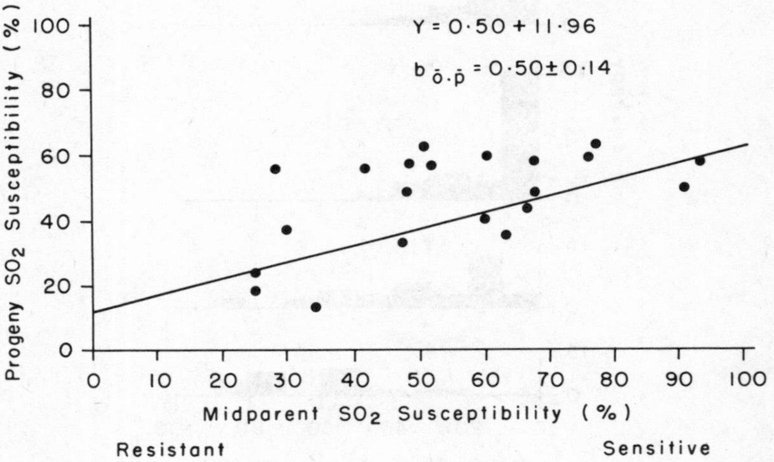

FIG. 3. Regression of the mean SO_2-susceptibility of progeny ($\bar{o}$) on the mean of their parents ($\bar{p}$), in *Geranium carolinianum.* The regression equation is shown, and the regression coefficient = $b_{\bar{o}.\bar{p}}$ ± standard error. Midparent SO_2 susceptibility is the average of the values for both parents. (From Taylor 1979.)

hundreds of hectares, although it was previously unrecorded there (Cox & Hutchinson 1981). These populations are highly tolerant to normally toxic levels of copper, nickel, cobalt and silver, compared with non-Sudbury Canadian and European populations of the same species. They are also tolerant to elevated concentrations of aluminium in solution. What could not be predicted from the analyses of the soils in which *D. cespitosa* grows at Sudbury is that the populations are also tolerant to lead, zinc and cadmium, which are not elevated in the soils (Cox & Hutchinson 1980). Coincidental metal tolerances thus also occur with the multiple-metal tolerances, confounding the idea of metal-tolerance specificity and suggesting some common mechanisms. We have found this also for other grass species and for two algal species (*Euglena mutabilis* and *Chlorella saccharophila*) from highly polluted habitats in the Canadian arctic (Havas & Hutchinson 1983). In addition, the multi-metal-tolerant Sudbury populations of *D. cespitosa* are also SO_2-tolerant, and tolerant of strong soil acidity. The evolutionary 'burst' of colonization is a remarkable one and suggests very rapid adaptation to a complex polluted environment. This rapid spread is undoubtedly helped by a striking lack of competitors in the inner smelter zone.

Pollutant interactions

Very rarely in the field do pollutants occur singly. Generally, mixtures of pollutants are the rule. Studies on gas mixtures have been rare but a number of interesting responses of vegetation have been described. SO_2 and nitrogen dioxide (NO_2) act synergistically on some crop plants, including corn, for four broad-leaved trees and several grasses, and for the garden pea (Horsman & Wellburn 1977, Ashenden & Mansfield 1978, Whitmore & Freer-Smith 1982). Ozone (O_3) and SO_2 have also been shown to be synergistic in some circumstances and for certain species. G. Hofstra (personal communication) has shown that O_3 sensitivity in bean and cucumber is affected by a pre-exposure to SO_2, as well as by the duration of that exposure. Pre-exposure sensitized the plants to greater O_3 injury.

SO_2 and acid-rain interactions are being studied actively, and synergistic effects have been reported for soybean (Irving & Miller 1981). Such effects can be of considerable significance in view of the spread of acid precipitation.

Among more exotic interactions, synergistic responses have been described from SO_2 and cadmium in garden pea in German studies (Grünhage & Jäger 1982), and in cress in Canadian work (Czuba & Ormrod 1981). In cress, injury was observed at lower cadmium and O_3 concentrations when the plant was exposed to them together rather than separately. In bean and tobacco, cadmium treatment increased the number of leaf lesions of tobacco mosaic virus, and this occurred at levels of cadmium which did not produce visible symptoms of phytotoxicity (Harkov & Brennan 1981). In another virus–pollutant interaction, SO_2 at 260 or 520 $\mu g/m^3$ increased infection in bean by southern bean mosaic and maize dwarf mosaic viruses (Laurence et al 1981).

Shriner (1977) described fungal infections that interacted with sulphuric acid-rain treatments and host–parasite relationships. The responses varied widely, from enhanced to reduced infections. Plant responses and adaptations will be governed by the genetic variability available to respond not only to the effects of single pollutants but also to the effects of the mixtures of pollutants, and to the induced interactions that involve microbes on the leaves and in the soil.

Effects on reproduction

The effects of pollutants on pollen germination and pollen-tube growth are of interest because they affect the ability of individuals and populations to reproduce themselves. Such responses to pollutants have generally been studied *in vitro*, and this has been challenged in relation to the validity of extrapolations to conditions *in vivo*. Nevertheless, decreased pollen germination, delayed germination and reduction in pollen-tube growth have been described for a number of species and for a variety of air pollutants. These latter include SO_2, O_3, fluorides and acid rain (Karnosky & Stairs 1974, Varshney & Varshney 1981, Dubay & Murdy 1983, Feder 1968, Sulzbach & Pack 1972). Interestingly, R. M. Cox (personal communication) has shown that for 13 boreal forest species, all pollen germination was significantly decreased by suspension pH values ranging from 5.6 to 2.6, and that broad-leaved trees were more sensitive to acidity than the conifers he examined. Pollen-tube elongation in sugar maple (*Acer saccharum*) was inhibited at pH 4.0, and in paper birch (*Betula papyrifera*) at 3.6, while jack pine (*Pinus banksiano*), white pine (*Pinus strobus*) and hemlock (*Tsuga canadensis*) tolerated pH 3.6, 3.0 and 2.6, respectively, before inhibition occurred. It is tempting to relate this directly to optimal pH conditions on the stigma surface at the time of pollination. It may also be of relevance in interpreting the declines and invasions of conifers and sugar maple in the limestone quarry study by Brandt & Rhoades (1972), described earlier.

Mechanisms of adaptation

A great variety of mechanisms appear to determine a plant's susceptibility or tolerance to air pollutants. Such mechanisms can involve morphological and anatomical features, such as size and angle of subtention of leaves, rooting depth of mature plants, position and number of stomates, whether or not leaves are maintained all year (evergreens), thickness of the leaf cuticle, and growth-form of the plant. Physiological and biochemical features are also often involved. The actual mechanism of damage from specific pollutants is generally rather poorly known. Pre-adaptational features often seem to be of considerable significance. For example, the buffering or neutralizing capacity of leaf surfaces may well play a critical role in determining sensitivity of leaf surfaces to acid-rain damage. The leaves of the highly tolerant herb *Artemisia tilesii*, from the Smoking Hills area of arctic Canada, are able to rapidly neutralize acidic droplets (pH 3.0–3.5) placed on their surface, and can raise the pH of such droplets to 5.6 or 7.5 in 30 minutes (C. Adams & T. C. Hutchinson, unpublished).

At the enzymic level rather little has been done. Population differences in peroxidase and acid phosphatase activities have been associated with differences in tolerance to SO_2 and acid rain. However, perhaps the most interesting report is that of Tanaka & Sughara (1980), who have studied the role of superoxide dismutase (SOD) in defence against SO_2 toxicity in leaves of poplar and spinach. Young poplar leaves are more tolerant to SO_2 than old leaves and were shown to have five times as much SOD activity. Activity was increased by fumigation with 0.1 p.p.m. SO_2. These 'induced' leaves were more tolerant to 2 p.p.m. SO_2 than control leaves. Spinach leaves sprayed with the SOD-inhibitor, diethyldithiocarbamate, lost a good deal of their SOD activity and this corresponded to a decrease in their resistance to the toxic effects of SO_2. It seems likely that SO_2 toxicity is in part due to the superoxide radical and that SOD participates in the defence mechanism against SO_2 toxicity.

In summary, plants have, on the whole, responded adversely to the onslaught of atmospheric pollution over the past 100 years. Many groups and species are particularly sensitive. Nevertheless, there is a great deal of genetic variability available for successful breeding programmes of pollution-resistant crop varieties. In natural populations, air pollution is acting as a selective force, with evolutionary changes already apparent.

REFERENCES

Ashenden TW, Mansfield TA 1978 Extreme pollution sensitivity of grasses when SO_2 and NO_2 are present in the atmosphere together. Nature (Lond) 273:142-144

Aycock MK 1972 Combining ability estimates for weather fleck in *Nicotiana tabacum* L. Crop Sci 12:672-674

Barkman JJ 1969 The influence of air pollution on bryophytes and lichens. In: Air pollution: 1st European conference on effects on plants and animals. Netherlands Institute of Agriculture, Wageningen, p 197-209

Bell JNB, Clough WS 1973 Depression of yield in ryegrass exposed to sulphur dioxide. Nature (Lond) 241:47-48

Bennett JP, Resh HM, Runeckles VC 1974 Apparent stimulation of plant growth by air pollutants. Can J Bot 52:35-41

Bleasdale JKA 1973 Effects of coal-smoke pollution gases on the growth of ryegrass (*Lolium perenne*). Environ Pollut 5:275-285

Bradshaw AD, McNeilly T 1981 Evolution and pollution. Edward Arnold, London

Brandt DJ, Rhoades RW 1972 Effect of limestone dust accumulation on the composition of a forest community. Environ Pollut 3:217-225

Briggs D 1972 Population differentiation in *Marchantia polymorpha* L. in various lead pollution levels. Nature (Lond) 238:166-167

Brodo IM 1961 Transport experiments with corticolous lichens using a new technique. Ecology 42:838-841

Coker PD 1967 The effects of sulphur dioxide pollution on bark epiphytes. Trans Br Bryol Soc 5:341-347

Cox RM, Hutchinson TC 1980 Multiple metal tolerances in the grass *Deschampsia cespitosa* (L) Beauv. from the Sudbury smelting area. New Phytol 84:631-647

Cox RM, Hutchinson TC 1981 Environmental factors influencing the rate of spread of the grass *Deschampsia cespitosa* invading areas around the Sudbury nickel–copper smelters. Water Air Soil Pollut 16:83–106

Crittenden PD, Read DJ 1979 The effects of air pollution on plant growth with special reference to sulphur dioxide. III: Growth studies with *Lolium multiflorum* and *Dactylis glomerata*. New Phytol 83:645-651

Czuba M, Ormrod DP 1981 Cadmium concentrations in cress shoots in relation to cadmium-enhanced ozone phytotoxicity. Environ Pollut Ser A Ecol Biol 25:67–76

DuBay DT, Murdy WH 1983 The impact of sulphur dioxide on plant sexual reproduction: in vivo and in vitro effects compared. J Environ Qual 12:147-149

Dunn DB 1959 Some effects of air pollution on *Lupinus* in the Los Angeles area. Ecology 40:621-625

Feder WA 1968 Reduction in tobacco pollen germination and tube elongation, induced by low levels of ozone. Science (Wash DC) 160:1122-1124

Ferguson P, Lee JA 1980 Some effects of bisulphite and sulphate on the growth of *Sphagnum* species in the field. Environ Pollut Ser A Ecol Biol 21:59-71

Freedman B, Hutchinson TC 1980 Long-term effects of smelter pollutants on forest litter decomposition near a nickel–copper smelter at Sudbury, Ontario. Can J Bot 58:2123-2140

Gilbert OL 1969 The effect of SO_2 on lichens and bryophytes around Newcastle upon Tyne. In: Air pollution: 1st European conference on effects on plants and animals. Netherlands Institute of Agriculture, Wageningen, p 223-235

Grünhage L, Jäger H-J 1982 Kombinationswirkungen von SO_2 und cadmium auf *Pisum sativum* L. 2: Enzyme, freie aminosäuren, organische säuren und zucker. Angew Bot 56:167-178

Harkov R, Brennan E 1981 Cadmium in foliage alters plant response to tobacco mosaic virus. Air Pollut Control Assoc J 31:166-167

Havas M, Hutchinson TC 1983 The Smoking Hills: a natural acidification of an aquatic ecosystem. Nature (Lond) 301:23-27

Hawksworth DL, Rose F 1976 Lichens as pollution monitors. Edward Arnold, London

Horsman DC, Wellburn AR 1977 Effects of SO_2 polluted air upon enzyme-activity in plants originating from areas with different annual mean atmospheric SO_2 concentrations. Environ Pollut 13:33-39

Hutchinson TC, Whitby LM 1974 Heavy metal pollution in the Sudbury mining and smelting region of Canada. I: Soil and vegetation contaminated by nickel, copper and other metals. Environ Conserv 1:123-132

Irving PM, Miller JE 1981 Productivity of field-grown soybeans exposed to acid rain and sulfur-dioxide alone and in combination. J Environ Qual 10:473-478

Karnosky DF, Stairs GR 1974 The effects of SO_2 on *in vitro* forest tree pollen germination and tube elongation. J Environ Qual 3:406-409

Laurence JA, Aluisio AL, Weinstein LH, McCune DC 1981 Effects of sulphur dioxide on southern bean mosaic virus and maize dwarf mosaic. Environ Pollut Ser A Ecol Biol 24:185-191

Leblanc F, Rao DN 1966 Réaction de quelques lichens et mousses épiphytique à l'anhydride sulfreux dans la région du Sudbury, Ontario. Bryologist 69:338-346

Murdy WH, Ragsdale HL 1980 The influence of relative humidity on direct sulphur dioxide damage to plant sexual reproduction. J Environ Qual 9:493-496

Ryder EJ 1973 Selecting and breeding plants for increased resistance to air pollutants. In: Naegle JA (ed) Air pollution damage to vegetation. American Chemical Society, Washington DC, Adv Chem Ser 122:75-84
Shriner DS 1977 Effects of simulated rain acidified with sulphuric acid on host-parasite interactions. Water Air Soil Pollut 8:9–14
Sinclair WA 1969 Polluted air: potent new selective force in forests. J For 67:305-309
Skelly JM 1980 Photochemical oxidant impact on Mediterranean and temperate forest ecosystems. Real and potential effects. In: Miller PR, Dochinger L (eds) Proc Symp effects of air pollution on Mediterranean and temperate forest ecosystems, June 22–27. Riverside, California (USDA Gen Tech Report PSW-43) p 38-50
Skye E 1968 Lichens and air pollution. Acta Phytogeogr Suec 52:1-123
Sulzback CW, Pack MR 1972 Effect of fluoride on pollen germination growth and fruit development in tomato and cucumber. Phytopathology 62:1247-1253
Tanaka K, Sugahara K 1980 Role of superoxide dismutase in defense against SO_2 toxicity and an increase in superoxide dismutase activity with SO_2 fumigation. Plant Cell Physiol 21:601-611
Taylor GE 1978 Genetic analysis of ecotypic differentiation of an annual plant species *Geranium carolinianum* L. in response to sulphur dioxide. Bot Gaz 136:362-368
Taylor GE, Murdy WH 1975 Population differentiation of an annual plant species *Geranium carolinianum* L. in response to sulphur dioxide. Bot Gaz 136:212-215
Varshney SRK, Varshney CK 1981 Effect of sulphur dioxide on pollen germination and pollen tube growth. Environ Pollut Ser A Ecol Biol 24:87-92
Whitby LM, Hutchinson TC 1974 Heavy metal pollution in the Sudbury mining and smelting region of Canada. II: Soil toxicity testing. Environ Conserv 1:191-201
Whitmore ME, Freer-Smith PH 1982 Growth effects of SO_2 and/or NO_2 on woody plants and grasses during spring and summer. Nature (Lond) 300:55-57

DISCUSSION

Clarke: Have the bare rocks at the Sudbury smelters in Canada always been bare or did they result from erosion after the vegetation had gone?

Hutchinson: They were originally vegetated, and covered with soil beneath a pine forest. Erosion took place rapidly after the vegetation had died as a result of smelter fumigations. Interestingly, a lot of the wood that died and was cut 50 or 60 years ago is still lying there intact at Falconbridge. It has not decayed as one would expect. The SO_2, metal emissions and acidity must have had strongly detrimental effects on the decomposing fungal populations.

Foy: How have human populations in this area been affected?

Hutchinson: A medical study was done by Federal researchers about 10 years ago, and it showed that the average life expectancy for males living in Sudbury was 10 years less than the national average. As well as high concentrations of sulphur dioxide and metals in the air, there are elevated nickel levels in the water supply, and various other problems too, as I mentioned, such as arsenic and selenium emissions which could affect health.

Bradshaw: In Sudbury, as in Prescot on Merseyside in the UK, a large number of species have not survived. Quite a number of those losses could be explained by the plants lacking the necessary pre-adaptations or, to put it the other way around, by their having the unnecessary or negative adaptations. Do you agree that a number of species did not persist there probably because they could not do the necessary evolving that plants like *Deschampsia* species, for instance, have been able to do?

Hutchinson: Yes. In studies of pollution effects scientists often like to compare two species within the same genus, and two such species happen to have been found in the Sudbury area. They exemplify your point. One of these is a very common acid-tolerant grass, *Deschampsia flexuosa*, which is found also all over British, European and North American woodland. In Europe this species has often evolved metal-tolerance and is found on mine spoil-heaps, but it has not developed in this way at Sudbury, despite having every opportunity. *Deschampsia cespitosa,* on the other hand, only came into the Sudbury area after 1974. The nearest population to Sudbury is 80 km away, growing on limestone, with no indication of metal tolerance. In Europe it has rarely evolved metal-tolerance. This problem of the 'wrong' species developing substantial multiple metal tolerances and now occupying hundreds of hectares of polluted soils is a great puzzle. If one were making predictions, this species is the last thing one would have predicted to survive at Sudbury, because of its habitat preferences elsewhere in Ontario.

Bradshaw: This is very interesting. There is a lot more that we ought to know about where such variability comes from and how it arises. The most comparable example we can grasp is resistance to warfarin in rats (see also p 17). In that case one is dealing simply with a mutation that does not arise very often. The mutant arose near Shrewsbury in the UK and led to a whole population of warfarin-resistant rats in the Shrewsbury region. This type of effect is easily identified. Each of the two species of *Deschampsia* that you mentioned can obviously mutate to give metal tolerance, but perhaps they do so at a very low frequency. So we could suggest that *D. cespitosa* has mutated somewhere and then started to colonize the area, but that *D. flexuosa* just hasn't done it yet, although, as you say quite rightly, it appears to have done so in Britain.

Hutchinson: I have to assume that at least in the Ontario populations of *D. flexuosa* the genetic potential for tolerance to copper and nickel is not present, whereas *D. cespitosa,* although it does not occupy acid environments in the field in Ontario, happens to have that right combination. If we screen control populations of *D. cespitosa* from limestone areas, we find that about 0.3–0.5% of that control seed population is either copper-tolerant or nickel-tolerant. This proportion would be similar to that for *Agrostis,* which you described in your paper (Bradshaw, this volume). How such plants can develop tolerance to nickel, copper and cobalt all at once, however, is a mystery.

Gressel: Do the populations of *D. cespitosa* at, say, 80 km from the Sudbury site interbreed? And will the population growing on the polluted site also grow on limestone? Are you sure they came from the same area?

Hutchinson: It would be nice to suppose that the Sudbury population of *D. cespitosa* came originally from the Bruce Peninsula area, 80 km away. We intend to look at the isoenzymes of both populations and to categorize plants involved in the Sudbury spread of several hundred hectares. We need to know whether the plants growing at Sudbury all form one subset of the control population. In reply to your second question we can grow the metal-tolerant population very well on limestone soils, and it grows vigorously, but the reverse is not the case. The control population grows poorly on acidic soils, irrespective of metal content.

Davies: Is hydrogen sulphide as well as sulphur dioxide produced under normal conditions of nickel smelting?

Hutchinson: In the final processing, all the sulphur is blown off by oxidizing it in a stream of oxygen. In those conditions not much H_2S would be produced. At Yellowstone, in the USA, *D. cespitosa* can be found growing in environments that do contain both H_2S and SO_2 (personal observations).

Bradshaw: In these relatively 'new' polluted environments of Sudbury and Prescot, one can't be sure that a genotype has not come in from somewhere miles away. There is some evidence for this in the spread of herbicide-resistant weeds. Since you don't know where the *D. cespitosa* came from it will be interesting to see whether you can pin it down by isoenzyme characterization.

Hutchinson: Yes. I have found another metal-tolerant population of *D. cespitosa* from the little town of Cobalt, Ontario which, as you might guess, has a cobalt mine. It is about 150–200 km from Sudbury. That population grows in the surrounding woodlands as well as on mine tailings and it is a much older population, which may have been there for a hundred years. The population at Sudbury may, in fact, be related to this population rather than to the population growing on limestone at the Bruce Peninsula. Seed dispersal could, just conceivably, have taken place via miners' muddy boots.

Gressel: Some of the fungal-produced toxicants such as cercosporin act by oxygen radical formation. Has anyone looked at the plants tolerant to ozone or SO_2 to see if they are more tolerant to some of these fungi?

Kuć: Not that I am aware of.

Gressel: It would be interesting to look into this. Some onion varieties are tolerant to ozone, and this is genetically well defined. In tobacco, too, there is a lot of genotypic variation.

Fowden: I am most interested in physiological and biochemical mechanisms of tolerance. When *multiple* tolerance is found, in both grasses and algae, is this likely to be because the plants are rejecting the metals (i.e. not accumulating them at all) or because the plants are accumulator species that package the metal inside the cells innocuously?

Hutchinson: I don't know but I have some information on this. *Agrostis* and *Deschampsia*, both grass genera, take up metals in quite large quantities. Exclusion is not a factor in their tolerance. W. Rauser, of Guelph University, and I are looking at some of the metal-binding proteins that seem to be produced in the root systems of the *Deschampsia* at Sudbury. His work to date suggests that a copper-binding protein, like metallothionein, is produced in the root systems. This would provide a fairly specific mechanism for copper tolerance. We know that the protein does not handle nickel, and we can separate nickel-tolerant and copper-tolerant plants by screening 'control' seedlings, whereas at Sudbury both tolerances are invariably present together because of soil pollution. The problem is that in preliminary experiments we are finding as much metallothionein-like substance in the controls as in the plants from the Sudbury area. Despite this apparent setback, I feel that some of the groupings of metals that we have come up with by simply taking one genotype, producing lots of clones, and looking at its simultaneous tolerance to many different metals, suggest that there are common mechanisms. Cell walls may suffice, in some plants, for taking certain metals out of circulation and preventing them from getting into the critical cytoplasm. Stokes (1975) at Toronto showed that copper is taken right across the cell, in an alga tolerant of copper, and dumped into the nucleus where it accumulates in large quantities. Some work on aluminium suggests that it also gets into the nucleus and there accumulates with the DNA (Matsumoto & Morimura 1980).

Foy: There is evidence that aluminium tolerance is associated with organic complexes (Foy 1983). Such things as citric acid, malic acid and oxalic acid in plants may act as deactivators. There is also evidence for compartmentalization. For example, in the tea plant one can find as much as 30 000 p.p.m. of aluminium in the old leaves and as little as 50 p.p.m. in the buds. Polyphenols in both the tea plant and the rubber plant can complex aluminium, and so a desolubilization or inactivation process may occur even though the metal remains inside the plant. Generally, aluminium-tolerant plants are accumulators, and in the varieties that I have studied we have not been able to relate aluminium tolerance to aluminium present in the top of the plant. We believe that *exclusion* is influential in the aluminium tolerance of wheat and barley at least.

Hutchinson: Plants that are selenium accumulators are known to produce specific amino acids. My impression is that these results cannot yet be extrapolated to the heavy metals. Chenery (1948) and Haridasan (1982) reported a very large number of aluminium accumulators amongst the flora of the cerrado of Brazil, where the ancient soils are strongly acidic, with high soluble aluminium levels.

Foy: Recent work by Siegel & Haug (1983a,b) indicates that aluminium binds to calmodulin and interferes with calmodulin-stimulated membrane-bound ATPase activity which is involved in maintaining trans-membrane potentials in plasma membrane preparations from barley roots.

Presumably sulphur dioxide has both a direct effect and an acid-rain effect. Is there any way to determine the relative importance of these two effects in different situations?

Hutchinson: The effects have tended to be lumped together as those of acidifying air pollutants. In Britain particularly, the humid atmosphere means that both effects of SO_2 will almost always operate together. In laboratory experiments where humidity is controlled, *Lolium perenne*, for example, has been studied purely for its tolerances to SO_2 (Bell & Clough 1973). Studies at Liverpool, Sheffield and Manchester have shown that charcoal-filtering of the atmospheric air will produce substantial enhancement of growth compared with growth in the polluted ambient air, as I mentioned. So, for many species in these environments, there is a general chronic suppression of growth, even where tolerance exists. There seems to be no relationship between tolerance to acute conditions and tolerance to these chronic conditions. Professor Jack Rutter has been working on that at Imperial College, London and found no such relationships between chronic and acute sensitivity to SO_2 in a whole series of different trees and grasses (see Garsed & Rutter 1982). One must suppose that different mechanisms are operating. Breeders have therefore to decide whether to breed for plants that can handle acute toxicity from smelters or to breed for tolerance to general ambient air-pollutants.

Bradshaw: It is very difficult to distinguish these two effects of SO_2; the tolerances are distinctly separate but the phenomena in nature are not (Ayazloo & Bell 1981). Even low chronic levels of SO_2 (about 200–300 $\mu g/m^3$) can contain within them periods of much higher concentration, so breeders must also bear this in mind.

Fowden: Charles Whittingham and his colleagues (Rothamsted Experimental Station Annual Report 1978) examined plants growing near one of the brick works in Bedfordshire, UK, which causes aerial pollution by SO_2 and fluoride. He tried an experiment on a specially designed chamber with filtered air, compared with a similar chamber with unfiltered air. Because the experiment was not set up down-wind, the pollution seemed to come in packages rather than being continuous. The effect in some years was to cause a marked reduction in barley growth, sometimes leading to 30% less production of dry matter. But the plant seemed to compensate in terms of its grain production, so that there was usually no statistically significant reduction in final grain yield. Only in one year was there a reduction in both total dry weight production and grain yield.

Graham-Bryce: Another truly gaseous pollutant that may not strictly be toxic but has pronounced growth-regulatory effects even after only short periods of exposure is ethylene. Distinct ethylene gradients can be detected from busy roads, giving levels above those that can produce physiological effects (see, for example, Abeles & Heggestad 1973). Is there any evidence of adaptive phenomena (related, perhaps, to premature ripening) resulting from ethylene exposure?

Hutchinson: I have no information about ethylene effects, but other hydrocarbons such as naphthalene have been studied. It is possible to select out tolerant individuals and varieties for this and other hydrocarbons (Treshow 1970).

REFERENCES

Abeles FB, Heggestad HE 1973 Ethylene—urban air pollutant. J Air Pollut Control Assoc 23:517-521

Ayazloo M, Bell JNB 1981 Studies on the tolerance to sulphur dioxide of grass populations in polluted areas. I. Identification of tolerant populations. New Phytol 88:203-222

Bell JNB, Clough WS 1973 Depression in yield in ryegrass exposed to sulphur dioxide. Nature (Lond) 241:47-48

Chenery EM 1948 Aluminium in plants and its relation to plant pigments. Ann Bot (Lond) 12:121-136

Foy CD 1983 Physiological effects of hydrogen, aluminum and manganese toxicities in acid soils. In: Adams F (ed) Soil acidity and liming, 2nd edn. Am Soc Agron, Madison, WI (Agronomy Monograph 12), in press

Garsed SG, Rutter AJ 1982 Relative performance of conifer populations in various tests for sensitivity to SO_2, and the implications for selecting trees for planting in polluted areas. New Phytol 92:349-367

Haridasan M 1982 Aluminium accumulation by some cerrado native species of central Brazil. Plant Soil 65:265-273

Matsumoto H, Morimura S 1980 Repressed template activity of chromatin of pea roots treated with aluminium. Plant Cell Physiol 21:951-959

Rothamsted Experimental Station Annual Report 1978 Part 1, p 43-44

Siegel N, Haug A 1983a Aluminum interactions with calmodulin. Evidence for altered structure and function from optical and enzymatic studies. Biochim Biophys Acta 744:36-45

Siegel N, Haug A 1983b Calmodulin dependent formation of a membrane potential in barley root plasma membrane vesicles: a biochemical model of aluminum toxicity in plants. Physiol Plant, in press

Stokes PM 1975 Adaptation of green algae to high levels of copper and nickel in aquatic environments. In: Hutchinson TC et al (eds) Pathways. Toronto, Ontario (1st Int Conf Heavy Metals in the Environment, Vol 2) p 131-154

Treshow M 1970 Environment and plant response. McGraw Hill, New York

Evolution of herbicide-resistant weeds

J. GRESSEL

Department of Plant Genetics, The Weizmann Institute of Science, Rehovot 76100, Israel

Abstract. The first generation of species-selective phenoxy herbicides went into use at the same time as modern antibiotics, chlorinated hydrocarbon insecticides and new fungicides and rodenticides. No resistance has appeared to the still widely used phenoxy herbicides. However, resistance has developed, in many isolated areas around the world, to *s*-triazine herbicides and, in a few small areas, to bipyridillium herbicides. The parameters of population genetics that govern herbicide resistance and those that govern resistance of microorganisms and fungi to other xenobiotics are the same: generations, selection pressure and fitness. Even though generation times are longer with plants, a sufficient number has passed for resistance to be apparent. The only special factor that controls the development of herbicide resistance in weeds is the spaced germination of seed throughout many seasons. The reason that resistance has not developed to the phenoxy herbicides is probably because of a low effective selection pressure; germination of susceptible seeds occurs late in the season after the herbicide is biodegraded. The highly persistent triazines, and the monthly used, highly ephemeral bipyridillium, paraquat, have exerted much stronger selection pressures. Different modes of tolerance and resistance seem to have evolved in the same species. Crop and herbicide rotation can considerably delay the possibility of resistance development until it is effectively precluded.

1984 Origins and development of adaptation. Pitman Books, London (Ciba Foundation symposium 102), p 73-93

Herbicides are the most widely used, but least discussed, pesticides in agriculture. Their usage provides the most cost-effective method of preventing weed competition with crops, far surpassing mechanical cultivation and without requiring the concomitant soil disturbance that causes erosion. Among all pesticides, herbicides have the lowest toxicity to mammals, thus conferring a large margin of safety on their use.

The chlorinated phenoxy herbicides were introduced in the mid 1940s. Like antibiotics and modern insecticides, their development was in large part the outcome of war-time work. Resistance quickly appeared to insecticides, fungicides and antibiotics (see papers by Sawicki & Denholm, Georgopoulos, and Datta, this volume). Researchers quickly predicted that resistance to herbicides would soon appear if agriculturalists were not careful. Yet farmers growing wheat in monoculture, who have used chlorinated phenoxy herbicides annually to control broad-leaf weeds for nearly 40 years, bestow only scorn on these predictions and on those who made them.

Fifteen years after the introduction of phenoxy herbicides the newer *s*-triazine herbicides replaced them (but only for maize cultivation) as they gave better control of the same weeds over a longer season. As recently as four years ago, triazine resistance was discounted as 'not being a serious problem, nor does it appear to be a major threat in the future' (Parochetti 1979). Resistance to the triazine herbicides has become widespread, usually over small areas and only in special circumstances of usage. The problem has become sufficiently alarming that a book has appeared summarizing what is known about the problem, what can be done to discover and prevent it, and even how to use resistant material positively for agriculture (LeBaron & Gressel 1982).

Types of selectivity and resistance

One of the predicted problems was not the appearance of resistant variant strains, or *biotypes*, in a previously susceptible species but the changing patterns of weed distribution. The susceptible species would be replaced by different species that were always tolerant or resistant. Frequent claims for such changing distribution patterns have not usually been based on good numerical ecology of agricultural systems. A notable exception, where there was an excellent data base *before* herbicide usage and good large-scale measurements after, produced ambiguous conclusions. In this study, Haas & Streibig (1982) directly correlated large-scale changes in weed distribution with the usage of phenoxy herbicides in Danish small-grain fields. Unlike most others dealing with this problem they also posed 'controls'; they compared the new patterns of weeds with other agricultural variables and found similar temporal correlations between the patterns and: increased area drained; increased nitrogen fertilization; and introduction of higher yielding varieties. Indeed, the evidence showed that some of these factors were clearly more important than the use of herbicides. Weed species that 'liked' soggy soils disappeared, as did those that did not respond to greater fertility. For the remainder of this paper, I shall discuss only cases where newly resistant plants appeared.

It is conceptually simple to design compounds that will differentiate between biologically unrelated organisms; i.e. killing insects, rodents and fungi in plants. It is far more formidable a problem to design compounds that selectively kill specific *plants* from within a group of plants, i.e. weeds within crops. Both the triazines and the phenoxy herbicides excel at controlling weeds in specific crops while the crops are standing and growing. Inherent biochemical characteristics confer these resistances; they are rather clearly understood for the triazines (Jensen 1982, Arntzen et al 1982) but less so for the phenoxy herbicides. Other modes of selectivity are less obviously biochemical; a contact herbicide can be used to kill weeds just before crop seeds germinate; a spray of contact herbicide can be directed away from the crop; the crop can be treated with 'protectants' (also called antidotes or safeners); or the crop may have special morphological features, such as thick cuticles or special leaf angles, which prevent herbicide effects.

One of the problems that the farmers and the chemical industry had with the phenoxy herbicides was their lack of season-long persistence, and considerable effort went into finding chemicals (such as the triazines) that had a longer soil life. Even now, all present-day herbicides are either progressively biodegraded or removed from the biosphere by other means. One mode of resistance to herbicides could be phenological: the seasonal timing of weed germination could be changed, through evolution, to a time when the herbicide would no longer be inhibitory. Putwain et al (1982) looked for such genetically determined phenological changes in *Senecio vulgaris* (groundsel), a species that seemed to be germinating at a season of minimal triazine concentration. They found, instead, that the seeds selected from plants that germinated during low triazine concentrations in the soil were not specifically enriched to germinate at that time; they had the same germination spacing as seeds from untreated areas.

Integrating the factors that enrich for resistant and tolerant individuals by the use of herbicides

Perceptive individuals began asking why resistance to phenoxy herbicides was so late in developing (L. G. Holm, personal communication 1960). In the mid 1970s, when I was having trouble isolating herbicide-resistant cells from suspension cultures of susceptible species, I began to enumerate the causes that might control the rate of resistance in the field. Many considerations of population genetics were borrowed from the seemingly analogous studies on heavy-metal tolerance by Bradshaw and his colleagues (see Bradshaw, this volume). These studies showed that genes for each type of resistance were found in at least a third of the species that were present before the appearance of the xenobiotics. It quickly became apparent that we had few data on many of the influential factors, especially the true selection pressure, of herbicides. The farmer and weed scientist can tell the 'kill'-rate of a herbicide but not the reduction that will be found in the number of propagules of the weed at the end of a season. This effective selection pressure can still only be guessed.

Now that we have herbicide-resistant biotypes, we know that they are usually highly 'unfit' in the classical (Haldanian) sense. When resistant biotypes are grown in competition with susceptible (wild-type) biotypes of the same species, without herbicides, their seed yield is about half that of the wild type (Radosevich & Holt 1982). The lack of fitness is meaningless when herbicides persist through the season; but it will have a considerable delaying effect on enrichment for resistance when non-persistent herbicides are used, because much of the season will be available for the remaining susceptible individuals to exert their superiority. This competition is especially fierce at the time of establishment (Aikman & Watkinson 1980), which was not studied by Radosevich's group. Thus, the fitness differential may be even greater than the factor of two that they reported. The more fit individuals will be more 'plastic'; they will be better able to follow the extrapolations of Parkinson's Law and 'expand to fit the space available'.

Persistence of herbicides interrelates not only with fitness but also with a special characteristic that differentiates weeds from crops as well as from other pests—their 'spaced out' germination. Weed seeds, because of a multitude of dormancy properties, germinate throughout the season. Susceptible seeds can germinate after the rapidly degraded herbicide has gone, and then can produce more seeds before the season is over, considerably lowering the effective selection pressure (Fig. 1).

Weed seeds can also germinate over a period of years. Every time we enrich for resistant individuals by the use of herbicides, the 'resistant' seeds are diluted by a seed bank of susceptible seeds from previous years which also exert a 'buffering effect' and delay resistance.

We need to know which of the factors affecting herbicide resistance are quantitatively more important, so as to modify agricultural practices specifically to delay the appearance of resistance. The effects of the selection pressure, the herbicide persistence and the seed bank on the rates of enrichment for resistance must be quantified, to allow us to see how modifying any of them will affect the rate at which resistant seedlings will appear. Such modelling has been done for the evolution of insecticide resistance by Georghiou & Taylor (1977), by Delp (1981) for fungicide resistance and by Goldie & Coldman (1979) for resistance of cancer cells to anti-tumour agents. These systems lack an equivalent for the large soil-seed reservoir. We have integrated the factors governing the rates of evolution of herbicide-resistant weeds, including the effects of the seed bank (Gressel & Segel 1978, 1982). A series of mathematical considerations has culminated in the simple expression:

$$N_{\mathrm{n}} = N_0 \left(1 + \frac{f\alpha}{\bar{\mathrm{n}}}\right)^n$$

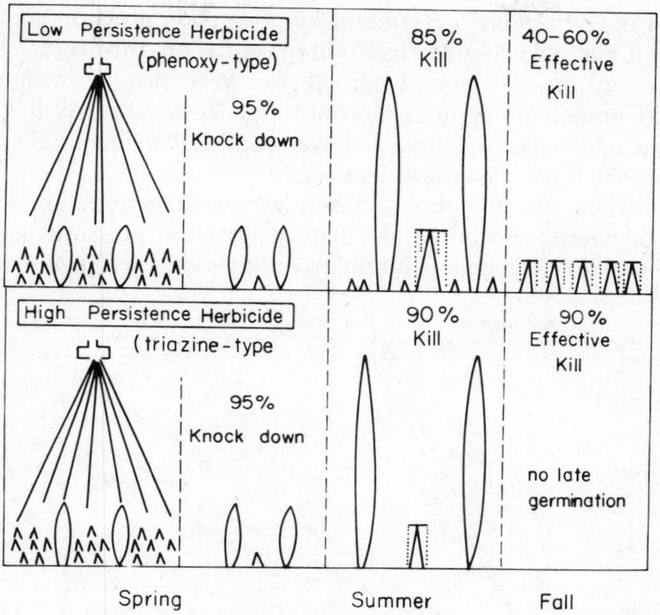

FIG. 1. Effect of herbicide persistence on the effective kill. In this (hypothetical) example two selective herbicides (phenoxy-type and *s*-triazine-type) kill 95% of the weeds (/\) without affecting the crop (0). In both cases the weed grows more than it would have, had it not been thinned by herbicides, because of an inevitable 'Parkinsonian' plasticity (see text). The effective kill, as measured by seed output, is about 90% by midsummer in both cases. However, the low persistent herbicide (e.g. phenoxy type) allows late summer germination of weeds growing in the layer underneath the crops, if viable seed is shed before winter; this further lowers the effective kill and hence the selection pressure. In both scenarios the crop yield may be essentially identical. Actual measurements of effective kill have not been reported.

that is, where n is the number of years of treatment; the proportion of resistants of a given species in the nth year of continued treatment of a given herbicide is N_n; and the initial frequency in the field prior to herbicide treatment is N_0. N_0 itself is a function of the frequency of natural mutation to the resistant biotype and the fitness of such a biotype. The factor in parentheses governs the rate of increase of resistance. It contains the overall fitness (f) of the resistant compared to the susceptible biotype, which in the known cases of herbicide resistance is between 0.3 and 0.5. The selection pressure (α) is defined as the proportion of the remaining resistants divided by the proportion of remaining susceptibles. For example, if no resistants are killed and 95% of the susceptibles are killed, $\alpha = 1/0.05 = 20$. In the case of heavy-metal resistance (see Bradshaw, this volume), α will approach infinity because of the extremely high rate of kill. Selection pressure and fitness are divided by n̄, the only complex factor in the equation, the average life span of the species in the soil seed-bank. If we were dealing with crops that germinated immediately, then n̄ would equal one year. With most weed species, n̄ would be between two and five years, and an increasing n̄ depresses the rate at which resistance will increase.

The interrelations are clearer when we use the equation to generate hypothetical lines from different effective selection pressures and from an average seed-bank life-span, with different fitnesses (Fig. 2). We have started

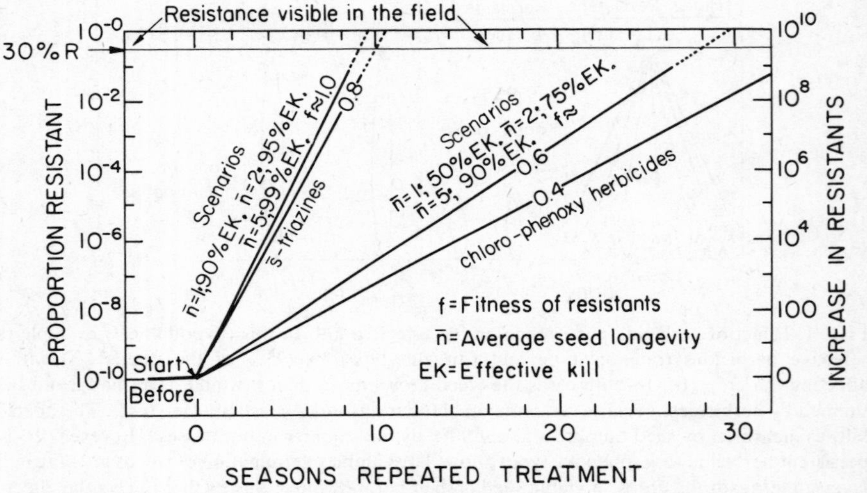

FIG. 2. Effects of various combinations of selection pressure α (measured as effective kill, EK) and of soil seed-bank longevity (n̄) on the rates of enrichment of herbicide-resistant individuals over many seasons of repeated treatment. The values are plotted for fitnesses which would be allowed to develop after the herbicide becomes degraded. With the persistent triazine-type herbicide the fitness ($f \approx 1.0$; $f \approx 0.8$) would be near to unity as the fitness differential has no time to become apparent. With the phenoxy type, fitness differentials ($f \approx 0.6$; $f \approx 0.4$) will have time to be influential. Resistance (R) would become apparent in the field only when more than 30% of the plants are resistant. The scale on the right indicates the increase in resistance from any unknown initial frequency of resistant weeds in the population, whereas the scale on the left starts from a theoretically expected frequency of a recessive monogene. Plotted from equations in Gressel & Segel (1978).

in year zero from a frequency of 10^{-10} (e.g. where resistance might be inherited by a single recessive gene in a diploid weed). It is possible to move the frequency scale in Fig. 2 to fit any other initial field frequency. From the slopes it is clear that we should be enriching yearly for herbicide-resistant individuals. If we follow the slopes, we see that it will take many years until we reach a frequency of resistant weeds that will be noticeable (i.e. more than the 1–10% that usually remain after a herbicide treatment). Thus, we will not realize that we are enriching for herbicide resistance until it is upon us. This rate of enrichment cannot be measured in simple laboratory experiments. Only Nosticzius et al (1979) have attempted to make such measurements in the field. They made weed counts from 1974 in maize fields in Hungary after the fields had been in maize/triazine monoculture for over five years (Fig. 3). In 1974 the level of infestation by *Amaranthus retroflexus* was the same as it

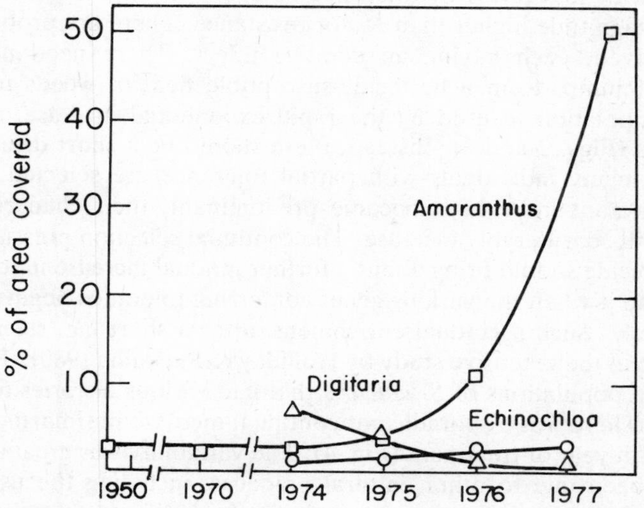

FIG. 3. Changes in weed populations in a monoculture maize field treated annually with atrazine. *Amaranthus retroflexus*, *Echinochloa crus-galli* and *Digitaria sanguinalis*, the foremost weeds, were counted. The field (in the Agricultural Combine of Babolna, Hungary) received atrazine in maize from 1970. (Data were plotted by Gressel & Segel 1982, from Table 1 in Nosticzius et al 1979, and are reproduced by permission.)

had been in previous years. Note the rapid five-fold increase in *Amaranthus retroflexus* between 1975 and 1976, and again between 1976 and 1977 at the expense of *Echinochloa crus-galli* and *Digitaria sanguinalis*, which follows the predictions of the model.

The cases in Figs. 1–3 best describe the appearance of a single-gene-inherited total resistance and not a partial tolerance where there would be greater than 90% kill of the normal strain and 50% kill of the partially tolerant strain. The selection pressure for partial tolerance is lower than that for resistance (i.e. $\alpha = 0.5/0.1 = 5$). Thus, the rate of increase of partial tolerance to herbicides should be rather slow, in seeming contradiction to the large number of reports of many tolerant biotypes (see LeBaron & Gressel 1982). In any genetic framework of resistance, it is to be expected that many more individuals would be partially tolerant than resistant, especially if quantitative polygenic inheritance is involved. The α value (selection pressure) will vary for each level of tolerance. The initial frequency will probably also vary; the greater the tolerance, the lower the frequency of individuals tolerant to a herbicide treatment. In the terms of the equation, this can be summarized as follows: although α is lower with tolerants, N_0 (the frequency of strains with low levels of tolerance before spraying) is probably a few orders of magnitude higher than N_0 for resistance. There are probably many genes involved, each having an additive effect. There need not be the exponential jump, from a herbicide-susceptible field of weeds to a highly resistant population, caused by the rapid exponential increase in resistant populations (Figs. 2 and 3). Instead, there should be a short delay of a few years until many individuals with partial tolerance are selected. As these partially tolerant individuals become predominant, their chances of interbreeding will considerably increase. The continual selection pressure exerted by the herbicide should bring about a further gradual increase in tolerance of the weeds to it when the various genes conferring tolerance begin to interact quantitatively. Such a gradual enrichment, after a short lag, seems to have taken place in the extensive study by Holliday & Putwain (1980). They tested tolerance in populations of *S. vulgaris* that had various histories of simazine treatment. There was a sporadic but continual increase in simazine tolerance after the fifth year of treatment (Fig. 4). The variability was great and may in part have been due to other cultural practices, including the use of other herbicides. Still, the regression was statistically significant.

No one can yet say for sure how different are the effective selection pressures for the phenoxy and the triazine groups. One mid-season survey of 17000 quadrants in 859 North Dakota wheat fields showed that the best control of *A. retroflexus, Chenopodium album* and *Brassica campestris* with a phenoxy herbicide gave 0.2 plants of each weed per m^2 (Dexter et al 1981). Considering the plasticity of these species, there would be sufficient seeds for a good stand of weeds the following year. If the herbicide were not used, these stands could be as high as 59, 46 and 32 plants per m^2, respectively. These data do not include any late germination that may give rise to seed in autumn. I have not found similar data for the triazines.

Incidence of and conditions leading to resistance

The mathematical models leading to Fig. 2 can be used to predict what happens when a mono-herbicide culture is *not* used (Gressel & Segel 1978,

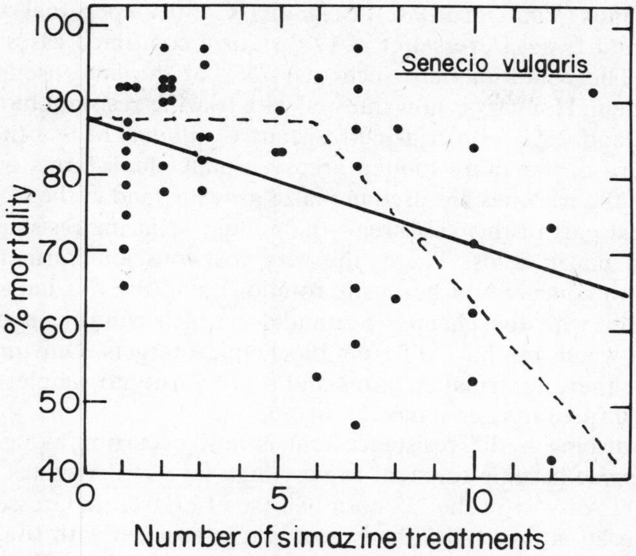

FIG. 4. Increased *Senecio vulgaris* tolerance to simazine, a triazine herbicide, as a result of repeated treatments. *S. vulgaris* seed was collected at 46 locations in England where the previous treatment history of the site was known. Almost all sites with more than three simazine treatments were also treated at various times with other herbicides. The last treatment of each population was with 0.7 kg/ha simazine under standardized conditions, yielding the results in this figure. The variance due to the regression was highly significant ($P < 0.01$). This figure was drawn by Gressel & Segel (1982) from data in Figs. 3 and 4 of Holliday & Putwain (1980). The solid line is the calculated regression and the dashed line fits the explanation in the text. (Reproduced by permission.)

1982). Simply stated, it is the number of weed generations which are treated that affects enrichment. If it would take 20 years in mono-herbicide culture to obtain resistance, it would take 40 or 60 years in a 1 in 2 or 1 in 3 rotation, respectively. Indeed, all *s*-triazine resistance confirmed to date has come from mono-herbicide culture: yearly treatments in maize (which has biochemical resistance); yearly treatments in trees and vineyards (having roots beneath the soil levels to which triazines are bound); and sterilant concentrations used on roadways. It seems as if the incidence and area have increased exponentially (as would be predicted by the model) from the first case in the late 1960s to the 30 species that were confirmed as triazine-resistant in November 1981 in 23 US states, four Canadian provinces and seven European countries (LeBaron & Gressel 1982). Indeed, to fulfil the apparent requirement that all 'developed' countries must prove themselves by having triazine resistance, the Netherlands (Van Dord 1982), Belgium (R. Bulcka personal communication 1983) and Israel (Gressel et al 1983a) have confirmed cases of triazine resistance. The conditions are such that 75% of the land used for monoculture maize in Hungary is now infested with triazine-resistant biotypes of *A. retroflexus* and 25% with triazine-resistant *C. album*. Most other cases of resistance are on far more limited areas—usually small farms or orchards. The bulk of the triazines are used in maize-growing land in the US corn belt. In the richest part of this vast area—the middle—triazine resistance has not appeared in maize fields. There, the very cost-conscious farmers use crop rotation (with concomitant herbicide rotation), and the triazines are usually used together with the cheaper acetanilides which control an overlapping spectrum of weeds but have different biochemical targets. Only in peripheral areas where there are smaller farms and a propensity to simpler operations (i.e. monoculture) has resistance occurred.

More disturbing is the resistance that is now occurring along roadways. *Kochia scoparia* became resistant to very high levels of triazines used along railroad rights-of-way in the US corn belt (see LeBaron & Gressel 1982). In Israel, the road authority has had most verges sprayed with triazines; first, *Brachypodium distachyon* (Gressel et al 1983a) and then six other newly evolved grass biotypes—*Phalaris paradoxa*, *Alopecurus utriculatus*, *Alopecurus myosuroides*, *Lophochloa phleoides*, *Lolium rigidum*, *Polypogon monspeliensis*—were found that were triazine-resistant (A. Nir and B. Rubin personal communications, 1983). These results are worrying, as moving vehicles are excellent disseminators of seed of the resistant biotypes, taking the seed directly into the fields of the farmers who had been so careful to exclude them.

There are four cases in England, Egypt and Japan of resistance and one of strongly increased tolerance to paraquat, a contact bipyridillium herbicide that has no biological persistence (see LeBaron & Gressel 1982, Watanabe et al 1982). This seemingly contradicts the models which place persistence as an important factor governing selection pressure and fitness. The lack of field persistence of paraquat is compensated for by the persistence of the farmers who treated with this compound 8–12 times per year!

Biochemistry and genetics of triazine tolerance and resistance

Maize is the major crop in which *s*-triazines are used as a biochemically selective herbicide. The resistance of the maize itself is due to a specific glutathione-*S*-transferase in maize, which conjugates with chloro-*s*-triazines. A nuclear gene is known to confer this resistance (Jensen 1982).

When the first triazine-resistant weeds were found, there was considerable surprise that they did not have the same mechanism as maize; they degraded atrazine slowly and, unlike maize, their chloroplasts were totally resistant to triazines, which usually interferes with photosystem II of photosynthesis (Radosevich & Holt 1982, Arntzen et al 1982). It was later shown that plastids of the resistant biotypes do not bind the triazines (Arntzen et al 1982), which prevents them from 'committing suicide'. This trait is maternally inherited (Souza-Machado 1982), and is presumably borne on the plastid genome. The ecological implications of this special inheritance are that the trait of resistance cannot be spread by pollen, only by seed. The appearance of plants with plastome mutations is rare in nature. J. Duesing has tried, mainly from theoretical considerations, to estimate the frequency of triazine-resistant individuals, and found a frequency of less than 10^{-10}. He has characterized a nuclear gene in a triazine-resistant *Solanum nigrum* accession that increases the frequency of chloroplast mutations by about 1000-fold (J. Duesing, unpublished paper, Weed Science Society of America, February 1983). Such a plastome-mutator gene may be a transient step in the appearance of cytoplasmically inherited resistances that are quickly bred out of the populations in nature. This property has not been seen in other triazine-resistant biotype accessions.

There is considerable evidence that the appearance of triazine resistance in the same species in different isolated areas is due to concurrent evolution and not to spread. In France, many different 'chemotypes' of *C. album* are to be found in the same field. Resistance has appeared in different chemotypes in a typical founder effect manner in each of the locales; that is, only one resistant chemotype was found in each locale with interlocale differences (Gasquez & Compoint 1981).

There are slight biochemical differences in the triazine-binding properties of the different biotypes; the ratio between binding constants of resistant and susceptible types varies among species (Arntzen et al 1982). There are also slight differences in the dose–response curves of different resistant biotypes of the same species (Fig. 5). The best explanation is that there are different amino acid substitutions at various points in the triazine-binding protein that may confer this difference. The binding protein is presumably a part of photosystem II, as triazines inhibit this pathway. Thus, the lesion responsible for non-binding of the herbicide probably causes the decrease in efficiency of photosystem II observed by Radosevich & Holt (1982), which is translated into the reduced fitness of these biotypes. Considerable effort is being expended in many laboratories in the isolation, cloning and sequencing of the gene for this protein.

Tolerance to *s*-triazines has also evolved in many species. These include species such as *Echinochloa crus-galli* (see Gressel et al 1982) and *S. vulgaris* (Fig. 4) which have also evolved resistance at the plastid level in other locales.

The tolerance in *S. vulgaris* is probably inherited polygenically on nuclear genes, as it has a low 'broad sense hereditability' (Holliday & Putwain 1980). An unsuccessful effort was made to ascertain if the tolerance in *S. vulgaris* was due to a more rapid degradation in a tolerant biotype; no differences in rates of degradation were found in material provided by P. D. Putwain (Gressel et al 1983b).

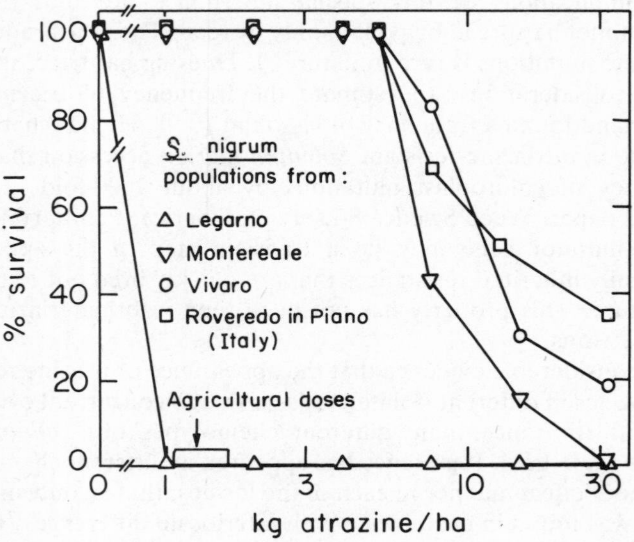

FIG. 5. Variable response of *s*-triazine-resistant accessions of *Solanum nigrum*. Seeds of the resistant biotypes were gathered from the four isolated places listed (in northern Italy) and were assayed in pot tests. Figure plotted by me from data in Table 3 of Zanin et al (1981).

There may have been a dual, concurrent evolution to a state of resistance and towards tolerance in *B. distachyon*. The plastids of the resistant biotype of this species have the immediate resistance response typical of other triazine-resistant weeds. The resistant plants were also found to degrade [^{14}C]atrazine much more rapidly than the susceptible biotype (Fig. 6). This suggests that while this dual evolution was occurring, the enrichment for total resistance may have become dominating, while the enrichment for tolerance had not yet arrived at a point where the plants were able to withstand high herbicide doses. If the tolerance is inherited on the nucleus and resistance is inherited on the plastome, we should be able to separate them in the offspring of the reciprocal crosses that are now in progress in my laboratory.

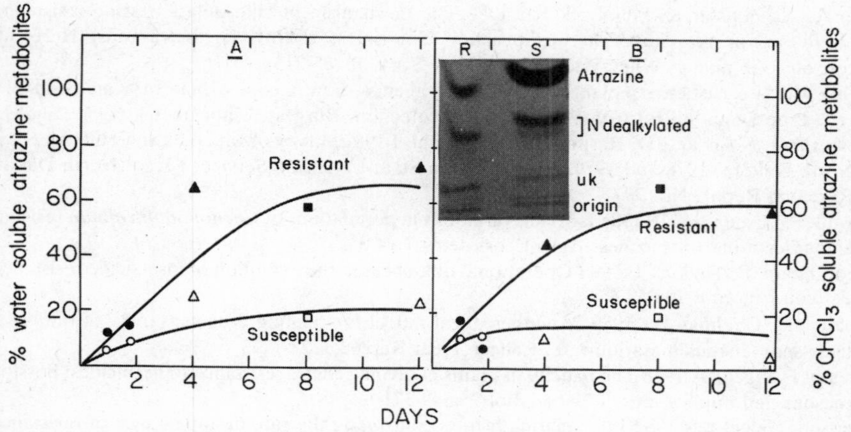

FIG. 6. Metabolism of [^{14}C]atrazine by resistant and susceptible *Brachypodium distachyon* biotypes. (A) Metabolism to water-soluble metabolites; the data are presented as percent of the total metabolites that partition into the aqueous phase. Water-soluble metabolites are all inactive. Each symbol shape represents a separate experiment. (B) Metabolism to chloroform-soluble components. The percent is plotted of radioactivity on thin-layer chromatographic plates *not* in parent atrazine (i.e. in the inactive *N*-dealkylated products and the unknown component). (Insert: thin-layer chromatographic separation of atrazine and its chloroform-soluble metabolites from resistant (R) and susceptible (S) plants. uk, unknown metabolite. (From Gressel et al 1983a; reproduced by permission.)

Conclusions

Resistance has appeared mainly to *s*-triazine herbicides around the world. Because of population enrichment considerations, it is deemed impossible to prevent the appearance of resistance. Nevertheless, wise agronomic practices, especially herbicide rotations, can considerably *delay* the problem for each herbicide. Resistance and tolerance have evolved by concurrent evolution, sometimes in the same species, in the same or in different locales.

Acknowledgements

The author is the Gilbert de Botton professor of plant sciences. Great indebtedness is due to Professor L. G. Holm who first whetted my interest in this subject and to the fruitful collaboration with Dr H. M. LeBaron which greatly enhanced my knowledge about resistance.

REFERENCES

Aikman DP, Watkinson AR 1980 A model for growth and self-thinning in even-aged monocultures of plants. Ann Bot (Lond) 45:419-427

Arntzen CJ, Pfister K, Steinback KE 1982 The mechanism of chloroplast triazine resistance: alterations in the site of herbicide action. In: LeBaron HM, Gressel J (eds) Herbicide resistance in plants. Wiley-Interscience, New York, p 185-213

Delp CJ 1981 Resistance to plant disease control agents—how to cope with it. In: Kommendahl T (ed) Proc Symp 9th Int Congress on Plant Protection, Burgess, Minneapolis, vol 1:253-261

Dexter AG, Nalewaja JD, Rasmusson DD, Buchli J 1981 Survey of wild oats and other weeds in North Dakota; 1978 and 1979. North Dakota State Extension Service, Fargo (North Dakota Research Report No. 79)

Gasquez J, Compoint JP 1981 Isozyme variations in populations of *Chenopodium album* resistant and susceptible to triazines. Agro-Ecosystems 7:1-10

Georghiou GP, Taylor CE 1977 Operational influences in the evolution of insecticide resistance. J Econ Entomol 70:653-658

Goldie JH, Coldman AJ 1979 A mathematical model for relating drug sensitivity of tumors to their spontaneous mutation rate. Cancer Treat Rep 63:1727-1733

Gressel J, Segel LA 1978 The paucity of plants evolving genetic resistance to herbicides; possible reasons and implications. J Theor Biol 75:349-371

Gressel J, Segel LA 1982 Interrelating factors controlling the rate of appearance of resistance. The outlook for the future. In: LeBaron HM, Gressel J (eds) Herbicide resistance in plants. Wiley-Interscience, New York, p 325-347

Gressel J, Ammon HU, Fogelfors H, Gasquez J, Kay QON, Kees H 1982 Discovery and distribution of herbicide-resistant weeds outside North America. In: LeBaron HM, Gressel J (eds) Herbicide resistance in plants. Wiley-Interscience, New York, p 32-55

Gressel J, Regev Y, Malkin S, Kleifeld Y 1983a Characterization of a *s*-triazine resistant biotype of *Brachypodium distachyon*. Weed Sci 31: in press

Gressel J, Shimabukuro RH, Duysen ME 1983b *N*-dealkylation of atrazine and simazine in *Senecio vulgaris* biotypes: a major degradation pathway. Pestic Biochem Physiol 19:361-370

Haas H, Streibig JC 1982 Changing patterns of weed distribution as a result of herbicide use and other agronomic factors. In: LeBaron HM, Gressel J (eds) Herbicide resistance in plants. Wiley-Interscience, New York, p 57-80

Holliday RJ, Putwain PD 1980 Evolution of herbicide resistance in *Senecio vulgaris*: variation in susceptibility to simazine between and within populations. J Appl Ecol 17:779-791

Jensen KIN 1982 The roles of uptake, translocation and metabolism in the differential intraspecific responses to herbicides. In: LeBaron HM, Gressel J (eds) Herbicide resistance in plants. Wiley-Interscience, New York, p 133-162

LeBaron HM, Gressel J (eds) 1982 Herbicide resistance in plants. Wiley-Interscience, New York

Nosticzius A, Muller J, Czimber G 1979 The distribution of *Amaranthus retroflexus* in monoculture cornfields and its herbicide resistance [in Hungarian]. Bot Kozl 66:299-307

Parochetti JV 1979 Herbicide resistance found in some weeds. Crops Soils Mag (July) p 9-10

Putwain PD, Scott KR, Holliday RJ 1982 The nature of resistance to triazine herbicides: case histories of phenology and population studies. In: LeBaron HM, Gressel J (eds) Herbicide resistance in plants. Wiley-Interscience, New York, p 99-116

Radosevich SR, Holt JS 1982 Physiological responses and fitness of susceptible and resistant weed biotypes to triazine herbicides. In: LeBaron HM, Gressel J (eds) Herbicide resistance in plants. Wiley-Interscience, New York, p 163-184

Souza-Machado V 1982 Inheritance and breeding potential of triazine tolerance and resistance in plants. In: LeBaron HM, Gressel J (eds) Herbicide resistance in plants. Wiley-Interscience, New York, p 257-274

Van Dord DC 1982 Resistentie van *Chenopodium album* tegen atrazin en *Poa annua* tegen simazin in Nederland. Meded Fac Landbouwwet Rijksuniv Gent 47:37-44

Watanabe Y, Honma T, Ito K, Miyahara M 1982 Paraquat resistance in *Erigeron philadelphicus*. Weed Res 27:49-54
Zanin G, Vecchio V, Gasquez J 1981 Indagini sperimentali su popolazioni di dicoteledoni resistenti all atrazina. Riv Agron 5:196-207

DISCUSSION

Davies: I understood that triazine-resistance was due to an altered membrane protein in the chloroplast.

Gressel: Arntzen et al (1982) have claimed this for weeds. The triazine-resistant chloroplasts do not bind the herbicide atrazine. This being so, then they don't 'commit suicide' (Arntzen et al 1982). Various ideas are circulating about this, and the membrane protein that you mention is believed to have a relative molecular mass of 32 000.

Clarke: Is it indeed true that, in the gene for triazine resistance on the chloroplast genome, a single amino acid substitution is responsible for the resistance?

Gressel: The evidence for single amino acid substitution is that in a single *Amaranthus hybridus* plant that is *resistant* to atrazine, there is one amino acid difference in the DNA sequence for the 32 000 protein from the sequence found in a single *susceptible* plant of the same species. This protein in *A. hybridus* has one amino acid difference from the same protein in spinach, which in turn has differences from maize. We don't yet know the levels of interspecies and intraspecies variability. The evidence for the particular protein comes from photoaffinity labelling of the protein but the photoaffinity site is on the exactly opposite side of the triazine molecule from the herbicidally active site. Because the electron transfer system is closely packed on lamellae, there is a possibility that a protein opposite the herbicide-binding protein is photoaffinity-labelled (Gressel 1982). Arntzen's group have probably isolated the gene but the evidence is not entirely convincing to me.

Wood: In your model for the evolution of resistance, do you consider gene flow from outside, or is this a closed population? This will obviously affect both the rate of evolution of resistance and also its rate of decline. In simulations of insecticide resistance done by Professor Georghiou's group and our own, both migration and escape of insects are potent factors in controlling the rate of evolution of any resistance.

Gressel: One might say that we have so few data that it isn't worth worrying about it until more become available. Selection pressure is probably an overriding feature here. Gene flow would explain interspecific differences in resistance much better than it would explain differences in resistance of a single species to two or more different herbicides. If we use two insecticides, one of which exerts much more selection pressure than the other, any given species will evolve more rapidly to the one that has a higher selection pressure and less rapidly to the one with a lower selection pressure. In a different genus, which has gene flow, the rate at which resistance appears might be faster, especially when polygenically inherited. We have so few species that are herbicide-resistant—only about 40—that we cannot yet predict anything about resistance based on gene flow. In the specific case of triazine resistance in weeds, gene flow is reduced because the trait is maternally inherited, and gene flow through pollen is precluded.

Georgopoulos: Do you ascribe the difference between the resistance to triazines and the lack of resistance to phenoxy herbicides entirely and solely to the difference in persistence of the herbicide, or do you think that the modes of action have something to do with this?

Gressel: The mode of action may well be important here, but we know nothing about the mode of action of the phenoxy herbicides. We know that monocotyledonous plants degrade phenoxy herbicides more rapidly than dicotyledonous plants, which is why monocotyledons are resistant to the phenoxys. But the selection pressure of phenoxys is so low, because of degradation in the soil, that one would not have expected resistance to have appeared in 50 years.

Wood: I do not understand how you were defining the two phenomena of tolerance and resistance in your paper. In studies of insecticide resistance we define these phenomena somewhat loosely but we do make a distinction between them, tending to restrict the term tolerance to small increases in LD_{50} towards a range of unrelated chemicals, and the term resistance to larger, more specific changes. What distinction do you make?

Gressel: After using any dose of herbicide that totally kills the wild biotype, one looks at the biotypes that one has selected for. If the pesticide has had little or no effect whatsoever on any aspect of growth or life-cycle then *resistance* is said to occur, as the ultimate effect (see Fig. 1). Anything between this example of no effect and partial sparing effect we would call *tolerance*. Part of the misunderstanding about terms is that entomologists usually talk only about resistant *populations*. In my paper I described how the rare resistant *individuals* are enriched for within a population, until the population becomes resistant.

Wood: So your definitions for herbicides are not parallel ones to ours for insecticides.

Georghiou: The term *tolerance* has been used in different ways in entomology. It is often used to denote changes in the LD_{50} that are small enough not to involve noticeable economic problems. In some cases a 5-fold increase has been used on the threshold between tolerance and resistance. This is a very

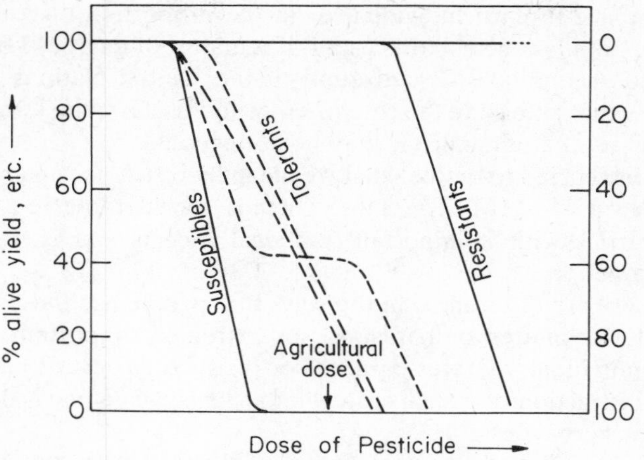

FIG. 1. (*Gressel*) Examples of susceptible, tolerant and resistant phenotypes. Note that tolerance can have many forms, but tolerant individuals are only partly affected at an agricultural dose. Resistance can be absolute, with saturating doses being ineffectual, but doses well above those commonly used can be lethal.

artificial distinction. *Tolerance* could be reserved for those cases in which the population—usually the entire species—is naturally endowed with the ability to survive significantly high doses of the chemical prior to the application of selection pressure. Several examples of such tolerance are known, e.g. in grasshoppers and the Mexican bean beetle for DDT, and in *Chrysopa carnea* for pyrethroids. Clearly there is a need for a term to describe this condition, and *tolerance* or *natural tolerance* appear to be appropriate.

Gressel: One does see *tolerant* genotypes, which may have longer generation times so that they live poorly but cope. I consider this to be *tolerance* because of slow degradation of the pesticide. But if degradation is fast enough for the pesticide to do nothing to the life-cycle, I would call it *resistance*. Because of the way that one enriches for resistant individuals by the use of herbicides, if one has 10% resistant individuals in one year, it is guaranteed that the next time the pesticide is used one would have 100% resistant individuals within the population (see Fig. 2 in my paper, p 78).

The species is not resistant prior to selection, but individuals or populations of it are. If a species will evolve resistance, susceptible populations must contain resistant individuals at some low, unnoticeable frequency.

Clarke: Since the limits of definition seem to be different for nearly everyone we might just treat the two terms as equivalent.

Georgopoulos: Independently of whether one speaks of differences in the level of resistance in each individual or in the proportion of resistant individuals in the population, in both cases it is a matter of degree. I believe that the Food and Agriculture Organization of the United Nations (1979) has recommended that at least in the case of fungi and bacteria the term tolerance should not be used at all, since it may be ambiguous.

Sawicki: I have tried to define what we mean by resistance and tolerance in my paper (see p 152-166). The subject needs considerable re-examination because it carries with it important practical aspects and is not a purely academic distinction.

Clarke: There are differences in the ways that people use these terms. It is evidently, in the matter of parasites, an extremely important distinction whether an individual tolerates parasites or resists them. But I am uncertain whether the distinction is of such critical importance when one is dealing with insecticides or herbicides.

Graham-Bryce: The question of fitness is likely to be a key issue in our discussions. Professor Gressel in his paper was tending to treat fitness as a fixed property, whereas we should also be interested in the *acquisition* of fitness. Resistant individuals are, almost by definition, less 'fit' or simply less 'competitive' with the others; otherwise these genotypes would be more numerous in the population. This can be shown not only for weeds, but also in insects and mites by their relative reproductive rates (e.g. Cranham 1982), and in pathogens by competition experiments (e.g. Hollomon 1975). When one applies a strong selection pressure such as pesticide treatment, one changes the environment such that these advantages are outweighed, and survival of resistant individuals is favoured. But if one then releases that selection pressure without having eliminated the normal (susceptible) individuals, presumably their greater competitiveness in the absence of the selection pressure would allow them to re-establish. While this situation prevails, resistance is manageable by a strategy of removing the particular selection pressure, and allowing the population to revert. However if selection pressure is maintained and the resistant individuals acquire the competitive ability of the normal susceptible population through genetic rearrangement, susceptibility will not drop back to the original level when selection pressure is relaxed, and there will then be an effectively permanent problem of established resistance. A crucial factor in contending with the threat of pesticide resistance would therefore appear to be to prevent the acquisition of this competitive ability. For weeds there has been less opportunity, compared with pathogens or insects, for genetic rearrangement; so are weeds still at that first stage, when one could take steps to manage these effects?

Gressel: The proportion of resistant individuals decreases in a population if one stops the herbicide treatments. Normally, repeated treatments with herbicide would enrich rapidly for the resistant individuals in the population.

Many years ago Crow (1960) showed that the proportion of resistant flies, retreated with insecticides a few generations after the initial treatment, immediately began to increase again. This is to be expected from the equation that I described in my paper. Another possibility, of course, is that interbreeding within the resistant population and between the resistant and the wild type raises the fitness to nearer that of the wild-type, so that the decay in resistance is then slower once treatment has stopped. Still, Professor Bradshaw's work (see p 4-19) shows fitness differentials in metal tolerance hundreds of years after the resistance evolved.

Clarke: We seem to be in a terminological mine field when discussing the *acquisition* of fitness. As I understand it, adaptation and fitness are the same thing in the sense that they both mean 'appropriateness'. We believe that natural selection acts in such a way as to make organisms appropriate to their environment, but it is not something that we generally can measure. We do not measure the 'fitness' or the 'adaptation' of an organism, except in relative terms, i.e. relative to something else. Fitness is therefore not an absolute property that one can acquire, but it is only a *state* relative to some other state. Adaptation comes into the same category. The implication is that if something has adapted it has done so *compared with* some other specified state, which is less adapted. As I said earlier (p 1), there are different degrees of propriety to the environment. But there isn't anything that we can call absolute propriety.

Foy: We should distinguish between inherent properties and stress-induced properties of plants. For example, one cannot simply develop plants for drought tolerance by selecting fine-rooted plants. If put under stress (such as aluminium toxicity), those plants might no longer be fine-rooted; morphologically and physiologically they would be different. In this case, the property selected for, to extract water, will not extract water efficiently if aluminium toxicity or another stress is present. More of our plant screening (e.g. for photosynthetic efficiency) should be done in the presence of anticipated stress factors rather than under ideal conditions.

Hartl: I don't think we can infer that a particular genotype has a *fitness* that is associated with it as, say, hair colour is associated with an individual. A natural environment is a mosaic of environments, and the same genotype in a natural environment will have many different fitnesses in which, for any generation, a sort of 'average' fitness is important in terms of changing gene frequency. This is just another way of saying that fitness is a complex trait determined by an interaction between the genotype and the environment. As Dr Foy said, some of these genotypes will be in stressful environments and will express a different fitness in that particular environment than in a less stressful environment. We need to know why certain species have pre-existing genetic variation that adapts them to extreme environments while other species do not. One plausible explanation is that the mosaic of natural environments for the polymorphic species is such as to maintain a non-negligible frequency of resistance polygenes which, under intense selection, can be brought together by segregation and be expressed as a more tolerant or more resistant genotype. A frequency of 1 in 300 copper-resistant individuals of *Agrostis* is surely much higher than can be accounted for by new mutations; those populations must be segregating.

Clarke: What Dr Foy said about stress and the need to do more screening is particularly true when we don't understand the genetics of the characteristics that we are looking at. As soon as we change the environment we change the expression, the heritability and all the quantitative parameters.

Datta: Don't you think that the word 'adaptation' implies a *change* from one state to another, whereas fitness doesn't?

Clarke: I would counter that by asking how could we ever define fitness?

Datta: Well we can't!

Clarke: It does not necessarily follow that because something is rarer it is less fit; it depends on the particular way in which it has become rarer. For example, at an equilibrium point between a rare and a common genotype, both genotypes may be equally 'fit'.

Graham-Bryce: In cases where this comparison has been made in competition experiments, such as those that Professor Gressel described, and in the measurements of reproductive rates of insects and mites, and the competitive studies with pathogens to which I referred earlier, the resistant individuals appear to be less successful. So, as far as one can *measure* the fitness, the resistant individuals do appear to be less 'fit'. Is it worth focusing more attention on the acquisition of a greater competitive ability by these resistant individuals?

Bradshaw: One should not use the term 'competitive ability' here because it is a specific term that relates to fighting for particular resources. You are here concerned with fitness under two conditions. Your term fitness is applying to fitness, or adaptation, under *normal* conditions when the herbicide or metal is not present, and you are comparing that with fitness under conditions where the insecticide (or equivalent) *is* being applied. A genotype resistant to insecticide or metals will obviously be fit in the *presence* of the stressful factor. But what we are interested in is its fitness in *normal* conditions, because that will then determine how fast it declines in frequency. The next point is whether the character of resistance is indissolubly associated with something that lowers the fitness of the individual that carries the character in normal conditions. This is a physiological and genetic problem. The likely explanation for metal-tolerant plants not being as fit in normal conditions as normal plants are is that the metal-complexing mechanism reduces the uptake of the essential trace quantities of the metal that they require. Copper-tolerant individuals grow better, in terms of rooting for instance, if they are given a small amount of copper; they have actually developed a copper *requirement*, presumably related to some chelating process whose operation is not stopped in normal soils, so putting them at a disadvantage.

Georgopoulos: The mechanism involved is important here. If the same gene that gives resistance also alters some other property at the cellular level, leading to reduced fitness in the absence of the chemical, for example the higher requirement for copper, then we *can* speak of a reduced fitness and this is unambiguous.

Clarke: Perhaps a general principle is involved here: in a complex system with interacting parts, a firm movement of any particular part of that system will inevitably shift many other parts which may be nicely adjusted to their environment. This inevitably would reduce the system's ability in other respects.

Datta: Yes; that is presumably what is happening with these copper-resistant plants. The organism *Staphylococcus aureus* is carried by about 50% of normal healthy people. Outside a hospital or any place where much penicillin is being used, about 50% of the strains are resistant to penicillin by virtue of being able to produce penicillinase, which doesn't seem to affect their fitness in any measurable way. They are certainly able to cause disease.

REFERENCES

Arntzen CJ, Pfister K, Steinback KE 1982 The mechanism of chloroplast triazine resistance—alternations in the site of herbicide action. In: LeBaron HM, Gressel J (eds) Herbicide resistance in plants. Wiley Interscience, New York, p 185-213

Cranham JE 1982 Resistance to organophosphates, and the genetic background, in fruit tree red spider-mite, *Panonychus-ulmi* from English apple orchards. Ann Appl Biol 100:11-23

Crow JF 1960 Genetics of insecticide resistance: general considerations. In: Research progress on insect resistance. Misc Publ Entomol Soc Am 2:69-74

Food and Agriculture Organization of the United Nations 1979 Pest resistance to pesticides and crop loss assessment. 2: report of the second session of the FAO panel of experts. (Rome, 28 August–1 September 1978)

Gressel J 1982 Triazine herbicide interaction with a 32000 M_r thylakoid protein—alternative possibilities. Plant Sci Lett 25:99-106

Hollomon DW 1975 Behaviour of a barley powdery mildew strain tolerant to ethirimol. Proc 8th Br Insectic Fungic Conf 1:51-58

General discussion

Relative fitness and the stability of resistance

Wood: In insecticide resistance, as far as heterozygous advantage is concerned (see also p 87-93), linked and balanced lethal genes can be found in association with resistance genes as, for example, in the spotted root maggot. Hooper & Brown (1965) found, when selecting a resistant population of this insect with dieldrin or DDT, that they ended up with only heterozygotes.

We have actually measured relative fitness (w), under conditions where resistant populations of mosquitoes in the field have been allowed to revert towards susceptibility (Curtis et al 1978, Wood & Bishop 1981). For *Anopheles stephensi* we considered five urban populations from India which had been relieved of DDT selection for between 20 and 79 months, and w was 0.91 for the *resistant* homozygote (where the *susceptible* homozygote had a fitness of 1.0—only a 10% difference). At the same time in a laboratory population, from an urban sample kept for 80 generations and left in the laboratory without exposure to DDT, the value of w was 0.96. So these are sufficiently close to suggest that there was not a substantial amount of immigration into the field populations which could have accounted for some of the decline in resistance. Fitness is measured here as $1 - s$, where s (the selection coefficient) is calculated from the change in the relative frequency of the resistance gene in the population over time (Wood & Cook 1983). We also looked at *Anopheles culicifacies* in India after relaxation of DDT treatment. In 28 villages we found relative fitness values of 0.62 to 0.97 for the RR homozygotes. So in some cases the relative fitness was almost as high as normal, but it could go down to 0.62, depending presumably on the degree of co-adaptation that had gone on in the original selection.

Gressel: Do you have any correlations between relative fitness and time since the last treatment of DDT? This might, for instance, show that the value of 0.97 came in 10 years after treatment (after interbreeding had occurred), while 0.62 applied to the time when resistance had just appeared.

Wood: We don't have such correlations because we looked at the whole process over the period of time. We did, however, look at the selection coefficient in relation to the initial frequency of the susceptible gene—in other

1983 Origins and development of adaptation. Pitman Books, London (Ciba Foundation symposium 102) p 94-99

words, to see whether there was any frequency-dependence. We did find such frequency-dependence; as the gene became rarer, so the rate of loss, or the selection against the resistance gene, slowed down.

Bradshaw: To what do you attribute that?

Wood: We were uncertain, and were unable to exclude immigration as a possible influence; immigration of susceptible individuals might explain such frequency dependence. We have also suggested various other possibilities including frequency-dependent selection, or heterosis, leading to a stable polymorphism. Another possibility is that there are two resistance mechanisms: a low-resistance factor with a high fitness and a high-resistance factor—a different gene—with a low fitness.

When we did similar studies with dieldrin, we found lower values for w (as low as 0.44). In other words, the dieldrin-resistance homozygote appeared to be relatively less fit than the DDT-resistance homozygote.

Georghiou: Some of our results suggest that the stability of resistance may also depend on the mechanism involved. Esterase-mediated phosphate resistance in *Culex quinquefasciatus* in our laboratory has shown a pronounced tendency to regress. We recently compared the rate of regression in three strains that had been similarly selected by temephos (esterase resistance), propoxur (oxidase), or permethrin (mainly *kdr*), respectively. Upon removal from selection, resistance to temephos declined rapidly, propoxur resistance slowly, and permethrin resistance declined at an intermediate rate. Since the three strains had originated from the same 'synthetic' population following a period of inbreeding, and were thus presumed to have similar fitness qualities, we suggested that the resistance mechanism itself may influence the rate of regression (Georghiou et al 1983).

Elliott: Which pyrethroids have you studied (see also p 132-137)?

Georghiou: Permethrin; but the cross-resistance to pyrethroids is so broad as to involve nearly all members of the group.

Hodgson: Resistance due to metabolism, whatever the mechanism, carries an energetic cost. For a mechanism that involves increased levels of several enzymes, the energetic cost is likely to be higher, even if we consider only protein synthesis. This may or may not be expressed as a lack of fitness under a particular set of circumstances, but it raises the possibility that the energetic cost may determine subsequent fitness.

Davies: Another problem is that one cannot generalize with respect to organisms. Microorganisms behave totally differently from plants. Resistance in microorganisms, regardless of the mechanism, is often extremely stable. For microorganisms there is no clear indication that a given metal or antibiotic resistance mechanism will make the organism more or less able to compete with the wild-type organism under conditions of non-selection.

Georghiou: It is interesting, though, to notice that with insects in a laboratory environment, where the conditions are steady and all colonies are treated in the same fashion, one still sees differential rates of decline of resistance, as I just described.

Sawicki: I disagree with this. We have kept strains of houseflies with single resistance genes that we have isolated from field populations, for 20 years or more. The question of contamination crops up, unfortunately, far too frequently but, provided one spends about 50% of one's time just making sure that this doesn't happen, one *can* maintain pure populations!

Hartl: I would like to elaborate on Dr Davies' point, with some data for microorganisms. This has to do with chromosomal resistance to streptomycin which, in *Escherichia coli*, is due to a mutation in the gene for a ribosomal protein. The mutation causes a significant fraction of misreading of the messenger RNAs and produces defective proteins because of amino acid substitutions. When two isogenic strains are grown under glucose limitation in chemostats, in the absence of streptomycin, there is no detectable difference in growth rate between the streptomycin-resistant strain and the normal strain, in spite of the abnormal proteins that the resistant strain is producing (D.E. Dykhuizen, unpublished observations). The limit of resolution of this method is a selection rate of less than 0.005.

Gressel: Is that general in microorganisms? Plants have been regenerated from parallel cell cultures: one was selected with streptomycin, to get resistance, and a parallel culture was kept without streptomycin. The streptomycin-resistant plants always appear to be defective, i.e. less fit.

Hartl: There is a complication in plants that has not yet been mentioned. Very intense selection in a small population involves a tremendous amount of inbreeding. As you are well aware, there is a pronounced inbreeding depression relative to traits associated with reproduction. So with a plant or an animal resistant to a pathogen or a heavy metal, if the population has a decreased reproductive performance relative to some standard population, one has no way of knowing whether that decrease is intrinsically a property of the mechanism of resistance or whether it has come about as a secondary effect of inbreeding due to the intense selection.

Bell: In higher plants, resistance of, for example, seeds to a mould or an insect may involve synthesis of, say, a trypsin inhibitor or a strange amino acid. Such a synthesis will involve specialized enzymes, and the diversion of nutrients from primary metabolic pathways, so producing a *cost* for resistance. This *cost* to the plant may however be less than the cost that would have been exacted by predators in the absence of these compounds.

Hutchinson: I mentioned in my paper (p 52-72) some of the plants in the Smoking Hills area of arctic Canada. One of the populations is an *Artemisia* species that turns out to be, perhaps, the most tolerant plant to sulphur dioxide and to acid-droplet damage that we have come across. Given 1000 years of selection at this site, this high degree of tolerance might be expected. But in another population of *Artemisia tilesii* from Alaska, which has had no contact with sulphur dioxide, the plants, when tested under the same conditions, were just as tolerant to SO_2 or 'acid rain'. My conclusion is that this species is pre-adapted to acidity and SO_2 by mechanisms present for quite different reasons, and there is *no* cost.

Hartl: And also there is no fitness differential.

Hutchinson: That's right.

Bradshaw: To return to metal tolerance, there must be some fitness differential that is fairly permanent. If one has a metal-contaminated area along side a non-contaminated area then, even after many generations which have provided many opportunities for genetic recombination, one does not see the gene moving out from the toxic area for which it is being selected. This suggests that the gene has a permanent disadvantage from which it cannot dissociate. We are used to pleiotropy as a normal aspect of gene action; there is no reason why genes cannot have disadvantageous pleiotropic effects.

Hartl: The discussion has strayed a little from Dr Davies' point, which was that one cannot assume that all of these things are detrimental, even though many of them are.

Georgopoulos: It is a matter of what is the mechanism of resistance in each case, I believe.

Clarke: I must take you up on your point, Dr Hartl, about the ribosomal protein that affects a lot of other gene products and does not show any fitness differential in the chemostat (p 96). We know that *E. coli* and other organisms have evolved extremely efficient mechanisms for preventing genetic mistakes. A striking general observation is that there are extremely good error-correcting systems of various kinds at work during DNA replication. If it were generally true that mutation didn't matter, we would not have such precise mechanisms. Perhaps your chemostat is the equivalent of the *Drosophila* bottle where mutants do quite well, even though, when taken into the harsh world outside, they do not survive.

Hartl: You are right. I have no doubt that if we change the conditions of the chemostat in an appropriate way we can show selection against the chromosomal streptomycin-resistance gene. My point was that a mutation that is detrimental under one set of conditions need not be detrimental under another set. One has to be careful to specify those conditions.

Davies: An interesting thing about ribosomal-ambiguity mutants is that the mechanism for degrading the unwanted mistranslated proteins (an ATP-dependent protease) requires energy, and under normal culture conditions this doesn't seem to matter in terms of growth.

Bowers: Doesn't stress of various kinds to plants cause chromosomal breakage here and there and allow a great deal of recombination in a shorter period of time? This recombination could eventually be expressed, for plants tolerant (or resistant) to a particularly nasty toxicant. If the stress is removed, the plant can then, so to speak, clean its genes and return to its more highly productive self. Isn't this built into all organisms, so that stress produces more opportunity for more chromosomal breakage and recombination? For example, cells dispersed from a tomato or potato leaf can be grown under various stresses. Cells that survive can then be selected and reconstituted to form plants that are different from the parent plant. It is suggested that this reconstitution allows the expression of genetic information that was always there but had never been expressed before. Other people disagree and say that these conditions of dispersion, propagation *in vitro* and reconstitution allow numerous opportunities for chromosomal breakage, exchange of material and recombination. In that case these reconstituted plants could be considered different *because* of the trauma. So where is the balance?

Clarke: Is not the tendency for chromosomal rearrangement a specific disorder that is suffered by cells as a consequence of tissue culture?

Gressel: Genetic problems in tissue cultures depend on the species and the medium. Some species are stable and some show drift.

Clarke: But general environmental traumas do not necessarily affect chromosomal rearrangement.

Bowers: So the germplasm needn't be affected?

Clarke: No.

Gressel: Well this depends on what you are calling environmental trauma, which can also be ultraviolet light and X-rays.

Clarke: There are, of course, specific things that do cause trauma, which we take for granted. But there is no reason to suppose that just because there is a stress there will necessarily be an increased frequency of chromosomal rearrangement; it depends on the nature of the stress, and that may have to be something very specific to the DNA or the chromosomes.

Kuć: The conditions for the regeneration of single-cell protoplasts into new plants are often mutagenic. In work with potatoes, plants regenerated from protoplasts of a single cultivar exhibited great variation. Some resembled egg plants or pepper plants; some were resistant to diseases to which the original cultivar was susceptible; some were differentially resistant to diseases. I would like to see the following experiment performed. Take a plant that has been regenerated and is now resistant to a particular disease, and go through that procedure again to see if the percentage of resistant plants in the regeneration process always remains the same or is higher. This should help determine whether the variability was genetic, in cells of the plant, or was introduced during protoplast formation and regeneration.

Gressel: Most of the selections done on plant-cell cultures could not be done in the field because of the sheer numbers involved: 10^7 plants in cultures can be contained in one bottle whereas 10^7 plants in the field would occupy a vast area. It is hard to treat fields uniformly with toxicants in the way one can treat a cell-suspension culture. With such a culture one can work at a kill rate of 99.99% in a selection whereas in the field the best one can obtain is about 95%, so that 'misses' occur.

Bradshaw: We must also remember that in cell culture the environment provided for the plant cells can nurse mutants over deficiencies of growth that would eliminate them straight away in the field. So mutants that are disadvantageous (i.e. which have low fitness under normal conditions) can be preserved and then assessed subsequently to see if they have adapted to the particular factor of interest.

Epstein: Before everybody becomes too enthusiastic we should be aware that selection at the cellular level does not necessarily give traits that can express themselves at the whole organism, or plant, level. This is certainly true, usually, for salt tolerance, my own interest, where there are many control points and many different processes that go into the making of salt-tolerant plants, including even morphology, sometimes. One cannot catch it all at the cellular level; one is forced to use the slower and more conventional selection and breeding at the plant level.

Gressel: What you say also refers to herbicide resistance. Not all cells that are resistant in culture give rise to resistant plants. The herbicide may affect more than one target in the whole plant and some of these targets may be absent in cell cultures. For example; most cell cultures are white and do not have the photosynthetic apparatus that is the target of many herbicides.

REFERENCES

Curtis CF, Cook LM, Wood RJ 1978 Selection for and against insecticide resistance and possible methods of inhibiting the evolution of resistance in mosquitoes. Ecol Entomol 3:273-287

Georghiou GP, Lagunes A, Baker J 1983 Effect of insecticide mixtures and rotations on the evolution of resistance. Pergamon Press, London (Proc 5th Int Congr Pestic Chem, Kyoto, Japan) IVd-11

Hooper GHS, Brown AWA 1965 Dieldrin-resistant and DDT-resistant strains of the spotted root maggot apparently restricted to heterozygotes for resistance. J Econ Entomol 158:824-830

Wood RJ, Bishop JA 1981 Insecticide resistance: populations and evolution. In: Bishop JA, Cook LM (eds) Genetic consequences of man made change. Academic Press, London

Wood RJ, Cook LM 1983 A preliminary note on estimating selection pressures on insecticide resistance genes. Bull W H O 61:129-134

Phytoalexins and disease resistance mechanisms from a perspective of evolution and adaptation

JOSEPH KUĆ

Department of Plant Pathology, University of Kentucky, Lexington, Kentucky 40546, USA

Abstract. Plants respond to cellular injury and infection by accumulating low molecular weight antimicrobial stress metabolites called phytoalexins. The accumulation of phytoalexins, together with lignification, suberization, callose formation and the production of agglutinins and inhibitors of extracellular microbial hydrolases, appears to be part of a multi-component response mechanism associated with disease resistance and wound repair. Compared to the antibody–antigen response in animals, the phytoalexin response in plants has low specificity for induction and activity of the phytoalexins. Plants also contain preformed antimicrobial chemical and physical barriers to infection in their external tissues. The successful pathogen has evolved to cope with preformed inhibitors and barriers and either avoids eliciting the response mechanism, or suppresses the mechanism, or detoxifies its antimicrobial components.

Annual plants can be systemically immunized against diseases caused by viruses, bacteria and fungi by limited infection with any one of the respective organisms. As with animals, disease resistance in plants depends on the rate and magnitude of response rather than on the ability to respond. The genetic information for disease resistance is found in all organisms, and disease resistance is the rule in nature. The interactions of plants with microorganisms in their environment are nature's example of diplomacy—compromise, adjustment to change and avoidance of deadly conflict.

1984 Origins and development of adaptation. Pitman Books, London (Ciba Foundation symposium 102), p 100-118

Adaptation to stress is a physiological process, and survival of a species testifies to the success of the process. An organism persists in an ecological niche because the environmental and nutritive conditions it encounters are adequate for it to develop and compete successfully with other organisms in the environment. The effect on an organism of change in a single environmental or nutritive factor, however, disrupts multiple chains of interdependent metabolic events. The organism must, therefore, not only adjust to stress due to a single change in a variable but must adjust also to all the metabolic changes resultant from that change. The speed and effectiveness with which an organism adjusts are vital to its survival.

It is axiomatic that plants have developed effective mechanisms for resistance to all infectious agents in their environment to survive the selection pressure of evolution. Mechanisms adequate for the survival of a species, however, may not satisfy the demands for high crop yield and quality imposed by modern agriculture. Nevertheless, susceptibility to disease is the exception in nature, and resistance is the common phenomenon. One hypothesis from the above observations is that all plants contain the genetic potential for resistance to fungal, bacterial and viral diseases. The determinants of resistance would then be the rate and magnitude with which this potential is expressed.

One mechanism for disease resistance in plants is their ability to accumulate low molecular weight antimicrobial compounds (phytoalexins) as a result of infection. The phytoalexins are localized in and immediately around the site of infection, and the rate and magnitude with which they are produced and accumulate appear to determine the disease reaction in some plant–microbe interactions.

It appears, however, that phytoalexin accumulation is only one component of the plant's mechanism for disease resistance. Other mechanisms, and their coordination with phytoalexin accumulation, determine disease resistance or susceptibility (Kuć 1982a,b,c, Kuć & Caruso 1977). Other components of the disease-resistance mechanism include lignification, suberization, and formation of callose, agglutinins and enzyme inhibitors. Many of these responses are typical of a wound response.

Are phytoalexins and other agents for resistance compatible with the hypothesis that all plants have the genetic potential for disease resistance to all infectious agents in their environment? Is there a metabolic pattern in evolution for development of disease-resistance mechanisms? Does disease resistance in plants depend on the activation of a mechanism in the host which is unique for its role in defence against disease and which, in turn, is elicited by structural components or metabolites unique to the infectious agent?

Phytoalexins—structural considerations

The accumulation of phytoalexins has been demonstrated in at least 15 plant families (Coxon 1982, Ingham 1982, Kuć 1982d). They have been reported frequently in angiosperms and dicotyledons, rarely in gymnosperms and monocotyledons, and not at all in non-vascular plants.

Similarities are evident among phytoalexins from plants within a family. Plants from the Leguminosae, Solanaceae, Compositae and Convolvulaceae

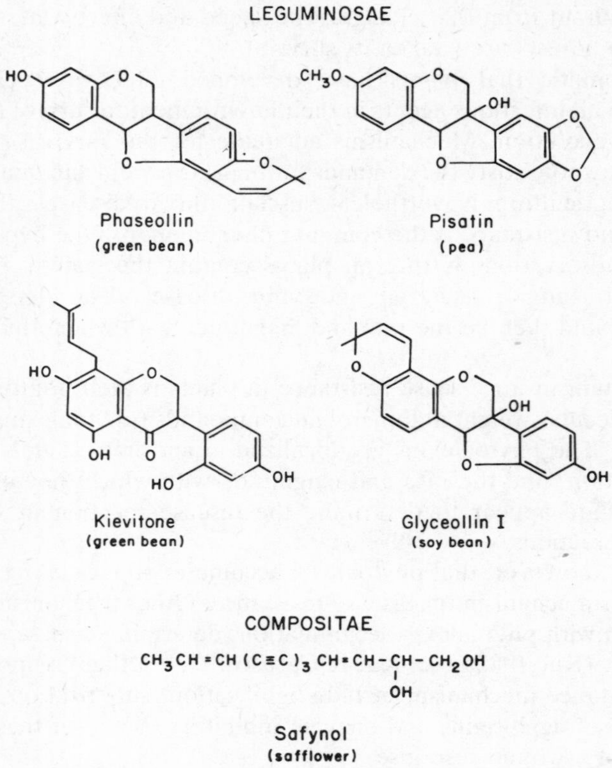

FIG. 1. Phytoalexins from members of the Leguminosae and Compositae.

produce predominantly isoflavonoids, carbocyclic sesquiterpenoids, polyacetylenes and furanosesquiterpenoids, respectively (Figs. 1 and 2). Of the 102 phytoalexins reported in the Leguminosae, 84 are isoflavonoid derivatives, eight are furanoacetylenes, four are stilbenes, three are benzofurans, two are chromones and one is a flavanone (Ingham 1982). Similarly, in the Solanaceae, 34 phytoalexins reported are terpenoid derivatives, six are phenylpropanoid phenols and three are polyacetylenes (Kuć 1982d). Though some phytoalexins are distributed among many families—e.g. caffeic acid derivatives accumulate in potato, carrot and sweet potato—plants in the Leguminosae have not been reported to produce sesquiterpenoid phytoalexins, and those in the Solanaceae have not been reported to produce isoflavonoids. Are these observations surprising? Clearly a potato is different from a bean; therefore, why should phytoalexins from two different plant

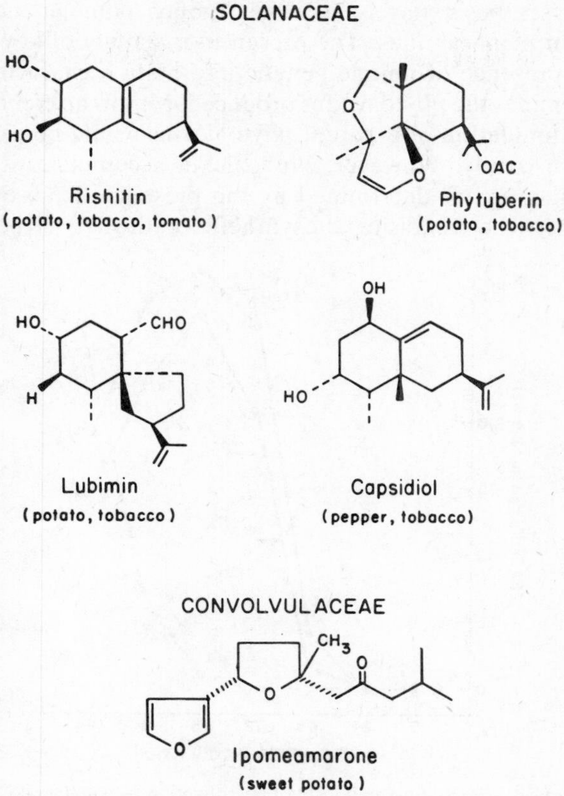

FIG. 2. Phytoalexins from members of the Solanaceae and Convolvulaceae.

families be in the same class of compounds that are synthesized by the same metabolic pathways? In animals of different families this happens. Antibodies are proteins, whether from dogs, cats or humans. The antibodies produced by dogs, cats and humans in response to a common antigen have relatively minor differences in amino acid content and sequence but all are rather specific with regard to their induction and interaction with the same antigen.

What are the metabolic implications of these observations in plants? The three basic pathways for biosynthesis in plants are the acetate–malonate, acetate–mevalonate and shikimate pathways. Some phytoalexins arise from precursors produced by all three, e.g. phaseollin, kievitone, the glyceollins; others are synthesized by only one, e.g. 6-methoxymellein, rishitin, ipomeamarone. Clearly the basic pathways are common to all plants, though animals lack the shikimate pathway.

Phytoalexin synthesis may be at a key branch point in the evolutionary development of plant families. The presence or activity of key enzymes, and therefore the presence of unique genetic information characteristic of plant families, determines the phytoalexins produced. Minor differences in structure or rates of accumulation of different phytoalexins would be expected among plants within a family. However, phytoalexin accumulation as a factor in disease resistance is not determined by the presence or absence of genetic information for the requisite biosynthetic pathways (Kuć 1972, 1976,

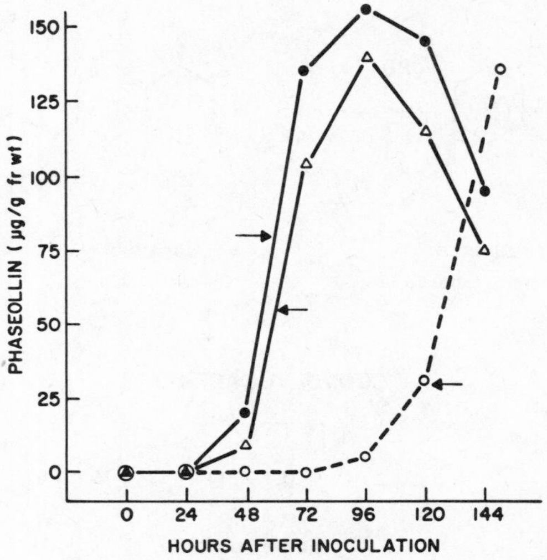

FIG. 3. Accumulation of phaseollin and onset of necrotization (→) in hypocotyls of a green bean cultivar inoculated with a compatible race (○ - - - ○) or incompatible race (△—△) of *Colletotrichum lindemuthianum* and in hypocotyl segments inoculated with a compatible race (●—●) taken from hypocotyls inoculated on both sides of the segments with an incompatible race or *Colletotrichum* non-pathogenic on bean 18 hours before inoculation with the compatible race.

1982a,b,c, Kuć & Caruso 1977), and would thus be compatible with my initial hypothesis that all plants contain the genetic potential for resistance mechanisms. This observation is consistent with the ability of dogs, cats and humans to produce antibodies, and all three, though susceptible to diseases, can be immunized for life against some of them.

The phytoalexins are not notably active as antibiotic agents though the quantity of phytoalexin that accumulates at sites of infection is sufficient to inhibit markedly the growth of some fungi and bacteria *in vitro*. Phytoalexins have a broad spectrum of activity against pathogens and non-pathogens of a host, and a broad range of pathogens, non-pathogens and physiological stresses can elicit accumulation of the same phytoalexin. A case for the involvement of phytoalexins in disease resistance depends on the rate and magnitude of their production (Figs. 3 and 4) and not on their selective toxicity *per se* or on their specific elicitation (Kuć 1976, Smith 1982). This is markedly different from the rather high level of specificity associated with the induction of antibodies and their interaction with antigens in animals, though

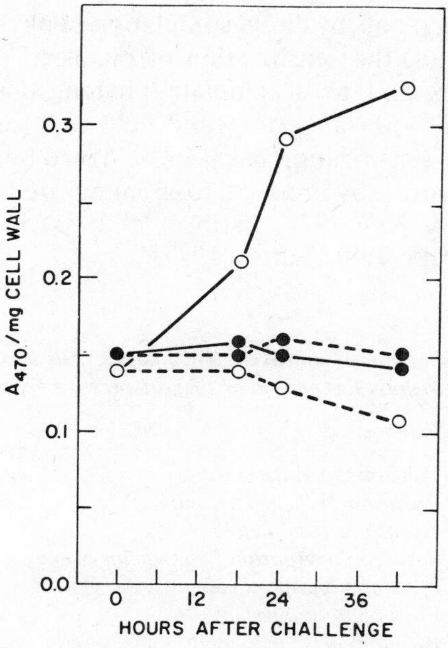

FIG. 4. Lignin content of Marketer cucumber apical tissue after challenge with *Cladosporium cucumerinum*. Lignin, extracted from cucumber cell walls with hot aqueous alkali, was measured colorimetrically. Systemic resistance was induced with *Colletotrichum lagenarium* and tissue was challenged with *C. cucumerinum* (○—○). Resistance not induced and tissue challenged, (●—●); resistance induced and tissue not challenged, (○ - - - ○); resistance not induced and tissue not challenged, (● - - - ●).

the efficacy of the antibody–antigen interaction in animals is also highly dependent on the rate and magnitude of antibody production. Unlike protein antibodies in animals, the phytoalexins have low relative molecular masses (M_r) and are generally lipophilic products of secondary metabolism. They are localized at the site of infection or stress. A signal which conditions plants to accumulate phytoalexins rapidly when infected, however, is translocated (Kuć 1982a,b,c, Kuć & Caruso 1977). Thus, phytoalexins accumulate rapidly after infection or stress in plants 'sensitized' to respond by prior infection (Figs. 3, 4). The energy and precursors required for phytoalexin synthesis are conserved until when and where they are needed. Many phytoalexins are phytotoxic at concentrations that inhibit the development of microorganisms. Attempts to protect plants against disease by stimulating production and maintaining high levels of phytoalexins systemically in foliage have proved unsuccessful. The frequent application of fungus-derived glucan elicitors to the foliage of green bean and soybean plants causes severe necrotization and stunting of growth.

Though protecting plants by direct stimulation of phytoalexin accumulation has been unsuccessful, the sensitization of plants to rapidly mobilize their defence mechanisms and to accumulate phytoalexins after infection has proved successful in both laboratory and field experiments (Figs. 3 and 4; Tables 1 and 2). The sensitization has been achieved by limited infection with fungi, bacteria or viruses, by products from immunized plants or by synthetic chemicals (Caruso & Kuć 1977, Farih et al 1981, Kuć 1982a,b,c, Kuć & Caruso 1977, Langcake 1981, Sutton 1979).

TABLE 1 The biological spectrum of effectiveness of systemic resistance induced by foliar infection of cucumber with *Colletotrichum lagenarium* or tobacco necrosis virus

Disease	*Pathogen*
Anthracnose	*Colletotrichum lagenarium*
Scab	*Cladosporium cucumerinum*
Gummy stem blight	*Mycosphaerella melonis*
Fusarium wilt	*Fusarium oxysporum* f. sp. *cucumerinum*
Downy mildew	*Pseudoperonospora cubensis*
Local necrosis	*Phytophthora infestans*
Angular leaf spot	*Pseudomonas lachrymans*
Bacterial wilt	*Erwinia tracheiphila*
Cucumber mosaic	Cucumber mosaic virus
Local necrosis	Tobacco necrosis virus

TABLE 2 Protection of cucumber plants in the field by immunization

Measurement	*Leaf 2*		*Leaf 3*	
	I[a]	*C*[b]	*I*	*C*
Number of lesions[c]	7	36	3	10
Diameter of lesions (mm)	1.4	4.1	0.4	3.5
Area of lesions (mm^2)	12	486	0.5	109
% Protection based on area of lesions	98%		99.5%	

[a] Plants immunized by restricted inoculation of the first true leaf (leaf 1) with spores of *Colletotrichum lagenarium*. [b] Control plants treated with water. [c] Plants challenged by inoculating leaf 2 with 40 and leaf 3 with 10 drops of a spore suspension of *C. lagenarium*.

Phytoalexins as stress metabolites

Though infection with fungi, bacteria and viruses elicits phytoalexin accumulation, other stress-inducing factors such as mercuric or cupric chloride, ethylene, ultraviolet irradiation, metabolic inhibitors, cell-wall components of fungi and bacteria, microbial metabolites, and insect and nematode damage are also elicitors (Kuć 1976). This is especially true for plants in the Leguminosae. This rather low level of specificity to elicitation neither supports nor detracts from the possible contribution of phytoalexins to disease resistance but should be noted when considering the molecular basis for elicitation and the evolution of the response.

A common feature of phytoalexin accumulation in plants after infection is a limited necrotization of plant tissues. Failure to accumulate phytoalexins is associated with the development of a pathogen without visible upset of plant metabolism, and without extremely rapid death of tissues, or rapid metabolism of the phytoalexins by pathogen or host.

In some susceptible reactions between fungi and plants, phytoalexins accumulate to higher levels than in resistance interactions. This is often evident when biotrophic growth terminates, sporulation occurs and lesions develop (Kuć 1972, Kuć & Caruso 1977). In this case phytoalexin accumulation occurs too late to contain the fungus and to restrict disease development.

Two distinct aspects of phytoalexin elicitation are apparent. The first is necrotization, or damage to plant cells by chemical or physical agents, which releases a non-phytotoxic substance. The second is the migration of the non-phytotoxic substance to healthy cells where it conditions increased synthesis of phytoalexins (Bailey 1982, Kuć 1972, 1982a,b,c, Kuć & Caruso 1977). The phytoalexins are then transported out of the producer cells and accumulate in the necrotized or damaged tissues. The actual elicitor of phytoalexin synthesis and accumulation is not the fungal glucan, the toxic chemical or the ultraviolet irradiation but, rather, the substance released by the damaged tissues. Thus, a major emphasis should be placed on *why* many fungal cell-wall glucans are highly phytotoxic rather than on how they elicit phytoalexin accumulation. The immediate elicitor of phytoalexin synthesis and accumulation may be this non-phytotoxic substance released from damaged cells.

Mechanical wounding or short-term injury, however, are generally inadequate stimuli for appreciable accumulation of phytoalexins (Kuć 1972, 1976, 1982a,b,c). A low-level, persistent stress, perhaps caused by the continued presence of the infectious agent or elicitor(s), seems to be required for high levels of phytoalexins to accumulate. This is consistent with the observation that when the infectious agent is contained, or with fungi and bacteria perhaps killed, phytoalexin levels drop towards those found in uninfected plants. Similarly, the hydrolysis of glucan elicitors and the dilution or sequestering of inorganic toxicants would, with time, remove the source of stress, and phytoalexin levels would drop unless the elicitors or toxicants caused the release of additional elicitors of host origin.

The antigen–antibody aspect of the disease-resistance mechanism in *animals* depends on the activation of a mechanism unique for its role in defence against disease, and which is elicited by structural components or metabolites unique to an infectious agent. Furthermore, the antibody is rather specific for the antigen. This uniqueness and specificity do not appear evident in the mechanisms for disease resistance reported in plants, though the reaction of plants to different pathogens and races of pathogens can be highly specific. The inflammation response and the production and action of interferon, however, are aspects of disease resistance in animals that are not unique for an infectious agent and have low specificity.

Susceptibility—a function of non-recognition of non-self or suppression of resistance?

Diseases initiated by obligate parasites depend on the maintenance of living host tissue for the development of the parasite. Rusts, smuts, powdery and downy mildews can persist in a plant without eliciting a host response that leads to necrotization or phytoalexin accumulation until reproduction has been initiated. Viral diseases require host participation in the replication of viruses, and in numerous bacterial diseases rapid multiplication of bacteria precedes the appearance of host damage. Some facultative fungi, e.g. *Phytophthora infestans, Colletotrichum lindemuthianum, Phytophthora megasperma* f. sp. *glycinea*, resemble obligate parasites during initial stages of pathogenesis, and some bacterial pathogens rapidly degrade host tissue during, or shortly after, infection.

The lack of host response during early stages in the development of biotrophic fungi may be due to the inability of the host to recognize the pathogen as *non-self*. With the onset of a shift from vegetative to reproductive development, physiological changes occurring in the pathogen may unmask recognition sites, or metabolites that are produced may cause physiological stress and host response. With viruses and bacteria, a physiological stress produced by a threshold amount of host or microbial metabolites, or by energy or substrate drain, can cause a host response. With some systemic plant viruses, bacteria or fungi this threshold value of non-self recognition may not be reached. A sizeable body of evidence shows that some pathogens suppress host responses, by *induced susceptibility*.

The subject of induced susceptibility has been thoroughly reviewed (Ouchi

& Oku 1982, Doke et al 1982, Kuć 1982d, Kuć et al 1983), and several reports indicate that it is associated with suppressed accumulation of phytoalexins. The germination fluids of *Mycosphaerella pinodes* were reported to contain low M_r suppressors of pisatin production (Ouchi & Oku 1982). The suppressors induced susceptibility of pea leaves to several non-pathogens and markedly reduced pisatin accumulation in the tissues. Ouchi & Oku (1982) also made the extremely interesting observation that the suppressors were active on other legumes that were susceptible (but not on those resistant) to the pathogen. The key to race specificity in the potato–*P. infestans* interaction may be the ability of compatible races of the fungus to produce extracellular glucans which suppress hypersensitive cell death, necrosis, and the accumulation of sesquiterpenoid phytoalexins (Doke et al 1982, Kuć 1982d, Kuć et al 1983).

Susceptibility—a function of toxin production or metabolism of phytoalexins?

Host-specific and non-specific microbial toxins have been characterized and these have been implicated in symptom development in plant disease (Daly 1982). Strong evidence exists that with the 12 documented host-specific toxins, the susceptibility of the plant to disease is determined by susceptibility to toxin. Both pathogen and toxin specifically damage host plants, and symptoms produced by either are often identical. Action of toxins may include their ability to suppress resistance mechanisms in a host. This could arise either because of rapid damage or death of host tissues in advance of the pathogen or because of more specific inhibition of host defence, which permits the development of the pathogen and the production of the toxin.

Detoxification of phytoalexins by fungi is well documented and appears to affect their virulence and host range (VanEtten et al 1982, Smith 1982). All phytoalexins reported to date are degraded by the plant tissue in which they are synthesized.

Induced resistance

Recent reports by our research group indicate that cucurbits can be systemically immunized against viral, bacterial and fungal diseases by infection with any one of the respective organisms or by fractions from immunized plants (Fig. 4, Tables 1 and 2; Caruso & Kuć 1977, 1979, Jenns & Kuć 1980, Kuć & Richmond 1977, Richmond et al 1979). In addition, green bean has been systemically immunized against anthracnose caused by *C. lindemuthianum* (Fig. 3), and tobacco against blue mould caused by *Peronospora hyoscyami* f.

sp. *tabacina* (Kuć 1982c, Kuć & Caruso 1977, Sutton 1979). Immunization followed by a booster inoculation protects plants through the period of flowering and fruiting. Controlled infection with *Colletotrichum lagenarium* or tobacco necrosis virus protected cucumber against disease caused by a broad spectrum of pathogens including obligate and facultative fungi, local lesion and systemic viruses, fungi and bacteria that cause either wilts or restricted and non-restricted lesions on foliage and fruit (Table 1). Induction of resistance is graft-transmissible and is effective against foliar as well as root pathogens. Mechanical injury or injury caused by dry ice, chemicals or components of fungi and uninfected plants did not elicit systemic protection. It appears that immunization in cucurbits involves sensitization of the plants to infection, and the response is rapid and multicomponent. One aspect of the response is rapid lignification (Fig. 4) at the sites of infection in immunized plants (Hammerschmidt & Kuć 1982).

Plants highly susceptible to all races of a pathogen, and therefore considered to lack genes for resistance to the pathogen, can be as effectively immunized against it as can plants that are resistant to some races. This is consistent with the hypothesis that all plants contain genetic information for highly effective resistance mechanisms, and that the difference between susceptibility and resistance is determined by the rate and magnitude with which genetic information is expressed. Our data (Kuć 1982a,b,c, Kuć & Caruso 1977, Jenns & Kuć 1980) indicate that plant immunization depends on the activation of multiple mechanisms for resistance and that the induction of the mechanisms is non-specific.

Discussion

Plants do not seem to have metabolic mechanisms that are unique for their role in disease resistance. The resistance mechanisms are not highly specific with regard to their induction, the products produced and the specificity of the products for inhibiting development of pathogens. In spite of this non-specificity, the mechanisms are extremely effective in protecting plants against disease and in permitting survival. Resistance, and not susceptibility, is the common phenomenon in nature.

A degree of specificity related to plant family is apparent in the structure of phytoalexins. Variations in substituent groups and in their location on molecules of a class of compounds are apparent between members of a family. The phytoalexins are generally lipophilic and are not translocated throughout the plant. They are not stable end-products of metabolism and they are degraded in host tissues even in the absence of infectious agents. Evidence suggests that phytoalexins can protect plants against diseases caused by fungi and bacteria, but neither phytoalexins, nor any other agent or mechanism, have been shown to influence the resistance of plants to viral diseases.

The highest specificity associated with resistance is the hypersensitive response observed in the interaction of plants with some fungi, bacteria and viruses. This response is associated with the very rapid but limited death of a few cells penetrated or immediately surrounding the infectious agent, and the response is characteristic of race-specific resistance and resistance of non-hosts. The hypersensitive reaction is also associated with rapid accumulation of phytoalexins. The rapid recognition of non-self may trigger the hypersensitive reaction but the mechanism for this recognition is unknown. Three components of a resistance mechanism involving phytoalexins are apparent: (1) necrotization; (2) induction of phytoalexin synthesis in healthy tissue adjacent to necrotic tissue by a chemical signal (the actual elicitor of phytoalexin accumulation); and (3) the movement of a second signal from infected tissues throughout the plant, which sensitizes cells to react rapidly when infected. This latter signal is not an elicitor of phytoalexin accumulation but it does *condition* the plant for resistance, including the rapid accumulation of phytoalexin, *in the presence of* a pathogen.

The successful pathogen has the ability to bypass or to degrade preformed physical and chemical barriers and to obtain adequate nutrition from the host. In addition, it must either avoid recognition by the plant as non-self, or actively suppress the host's resistance response mechanism, or cope with the mechanism.

The ability to immunize plants against some diseases is strong support for a systemic signal that sensitizes plants to recognize infectious agents and to respond rapidly. In both plant and animal systems, the genetic information is present for resistance and the key is its rapid expression.

In evolution, the plant's ability to respond to stress and injury are of principal importance. Plants respond by producing stress metabolites (some of which act as phytoalexins), by activating mechanisms for wound repair and by mobilizing mechanisms for isolating the affected tissues. As organisms have evolved and interacted with each other, some have developed the means to cope with the stress mechanisms of other organisms and have become pathogens. This has seldom resulted in the complete destruction of either of the interacting organisms, and susceptibility generally requires the fine tuning of the metabolism of both.

Immunization in animals after recovery from some diseases is well documented. The immune mechanism in animals is vital for survival, but it did not prevent smallpox and other diseases from repeatedly devastating humans for centuries. A resistance mechanism that permits the survival of humans or plants is effective from the long-term view of perpetuating the species. An apple that is covered with scab lesions is unacceptable in today's consumer market, though the seeds of that apple may be capable of germinating and producing trees that bear more apples. A human defence mechanism that allows for the death of multitudes periodically is no longer effective by our standards, but the manipulation of this same mechanism has produced the highly effective modern immunization which has eliminated smallpox and controlled other serious human diseases. The practical control of disease in plants by immunization is now a distinct possibility.

Acknowledgements

The author's research reported in this paper has been supported in part by grants from the Ciba-Geigy Corporation, Rockefeller Foundation, USDA Cooperative Agreement 58-7B300-185 and USDA/SEA Competitive Grant 78-59-2211-0-1-063-1. The investigation reported in this paper (no. 83-11-5) is in connection with a project of the Kentucky Agricultural Experiment Station and is published with approval of the Director.

References

Bailey JA 1982 Mechanisms of phytoalexin accumulation. In: Bailey JA, Mansfield JW (eds) Phytoalexins. Blackie, Glasgow, p 289-318

Caruso F, Kuć J 1977 Field protection of cucumber, watermelon and muskmelon against *Colletotrichum lagenarium* by *Colletorichum lagenarium*. Phytopathology 67:1285-1289

Caruso F, Kuć J 1979 Induced resistance of cucumber to anthracnose and angular leaf spot by *Pseudomonas lachrymans* and *Colletotrichum lagenarium*. Physiol Plant Pathol 14:191-201

Coxon DT 1982 Phytoalexins from other families. In: Bailey JA, Mansfield JW (eds) Phytoalexins. Blackie, Glasgow, p 106-132

Daly J 1982 The host-specific toxins of *Helminthosporia*. In: Asada Y et al (eds) Plant infection. Japan Scientific Soc Press, Tokyo, and Springer-Verlag, Berlin, p 215-232

Doke N, Tomiyana K, Furichi N 1982 Elicitation and suppression of the hypersensitive response in host–parasite specificity. In: Asada Y et al (eds) Plant infection. Japan Scientific Soc Press, Tokyo, and Springer-Verlag, Berlin, p 79-94

Farih A, Tsao P, Menge J 1981 Fungitoxic activity of efosite aluminum on growth, sporulation and germination of *Phytophthora parasitica* and *P. citrophthora*. Phytopathology 71:934-936.

Hammerschmidt R, Kuć J 1982 Lignification as a mechanism for induced systemic resistance in cucumber. Physiol Plant Pathol 20:61-71

Ingham JL 1982 Phytoalexins from the Leguminosae. In: Bailey JA, Mansfield JW (eds) Phytoalexins. Blackie, Glasgow, p 21-80

Jenns A, Kuć J 1980 Characteristics of anthracnose resistance induced by localized infection with tobacco necrosis virus. Physiol Plant Pathol 17:81-91

Kuć J 1972 Phytoalexins. Annu Rev Phytopathol 10:207-232

Kuć J 1976 Phytoalexins and their specificity in plant–parasite interactions. In: Wood R, Graniti A (eds) Specificity in plant disease. Plenum Press, New York, p 253-271

Kuć J 1982a Plant immunization-mechanisms and practical implications. In: Wood R, Tjamos E (eds) Active defense mechanisms in plants. Plenum Press, New York, p 157-178

Kuć J 1982b The immunization of cucurbits against fungal, bacterial and viral diseases. In: Asada Y et al (eds) Plant infection. Japan Scientific Soc Press, Tokyo, and Springer-Verlag, Berlin, p 137-153

Kuć J 1982c Induced immunity to plant disease. Bioscience 32:854-860

Kuć J 1982d Phytoalexins from the Solanaceae. In: Bailey JA, Mansfield JW (eds) Phytoalexins. Blackie, Glasgow, p 81-105

Kuć J, Caruso F 1977 Activated coordinated chemical defense against disease in plants. In: Hedin P (ed) Host resistance to pests. Amer Chem Soc Press, Washington DC (Amer Chem Soc Symp Ser No 62), p 78-89

Kuć J, Richmond S 1977 Aspects of the protection of cucumber against *Colletotrichum lagenarium* by *Colletotrichum lagenarium*. Phytopathology 67:533-536

Kuć J, Tjamos E, Bostock R 1983 Metabolic regulation of terpenoid accumulation and disease resistance in potato. In: Ness D et al (eds) Biochemistry and function of isopentenoids in plants. Marcel Dekker, New York, in press

Langcake P 1981 Alternative chemical agents for controlling plant disease. Philos Trans R Soc Lond B Biol Sci 295:83-101

Ouchi S, Oku H 1982 Physiological basis of susceptibility induced by pathogens. In: Asada Y et al (eds) Plant infection. Japan Scientific Soc Press, Tokyo, and Springer-Verlag, Berlin, p 117-136

Richmond S, Kuć J, Elliston J 1979 Penetration of cucumber leaves by *Colletotrichum lagenarium* is reduced in plants systemically protected by previous infection with the pathogen. Physiol Plant Pathol 14:329-338

Smith DA 1982 Toxicity of phytoalexins. In: Bailey JA, Mansfield JW (eds) Phytoalexins. Blackie, Glasgow, p 218-252

Sutton D 1979 Systemic cross protection in bean against *Colletotrichum lindemuthianum*. Australas Plant Pathol 8:4-5

VanEtten HD, Matthews DE, Smith DA 1982 Metabolism of phytoalexins. In: Bailey JA, Mansfield JW (eds) Phytoalexins. Blackie, Glasgow, p 181-217

DISCUSSION

Clarke: If the attacking fungus primes the plant so that it is ready to produce the phytoalexin response when it is attacked again, why does the plant not get itself into that 'ready' state all the time? What is the evolutionary explanation of the need to be primed?

Kuć: That is difficult to answer. To respond at all times to all things is a tremendous drain on the energy and mobilization of carbon compounds of the plant. People often ask a related question: if what I have said is true, and the initial inoculation induces resistance, then how is it that we have epidemics? In the field, epidemics do not arise by a gentle inoculation on a single leaf by a few spores, and then by a second inoculation with the pathogen after a waiting period. Often the build-up of inoculum to a high level on one plant will affect adjacent plants, and so on. Within that population will be a few plants that are protected. But such high levels of inoculum are thus generated, over a continuous period of time, that the plants subjected to these high levels don't have the time to develop a sensitization response. One could argue that it would be far better for them to be sensitive at all times. On the other hand, this might produce problems. In experiments with immunized cucurbits and tobacco, we do not see deleterious effects, but these might be possible. If the plants are sensitized to respond to many stimuli this may actually be metabolically harmful; they may be over-responding to stimuli continuously, which might restrict growth.

Hodgson: Your description of the mechanism suggests that when it is finally turned on one is seeing gene expression. In other words, a wide variety of proteins must suddenly be synthesized. Do inhibitors of events at the nucleic acid level shut off the whole mechanism, perhaps by preventing messenger RNA formation?

Kuć: Yes; this may be true in cucumber.

Hodgson: Does this approach offer some hope for manipulating the system rather than having to isolate a variety of rather complicated organic molecules?

Kuć: What intrigues me is that if my hypothesis is correct, the mechanisms are present for very effective resistance. The genetic information is there. We are talking about regulation. The alarm signals (immunization factors) we are considering may be simple molecules that are not acting as inhibitors of enzymes or metabolic pathways *per se*. They may regulate metabolism on a hormonal level which results in a cascade of metabolic events leading to resistance. It is encouraging that we can obtain both natural and synthetic compounds that lead to the expression of resistance.

Davies: Is it possible to isolate plant mutants that produce phytoalexins constitutively?

Kuć: I don't know of any such work. It might not be to the advantage of the plant to continuously have a high level of phytoalexins because at high levels these compounds themselves become phytotoxic, and their production and degradation would involve a tremendous expenditure of energy. I can see advantages to the way nature is doing this: the phytoalexin accumulates at the site at high levels and, at least with bacteria and fungi, the organism is contained, and then the phytoalexin levels drop.

Bailey: Most phytoalexins are phytotoxic to many organisms. In answer to Dr Davies' question, there is one example. Mutant pea plants were obtained which constitutively produced pisatin. Such plants were extremely dwarfed (Hadwiger et al 1976).

Datta: You said that the response is not specific for a particular attacker, and also that the response lasts for a life-time. In that case, surely, nearly all plants must be ready for attack because each must have been attacked by an invader early in its life and have responded. What is disadvantageous about that?

Kuć: I am not sure that all plants can really be considered diseased as a result of many organisms in the course of their development. A spore might land on a leaf of the plant and might not germinate if the atmospheric conditions (moisture, temperature) are not appropriate, or the germ tube from a spore might not penetrate the cuticle, or it might encounter pre-formed inhibitors. It is only when all environmental conditions are suitable and the other defences have been breached that the pathogen would enter. We put a sizeable number of spores on a very small spot under ideal conditions for penetrations, to sensitize the plant. In animal systems there is not only specificity of induction, but the antibody also has a rather high specificity with respect to the antigen. It would be difficult to say that we humans have avoided any particular pathogen, and yet we might not naturally become immunized to some.

Datta: But most of us are immune to lots of things.

Kuć: Yes, but not to *all* things.

Datta: But we are not plants!

Clarke: The second set response in the vertebrate immune system is, however, the consequence of its specificity; it has to build up the specificity in the 'memory', and that takes time. So I wonder why you saw a second set response in plants when there was no specificity.

Kuć: The sensitization (immunization) in plants appears to depend on the plant's rapid recognition of an infectious agent as non-self rather than as a particular infectious agent. Thus the plant responds with multiple non-specific resistance mechanisms effective against a broad range of infectious agents.

Bell: Is it correct that in certain circumstances phytoalexins are produced in response to mechanical damage rather than to invasion by an organism? How does that fit in with the specificity?

Kuć: This further supports what I said about phytoalexins not being specific with respect to induction and having a low specificity with respect to the organisms that they will inhibit. The phytoalexins are one aspect of the defence reaction of the plant, and I look upon them as a very primitive response to stress. In evolutionary development plants developed a rather non-specific but effective response to stress which included phytoalexin accumulation. This has become more sophisticated as we have progressed, through evolution, to higher forms of life.

Bell: Does this mean that if you were to damage the tissue mechanically, say with a pair of scissors, you would then subsequently be protecting it against further attack?

Kuć: No; we have never been able to immunize tobacco, potato or cucurbits systemically by injuring the tissue either mechanically, or by using toxicants of various sorts. Injury can, however, result in localized protection at or immediately around the site of injury.

Bell: This suggests that we are dealing with different mechanisms.

Kuć: No; it may be that we have not yet been clever enough to produce a suitable low-level and persistent stress in plants.

Georghiou: You indicated that grafting would transmit the protection. But would an insect feeding on a protected plant transmit the protective mechanism to another plant, perhaps mechanically by sap transfer?

Kuć: I have never tested that. With cucurbits, however, we can not only protect them against many diseases but we can also protect them against aphids by the techniques for immunization that I have described.

Sawicki: You mentioned the question of protection against viruses. How do you achieve this? Is an aphid-transmissible virus involved?

Kuć: We can immunize against mechanically or aphid-transmitted cucumber mosaic virus.

Sawicki: So you can protect against both modes of transmission?

Kuć: Yes. We believe that we are not just protecting against the aphid because we can protect the cucumber plant by the mechanically transmitted virus as well.

Gressel: Do you know which you are protecting against?

Kuć: The mechanisms for protection against viruses are under investigation. I do not know the molecular basis for the success of immunization in plants against virus diseases.

Sawicki: We are inoculating viruses onto very young seedlings of sugar beet, and we have always succeeded in inoculating with these viruses provided we use enough viruliferous aphids (A.D. Rice, M. Stribley, unpublished results). Could you suggest why this is so?

Kuć: If this is a virus that produces systemic symptoms and local lesions on the young seedlings, then your success wouldn't surprise me. On the other hand, it might be interesting to see what would happen if you stressed the sugar beet by inoculating it with another 'local-lesion type' pathogen first, and then inoculated it with the virus. This might be far less successful.

Bailey: You used the word 'stress', and I have some comments about cell injury or death, which relate to one or two of the questions. Cell death is clearly important in the systemic protection studies and it also has a crucial role in the way in which phytoalexins are formed. Cutting tissue is not very effective, but by other ways of killing or injuring cells, e.g. by using mercury, copper, various natural products, fungi, bacteria or viruses, one can see a relationship between the timing of cell death, the timing of phytoalexin formation, and the extent of cell death (Bailey 1982). If only part of a tissue is killed, one obtains phytoalexins, but if all the tissue is killed, one does not obtain them. There seems to be an interaction between the dying cells and the cells around them. Materials emanating from dead cells cause synthesis of phytoalexins around the dead cells. The synthesis of phytoalexins requires new messenger RNA and new synthesis of enzyme; it may also require activation of enzymes.

Kuć: One must differentiate between the signal that conditions cells to respond in one of many ways, some still unknown, from the actual response itself—the mechanism that leads to an end-product that contains the infectious agent.

Graham-Bryce: Is there any more to be learned from the pathogens that successfully overcome the systemic cross-protection that one sees by inoculation? You indicated that this cross-protection is not complete, and yet the systemic inoculation produces a capability for a whole range of responses. Do the pathogens that overcome those responses have any special characteristics?

Kuć: We are very interested in this in our potato work. We want to know why a branched-chain carbohydrate can apparently interfere with the phytoalexin-eliciting activity of a polyunsaturated fatty acid. We seem to have here a fascinating system for regulation.

Gressel: Isn't it true that different biotypes of plants make carbohydrases that keep fungi off roots?

Kuć: Plants do produce carbohydrases, but whether they are important in disease resistance to foliar or root pathogens is debatable.

Davies: You mentioned that arachidonic acid was one of the signals involved in phytoalexin production and breakdown. Do prostaglandins also work? Proteases are activated and phytoalexins are degraded, and something similar to arachidonic acid is involved. So a whole cascade of phosphorylation reactions could lead to synthesis and breakdown. Has this been studied?

Kuć: We are looking at this. Arachidonic acid is transformed, and lipoxygenases and prostaglandin-like compounds may be involved with the elicitation of phytoalexin accumulation by sesquiterpenoid in potato.

Gressel: Arachidonic acid causes earlier flowering in *Pharbitis* grown in a non-inductive day-length (Groenwald & Visser 1978). This might be considered as an adaptive response to attack in order to make seed earlier.

Bailey: When one sees a phytoalexin response one also finds that other compounds are simultaneously produced which are not antifungal (Bailey 1982). We call the antifungal ones phytoalexins, and we tend to forget the others.

REFERENCES

Bailey JA 1982 Mechanisms of phytoalexin accumulation. In: Bailey JA, Mansfield JW (eds) Phytoalexins. Blackie, Glasgow, p 289-318

Groenwald EG, Visser JH 1978 The effect of arachidonic acid, prostaglandins and inhibitors of prostaglandin synthetase on the flowering of excised *Pharbitis nil* shoot apices under different photoperiods. Z Pflanzenphysiol 88:423-429

Hadwiger LA, Sander C, Eddyreau J, Ralston J 1976 Sodium azide-induced mutants of peas that accumulate pisatin. Phytopathology 66:629-630

Insect–plant interactions: endocrine defences

W. S. BOWERS

Department of Entomology, New York State Agricultural Experiment Station, Cornell University, Geneva, NY 14456, USA

Abstract. It is the inevitable consequence of evolution that competitive species living together in a restricted space must try to exclude each other. Plants and insects are prime examples of this eternal competition, and although neither of these is in danger of extinction, their mutual defensive strategies are of compelling interest to the human race. Plant defences based on the insecticidal activity of certain of their secondary chemicals are readily apparent. Only through research into the fundamentals of insect physiology and biochemistry are more subtle defensive mechanisms revealed, linked to the disruption of the insect endocrine system. A diverse number of chemical structures are found in plants, which interfere with hormone-mediated processes in insects. Examples include: mimics of the insect's juvenile hormones such as juvabione from the balsam fir and the juvocimenes from sweet basil, which lethally disrupt insect development, and the precocenes found in *Ageratum* species, which act as anti-juvenile hormonal agents. The latter appear to serve as 'suicide substrates', undergoing activation into cytotoxins when acted on by specialized enzymes resident in the insect endocrine gland (corpus allatum) that is responsible for juvenile hormone biosynthesis and secretion. Consideration of these plant defensive strategies, which have been reached through aeons of evolutionary experimentation, may assist the human race in its defences against its principal competitors for food, fibre and health.

1984 Origins and development of adaptation. Pitman Books, London (Ciba Foundation symposium 102), p 119-137

The preservation of a natural equilibrium in ecosystems is well established. The interactive forces often referred to as 'the balance of nature' depend on the constantly adjusting evolutionary pressures in which organisms find momentary advantage against competitors, through specialized adaptations, so that for a time they flourish. Eventually, natural selective pressures supervene and other organisms achieve advantages that permit their momentary dominance or transitory success. Life on earth is dominated by vegetative plant life, and intense competition between plant species is recognized. Superimposed on the interactive forces of plant–plant interaction is the competition for space and sustenance by the dominant animate life, the insects.

The spatial and temporal distribution and success of insects and plants depend on their mutual competition and cooperation. For example, efficiency in the use of space and resources, and hardiness to climatic extremes, play an important part in the ultimate success of any species. Inevitably, since life itself is chemistry, much of coevolution can be viewed as a survival of the most chemically qualified species. Indeed the basic biochemistry of all living organisms (i.e. metabolism of carbohydrates, lipids and proteins) is similar, and variations on the fundamentals of this scheme evolve very slowly. Chemical mechanisms that regulate changes in form, size, rate of development, reproduction, movement in space and the ability to utilize available resources more efficiently than competitors are subject to constant change (i.e. evolution via genetic innovations). Apparent among these changes are the ways in which insects have modified their life-stages in order to exploit a variety of niches during their life-cycle. The immature larval forms are limited in movement and appear as highly simplified feeding machines whose principal function is to sequester as much energy as possible. The adult stages, as breeding machines, are far more complex functionally, highly mobile, often aerial, and are able to explore and exploit large spatial areas and distant resources for food and reproduction. The ability to assign specific functions to specific life-stages allows maximum utilization of resources and maximum efficiency in reproduction. It is little wonder, then, that insects are the most abundant and successful life-form on earth or that they are our principal competitors for space, food and fibre. The chemical regulators of such complex life processes are biochemical innovations arising from coevolutionary pressures.

Plants, on the other hand, are the most abundant and conspicuous life-form, and the ultimate energy collectors and food resource of life on earth. Since plants are relatively sessile and insects are often highly mobile, the competitive artifices used by insects and plants for survival in space and time are especially interesting. Plants have developed a variety of mechanisms to exploit the mobility of insects, through their production of flowers, nectar, pollen and a plethora of chemical attractants designed to influence insect behaviour favourable to cross-pollination. Chemists and biologists have long recognized that many plants produce a superabundance of chemicals that appear to confer little direct benefit on the growth, development and reproduction of the plant. Various explanations for these chemicals, including their storage excretion, or presence as end products of metabolism, are rendered unsatisfactory by any cost accounting. These chemicals, often called secondary plant compounds, represent significant energy investments by plants but do not seem to fit into the known framework of plant metabolism.

An explanation for the abundance and diversity of plant secondary chemicals must depend on external aspects of plant chemical ecology, unrelated to the nutritional, developmental or reproductive needs of the plant. Some of the morphological modifications assumed by plants to limit predation by herbivores are more apparent to the observer than others. Other factors that limit predation are not so casually defined. It is obvious that all insects do not eat all plants and that insect species which voraciously attack one species of plant will often starve to death in the presence of abundant plant material of a different kind. Simple human experience teaches that plants taste good or bad, are palatable or unpalatable, toxic or harmless. Taste, palatability and toxicity depend on plant secondary chemicals. Clearly, plants have also elaborated a vast arsenal of secondary chemicals for protection from the predation of herbivores. Certain of these defensive mechanisms, including repellents, 'antifeedents' and poisons, are well recognized and understood. We have profited immensely from using some of these plant defensive strategies to protect our agricultural plants and to control insect vectors of disease. A few of the natural insecticides derived from plants are intensely toxic to insects but minimally hazardous to human populations.

Fundamental studies of insect physiology and plant chemistry have revealed other, more subtle defensive strategies by plants against insects, which do not rely on intrinsic toxicity to fundamental biochemical processes common to all organisms. These strategies are, instead, directed to interrupt specific mechanisms that have made insects so successful—aspects of insect metamorphosis, reproduction, diapause and behaviour—by interfering with the insect endocrine system. Thus, secondary compounds that mimic the insect juvenile hormones have been identified in several plant species, and anti-juvenile hormonal compounds have now also been found in plants. Since the presence or absence of the insect juvenile hormone regulates embryogenesis, metamorphosis, reproduction, diapause, caste/morph determination, and even influences sex determination, disruption of these processes by plant secondary compounds can be seen as a powerful defensive strategy. It is perhaps not surprising that plants have engineered a great deal of economy into their defensive substances: some of these chemicals can have actions against both insects and plant pathogens (Bowers & Evans 1983).

Natural insecticides

Insects are the primary consumers of plants, and when these plants are also important sources of food and fibre to human populations, insect control becomes necessary. Insects are colourful, mobile and highly visible, and the earliest agrarians were certainly aware of the predation by insects on desirable plants. Equally, these early farmers were aware of plants that were not readily susceptible to insect attack and which could easily be demonstrated to contain substances toxic to insects. It seems only a small step from this discovery to the preparation of a simple plant extract and its application to desirable plants to prevent their predation by insects. Another way that natural insecticides were discovered came from observations of primitive peoples that certain plants, when cast into water, caused fish to die. Such

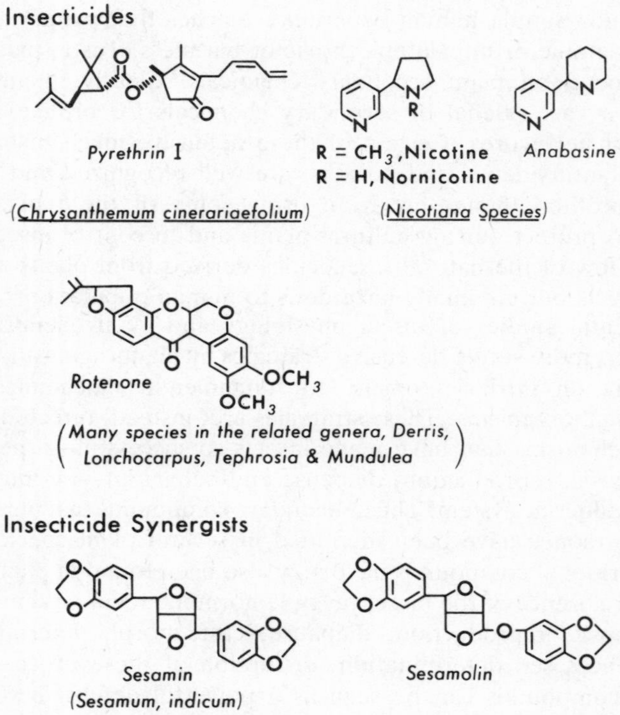

FIG. 1. Plant-derived insecticides and synergists.

fish-killing plants were quickly seized upon as possible insecticides. Rotenone (see Fig. 1) and deguelin are excellent examples of fish poisons that were not only strikingly low in toxicity to warm-blooded animals but were also excellent insecticides and were, for a time, important commercial products. Perhaps the most outstanding natural insecticides were those discovered in the chrysanthemum plant and which came to be called the pyrethrins (Fig. 1). The powdered flowers of these plants were found to be excellent insecticidal preparations that induced very rapid 'knockdown' of most insects and had significant follow-up toxicity. The advantages of such compounds as the pyrethrins for the chrysanthemum plant seem extraordinarily obvious. Attempts to consume the flowers result in rapid loss of sensibility, and potential predators are either knocked out for a time or killed. In any event, these interesting secondary compounds rapidly stopped predation by insects on the plant. Researchers in the UK (Elliot et al 1974) and Japan (Nakayama et al 1978) have taken full advantage of the sublime properties of these secondary compounds and, through selective molecular modification, have produced some of the most outstanding, safe and effective insecticides. Thus, the chrysanthemum plant has served not only to protect itself but also to provide inspiration for the development of chemicals of outstanding service to the human race. Eagelson (1942) discovered that the coformulation of the natural pyrethrins with sesame seed oil resulted in a more intensely toxic preparation. Haller et al (1942) found that the sesame oil contained certain compounds that greatly increased the toxicity of pyrethrins to many insects, and identified one active component as sesamin. A second highly active synergistic compound in sesame oil, sesamolin, was later identified by Beroza (1954). However, sesamin and sesamolin (Fig. 1) had no intrinsic toxicity apart from their combination with pyrethrins, and the increased toxicity resulting from the combination seemed to be only a fortuitous discovery. Later Doskotch & El-Feraly (1969), on re-examination of the crude pyrethrum extract, were able to identify sesamin as a natural component of it. This discovery vigorously supports the thesis that the pyrethrins are a superlative defensive component of the chrysanthemum, and it reveals the complexity of the defensive mechanisms that have evolved against insect attack. In independent studies, I found (Bowers 1968) that sesamin and sesamolin possessed juvenile-hormone activity against several insect species. This discovery led to the synthetic combination of structural moieties of the synergists with portions of the natural insect-juvenile hormone (Bowers et al 1965), culminating in hybrid hormonal 'mimics' that had hormonal activity against insects (Bowers 1968, 1969). The hybrid hormonal compounds were extremely effective at disrupting morphogenesis in insects, and several were 10 000-fold more active than the insect's own natural juvenile hormone. Interestingly, certain of these analogues were found to possess significant synergistic activity for pyrethrum against house flies (Fales et al 1970).

Synergists are compounds which, although lacking any direct toxic actions by themselves, are able to enhance the toxicity of other poisons when combined with them. We now know that the synergists such as sesamin and sesamolin are inhibitors of certain mixed-function oxidase enzymes present in the cellular microsomal fractions in animals. The mixed-function oxidases are important safeguards against a variety of poisonous compounds, which they degrade, so minimizing their toxicity. Sesamin and sesamolin inhibit these enzymes, and thus prolong and intensify the toxic action of natural or synthetic insecticides. The benefit of such mixed-function oxidase inhibitors to plants seems clear. Plants may store toxic compounds in their tissues to protect themselves against attack by pathogens or herbivores, but these components must not be toxic to the plant itself. Thus, compounds of only moderate toxicity may be stored in the plant tissues and, if consumed together with a synergist, may then prove sufficiently toxic for the plant's defensive purposes.

Alkaloids are produced in abundance by plants, and many are frankly toxic to animals generally. Thus, it is not surprising that farmers were using the powdered leaves of tobacco, or water extracts of them, to protect crops from insects long before chemists isolated and identified the principal toxic components. Schmeltz (1971) has referred to the use of tobacco as an insecticide and its importance in trade between the American colonies and Europe. Although they are not as widely used as in the past, nicotine, anabasine and nornicotine (see Fig. 1) are important natural products for insect control. Nicotine as a toxic defensive substance used by plants is rather widespread in the plant kingdom among many unrelated plants, including *Asclepias syriaca* (milkweed), *Atropa bella-donna* (deadly nightshade), *Equisetum arvense* (horsetails), and *Lycopodium clavatum* (club moth), testifying to its frequent independent evolution.

Thousands of insecticidal components of plants are known and documented in the literature. Few have achieved the commercial success of those already discussed here. Nevertheless, these toxins, seen so frequently in plants, are ample evidence of the evolutionary adaptation of many obscure biosynthetic pathways—obviously of little direct participation in the life and times of the plant—which have been retained and refined to produce toxic substances of defensive utility to the plant.

The structural diversity of the secondary compounds used in plant defence, and their effects on widely disparate biological mechanisms, provides an exciting and complex world for the chemical ecologist, but can also be viewed as a vast natural resource of chemical and biological activity awaiting imaginative optimization and utilization by mankind.

Plant secondary compounds affecting insect endocrine processes

Juvenile hormone mimics

Basic research into insect physiology and endocrinology has shown that plants produce secondary products which, although essentially non-toxic, can selectively perturb important aspects of the insect endocrine system. Two important hormonal regulators in insects are the juvenile hormones and the moulting hormones (ecdysones). The juvenile hormones, by their presence or absence, regulate many aspects of insect physiology. Slama & Williams (1965) discovered that a European insect *Pyrrhocoris apterus* failed to develop normally in the biological laboratories at Harvard University under their culture conditions and persisted in moulting into forms intermediate between the nymphal and adult stages, giving evidence of having been exposed to an excess of juvenile hormone. They ascertained the source of the hormonal

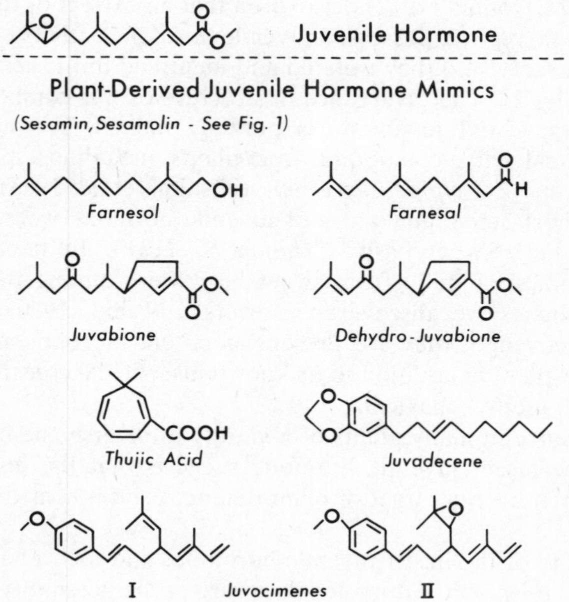

FIG. 2. Juvenile-hormone mimics found in plants.

influence as coming from the paper they had used to line the rearing jars, and finally located its ultimate source in the pulpwood of the balsam fir, *Abies balsamea*. We (Bowers et al 1966) were able to isolate from this tree a compound with extraordinarily high juvenile hormone activity, and we called it juvabione. Its relationship to the natural hormones is apparent in Fig. 2. Subsequently, a less active analogue (dehydro-juvabione) was found in a fir tree in Czechoslovakia (Cerny et al 1967). These compounds, applied to developing insects, exerted the same morphogenetic activity as the natural juvenile hormone. The resulting disruption of insect metamorphosis demonstrated the startling potential these compounds could have for a sensitive insect predator of the balsam fir. This fortuitous discovery greatly influenced chemical ecologists, because of the exciting prospect of a non-toxic defensive strategy linked to specific hormone receptors in insects. It also influenced economic entomologists, who viewed these discoveries as a potential resource of new, specific, and perhaps safe chemicals for insect control. Although the juvabiones possessed too narrow a spectrum of activity to be used commercially, they made scientists aware of hormone 'mimics' in plants, and stimulated investigations of other plants for hormonally active secondary constituents.

Barton & MacDonald (1972) discovered that an extract of the western red cedar tree, *Thuja plicata*, possessed juvenile hormone activity when tested on a number of insects, and they isolated and identified thujic acid as the active component (Fig. 2). The divergence in structure of this compound from the natural hormones and juvabione is puzzling indeed. We have identified several additional active compounds from plants, including a methylenedioxy compound from *Macropiper excelsum* (Nishida et al 1983); see Fig. 2. Perhaps the most active analogues of juvenile hormone were isolated from the common herb sweet basil, *Ocimum basilicum*. In fact, the juvenile hormonal 'mimics' (Fig. 1) from sweet basil were among the most active hormone analogues ever discovered (Bowers & Nishida 1980). Active in the picogram/insect range, they are present as potent morphogenetic agents in the sweet basil plant in addition to its known insecticidal constituents, methyl cinnamate and methyl chavicol.

The appearance in many plants of a variety of diverse chemical structures that mimic the insect juvenile hormone suggests that the insect endocrine system has often been a target of plant defences and is sensitive to chemical manipulation.

The discovery of the insect juvenile hormones and their analogous mimics in plants made economic entomologists aware of the possibility of developing safe and selective methods for controlling noxious insect pests by interrupting their endocrine-mediated processes. Industrial success was rapid and several preparations of juvenile hormone analogues are now commercially available for the control of insects that are pests to public health, including flies, fleas and mosquitoes. These insects are notably innocuous during their immature stages but are economically important as adults, and the juvenile hormone analogues are thus useful principally in controlling the insects that are important during the adult stages. Since the lethal morphogenetic derangement induced by juvenile hormone occurs only during the ultimate or penultimate instars of insect development, juvenile hormone and its analogues are of little consequence for controlling economically important insects during their early immature stages, and particularly during the agriculturally damaging immature feeding stages. Soon after the juvenile hormones were discovered and their activity clearly established, it became apparent that a more useful strategy for controlling plant-feeding insects would be to interfere somehow with the biosynthesis, secretion, transport or action of the juvenile hormone: classical endocrinological experiments demonstrated that, in the absence of juvenile hormone, insects would undergo a lethal precocious metamorphosis, or would be sterilized, or both.

Anti-juvenile hormonal compounds in plants

Because of the abundance of juvenile hormone mimics in plants, we wondered whether plants might also have adopted anti-hormonal strategies to protect themselves from predacious insects. After developing suitable bioassays, we began to extract plants and to test them for the ability to induce precocious metamorphosis. We discovered two compounds in the *Ageratum* plant that possessed the full anticipated anti-juvenile hormonal actions (Bowers 1976, Bowers et al 1976). These compounds were isolated and identified as simple substituted chromenes (Fig. 3). Structural optimization by chemical synthesis of analogues revealed the functional necessity of the 3,4-double bond and the substitution at the C7 position by small alkoxy substituents (Bowers 1977). We were also able to show that the corpora allata (the glands that secrete juvenile hormone) of sensitive insects that were

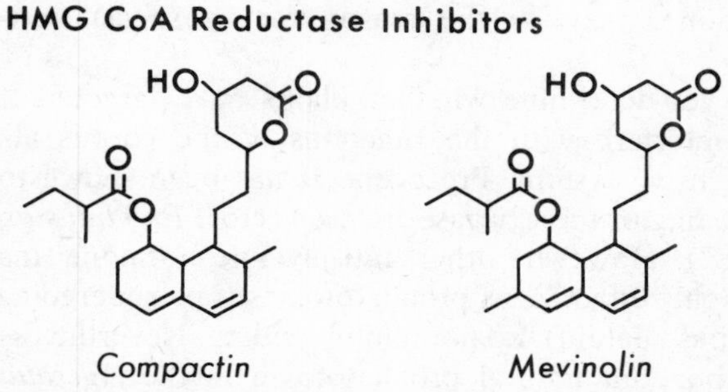

FIG. 3. Anti-juvenile hormonal compounds from plants. HMG-CoA reductase, hydroxymethylglutaryl-CoA reductase.

treated with these compounds, now called precocenes, were unable to undergo normal development (Bowers & Martinez-Pardo 1977). Subsequently, we found that the precocenes caused a cytotoxic disruption of the corpora allata (Unnithan et al 1977). Studies of metabolism (Ohta et al 1977, Soderlund et al 1980, and others; for review see Bowers 1982) revealed that the precocenes were undergoing an oxidative activation (via epoxidation) into highly reactive alkylating agents which destroyed the corpus allatum of the insect. The evidence is reasonably clear that these plant compounds are intrinsically non-toxic in their chromene form but, on oxidation by specialized enzymes in the insect corpora allata, they become lethal cytotoxins. Thus, the strategy that the plant adopts, as revealed by these studies, is startling. The *Ageratum* plant can synthesize and sequester compounds that are non-toxic to the plant itself but which, on consumption by an insect predator, are activated into lethal cytotoxins by the specialized enzymes located in the insect corpus allatum. Through this mode of activation the precocenes seem to be acting as suicide substrates for these unique enzymes in the corpus allatum. Since insect or other animal guts may have regions that contain mixed function oxidases or similar enzymes, one cannot discount the possibility that the precocenes could also act widely against a variety of other predacious animals whose mixed function oxidases could activate the precocenes. Alkylation of such gut tissues would cause necrosis of the epithelial intima, perhaps causing pain and other kinds of gastric distress which would signal the organism to stop feeding on *Ageratum*. Our principal findings, so far, have been a specific and total destruction of the insect corpus allatum by these compounds as a consequence of its extraordinarily high concentration of oxidase enzymes. We still need to know whether the enzymes involved in precocene activation are, in fact, those involved in juvenile hormone biosynthesis. The terminal step in this biosynthesis is the oxidation of methyl farnesenate to juvenile hormone III. The high concentration of oxidase enzymes in the corpus allatum, as a consequence of its singular function in the biosynthesis and secretion of juvenile hormone, may be the reason for its special sensitivity to the precocenes.

We cannot yet determine whether plants have targeted the precocenes especially to interfere with the functions of the corpus allatum or as a secondary defensive system. Precocene II has been shown to inhibit significantly the feeding of the Chagas' disease vector, *Rhodnius prolixus* (Azambuja et al 1982). However, other anti-juvenile hormone analogues of the precocenes, highly effective as proallatotoxins (i.e. undergoing lethal activation in the corpus allatum), do not inhibit feeding. Nevertheless, precocene II is the most abundant natural proallatotoxin in the *Ageratum* species. Its major function could be as a unique cytotoxin, activated, as described above, in regions of herbivore guts that contain mixed function oxidases, and producing necrosis that induces rejection of *Ageratum* as a food source. Precocene substrates may thus be devised to defeat the nominally protective mechanisms against xenobiotics offered by the mixed function oxidases in the alimentary system of many herbivores. The variety of toxic substrates that plants can produce to protect themselves against predation may have stimulated the herbivore to develop enzymes capable of degrading these xenobiotics. The *Ageratum* plant, in turn, has responded to this challenge by developing substrates which, on oxidation, are transformed into alkylating agents capable of destroying the herbivore's oxidizing enzymes themselves and/or other nucleophiles that the agents encounter in the herbivore's cells. The end result is destruction of tissue (i.e. destruction of the corpus allatum, in which anti-juvenile hormone results, or destruction of gut tissues, causing inhibition of feeding). It would be interesting to see if a further counter-defence by certain herbivores has been devised against these suicide substrates. In areas where mixed function oxidase enzymes are highly concentrated, disposable nucleophiles may coexist with the enzymes and may safely absorb any cytototoxins generated by the oxidative processes. Not all insects, especially many holometabolous species, are susceptible to the precocenes, and studies of precocene metabolism in insect guts should therefore be interesting.

Thallophytic plants, including the fungi *Penicillium brevicompactin* and *Aspergillus terreus*, produce inhibitors of hydroxymethylglutaryl-CoA reductase (Endo et al 1976, Alberts et al 1980), an important enzyme in the biosynthesis of sesquiterpenoids and triterpenoids. Monger et al (1982) have demonstrated that compactin inhibits biosynthesis of juvenile hormone *in vitro*. We have confirmed this result with compactin and with an analogous antibiotic, mevinolin (Fig. 3). Moreover, we also find that exposure of female milkweed bugs, in the teneral adult stage, to a residue of mevinolin, by contact, induces sterilization, although normal reproduction supervenes quickly if the bugs are removed from the treated substrate. It is not anticipated that the inhibitors of hydroxymethylglutaryl-CoA reductase have developed as a response to insect predation. Nevertheless, they are novel examples of plant defence, and such inhibitors may also be present in higher plants.

Synthetic proallatotoxins

The discovery of the precocenes and their activation by corpus allatum enzymes stimulated us to consider the overall utility of a suicide substrate as a plant-defensive mechanism. Plants must often be careful to store their toxins in ways to avoid autotoxicity. This process certainly requires extra precautions and significant energy expenditure. However, the storage of compounds like the precocenes, which are intrinsically non-toxic until activated, should be a much more efficient process. Jurd & Manners (1980) previously proposed that a variety of natural products may be activated into alkylating agents on oxidation. The oxidative activation of the precocenes (Fig. 4) seemed to parallel the activation of other plant-derived compounds, including obtusaquinone and obtusastyrene, which form quinone methides and are destructive alkylators that are presumably used for protection against plant pathogens. The close relationship between these modes of action prompted us to synthesize some simple analogues of the precocenes that lack the chromene ring, by substituting instead *o*-isopentenylphenols. Treatment of

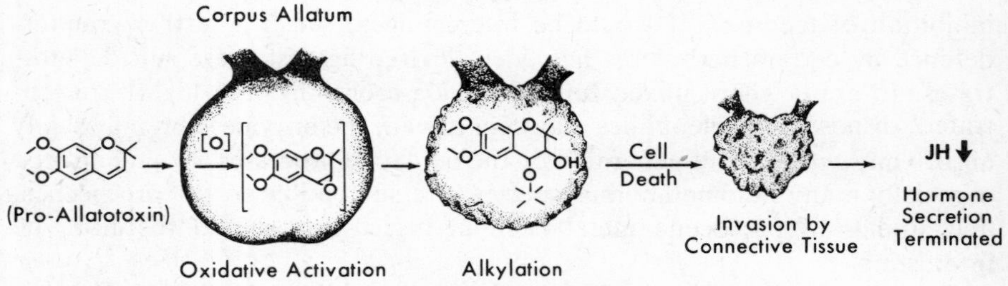

FIG. 4. Proposed mode of action for the precocenes. JH, Juvenile hormone.

insects with 5-ethoxy-4-methoxy-2-(3′-methylbut-2′-enyl)phenol resulted in anti-juvenile hormone actions identical with those produced by the precocenes (Bowers et al 1982). If the isopentenylphenols, commonly found among plant secondary chemicals, are activated by corpus allatum enzymes into the corresponding quinone methides, then they could be envisaged as another form of suicide substrate and another potential plant-defensive strategy against insect herbivores.

These studies reveal only a first insight into the subtle defence capabilities developed by plants in response to the activities of herbivores. It is impossible to distinguish yet between the importance of a specific defence (e.g. anti-juvenile hormone) and that of a general defence (e.g. general feeding deterrents).

Plants are superb chemists and from their roots to their meristematic tips they are loaded with chemical warfare agents. Insects, on the other hand, are possessed with immense vitality, plasticity and reproduction potential, able in a few generations to overcome most chemicals used by plants for defence. In this see-saw battle of the planetary dominants, we are privileged to be confounded, amused and, above all, instructed in the facts of organic evolution.

REFERENCES

Alberts AW, Chen J, Kuron G et al 1980 Mevinolin: a highly potent competitive inhibitor of hydroxymethylglutaryl-coenzyme A reductase and a cholesterol-lowering agent. Proc Natl Acad Sci USA 77:3957-3961

Azambuja PD, Bowers WS, Ribeiro JMC, Garcia ES 1982 Antifeedant activity of precocenes and analogs on *Rhodnius prolixus*. Experientia (Basel) 38:1054-1055

Barton GM, MacDonald BF 1972 Juvenile hormone-like activity of Thujic acid, an extractive of western red cedar. Bi-monthly Res Notes: Environ Can 28:22-23

Beroza M 1954 Pyrethrum synergists in sesame oil. Sesamolin, a potent synergist. J Am Oil Chem Soc 32:302-305

Bowers WS 1968 Juvenile hormone activity of natural and synthetic synergists. Science (Wash DC) 161:895-897

Bowers WS 1969 Juvenile hormone activity of aromatic terpenoid ethers. Science (Wash DC) 164:323-325

Bowers WS 1976 Discovery of insect antiallatotropins. In: Gilbert LI (ed) The juvenile hormones. Plenum Press, New York, p 394-408

Bowers WS 1977 Anti-juvenile hormones from plants: chemistry and biological activity. In: Marini-Bettolo GB (ed) Natural products and the protection of plants. Elsevier, New York. (Proc Pontifical Acad Sci, Vatican City, Italy)

Bowers WS 1982 Endocrine strategies for insect control. Entomol Exp Appl 31:3-14

Bowers WS, Evans PH 1983 A rapid sensitive assay for the discovery and isolation of naturally occurring fungicides. Int J Trop Plant Dis, in press

Bowers WS, Martinez-Pardo R 1977 Antiallatotropins: inhibition of corpus allatum development. Science (Wash DC) 197:1369-1371

Bowers WS, Nishida R 1980 Juvocimenes: potent juvenile hormone mimics from sweet basil. Science (Wash DC) 209:1030-1032

Bowers WS, Thompson M, Uebel EC 1965 Juvenile and gonadotropic hormone activity of 10,11-epoxyfarnesenic acid methyl ester. Life Sci 4:2323-2331

Bowers WS, Fales HM, Thompson MJ, Uebel EC 1966 Juvenile hormone: identification of an active compound from balsam fir. Science (Wash DC) 154:1020-1022

Bowers WS, Ohta T, Cleere JS, Marsella PA 1976 Discovery of insect anti-juvenile hormones in plants. Science (Wash DC) 193:542-547

Bowers WS, Evans PH, Marsella PA, Soderlund DM, Bettarini F 1982 Natural and synthetic allatotoxins: suicide substrates for juvenile hormone biosynthesis. Science (Wash DC) 217:647-648

Cerny V, Dolys L, Labler L, Sorm F, Slama K 1967 Dehydrojuvabione, a new compound with juvenile hormone activity from balsam fir. Collect Czech Chem Commun 32:3926-3933

Doskotch RW, El-Feraly FS 1969 Isolation and characterization of (+)-sesamin and β-cyclopyrethrosin from pyrethrum flowers. Can J Chem 47:1139-1142

Eagelson C 1942 Sesame oil as a synergist for pyrethrum insecticides. Soap Sanit Chem 18:125-127

Elliott M, Farnham AW, Janes NF, Needham PH, Pulman DA 1974 Synthetic insecticide with a new order of activity. Nature (Lond) 248:710-711

Endo A, Juroda M, Tsiyita Y 1976 ML-236A, ML-236B, and ML-236C, new inhibitors of cholesterogenesis produced by *Penicillium citrinum*. J Antibiot (Tokyo) 29:1346-1348

Fales JH, Bodenstein OF, Bowers WS 1970 Seven juvenile hormone analogs as synergists for pyrethrins against house flies. J Econ Entomol 63:1379-1380

Haller HL, McGovran ER, Goodhue LD, Sullivan WN 1942 The synergistic action of sesamin with pyrethrum insecticides. J Org Chem 7:183-184

Jurd L, Manners GD 1980 Wood extractives as models for the development of new types of pest control agents. J Agric Food Chem 28:183-188

Monger DJ, Lim WA, Kezdy FJ, Law JH 1982 Compactin inhibits insect HMG-CoA reductase and juvenile hormone biosynthesis. Biochem Biophys Res Commun 105:1374-1380

Nakayama I, Ohno N, Aketa K, Suzuki Y, Kato T, Yoshioka H 1978 Chemistry, absolute structures and biological aspect of the most active isomers of fenvalerate and other recent pyrethroids. In: Geissbuhler H (ed) 4th Int Cong Pest Chem, Zurich. Pergamon, New York. Advances in Pest Sci (No 2): 178-181

Nishida R, Bowers WS, Evans PH 1983 Juvadecene: discovery of a juvenile hormone mimic in the plant *Macropiper excelsum*. Arch Insect Biochem Physiol 1, in press

Ohta T, Kuhr RJ, Bowers WS 1977 Radiosynthesis and metabolism of the insect anti-juvenile hormone, Precocene II. Agric Food Chem 25:478-481

Schmeltz I 1971 Nicotine and other tobacco alkaloids. In: Jacobson M, Crosby DG (eds) Naturally occurring insecticides. Marcel Dekker Inc., New York, p 99-136

Slama K, Williams CM 1965 Juvenile hormone activity for the bug *Pyrrhocoris apterus*. Proc Natl Acad Sci USA 54:411-414

Soderlund DM, Messeguer A, Bowers WS 1980 Precocene–II metabolism in insects: synthesis of potential metabolites and identification of initial *in vitro* biotransformation products. J Agric Food Chem 28:724-731

Unnithan GC, Nair KK, Bowers WS 1977 Precocene-induced degeneration of the corpus allatum of adult females of the bug *Oncopeltus fasciatus*. J Insect Physiol 23:1081-1094

DISCUSSION

Clarke: Is it possible to make an anti-ecdysone i.e. an anti-moulting hormone?

Bowers: This has been attempted, but not successfully as far as I know. One can stop ecdysone biosynthesis by using drugs such as azasteroids on the insects. This will inhibit enzymes that normally convert plant sterols to ecdysone. But azasteroids are themselves poisonous. It would be very useful to be able to do this because the moulting hormone is intrinsic to almost every developmental stage whereas the juvenile hormone is only necessary periodically to certain stages. If one could safely and selectively interfere with the dealkylation of plant sterols to ecdysones this would provide an ideal method of insect control. Insects depend on obtaining their steroid nucleus from their food; they are unable to synthesize the cyclopentano-perhydrophenanthrene ring system.

Southwood: I am not aware of many accounts of how materials such as juvabiones, and other analogues of insect hormones, which are actually *in the plants* provide protection against other insects. Such materials present in ferns, for example, are in low concentrations and seem to be destroyed by the insect's digestion. I am particularly interested in the extent to which *Ageratum* is protected from herbivores by this device (Slama 1979).

Bowers: Plants such as sweet basil contain some very active compounds. If one can induce an insect to feed on enough sweet basil it will show morphogenetic damage in its development. The problem is that sweet basil also has a number of other compounds that are reasonably toxic to most insects and so the insects don't feed on it for very long anyway. *Ageratum* provides a better example. It is a popular garden plant because it is not susceptible to much insect damage, although white flies will attack it in the greenhouse. I once saw about 10 'woolly-bear' caterpillars feeding on an *Ageratum* plant, and they consumed several leaves each. About 30 minutes later all the caterpillars had moved off the plant. I therefore tried to rear the caterpillars on *Ageratum* plants in the laboratory but they repeatedly crawled off the plant when placed on it. By contrast, they fed quite happily on a bean plant. It seemed that the caterpillars had learnt a lesson from feeding initially on the *Ageratum*. We now know that the precocenes present in *Ageratum* are activated by oxidative enzymes into alkylating agents. We also know that insects are constantly feeding on plants that contain various toxins which can be broken down by enzymes localized to the insect gut, including oxidative enzymes. I would imagine that some of the precocene consumed by these particular caterpillars was oxidized and that it destroyed areas of the intestinal lining containing oxidizing enzymes. This may have caused an intestinal upset, so that the insects learned not to feed on it subsequently. I believe that grasshoppers, which do not like to feed on *Ageratum*, can be fed it unknowingly in a sandwich of leaves from other plants. When this happens they undergo precocious metamorphosis. So these provide examples of precocenes offering protection to the plant against insects. In early evolutionary history these compounds may have been much more effective than they are now.

Bell: With respect to the work of Slama & Williams (1965) that you mentioned, in which a hormonal influence was detected in the paper lining of the rearing jars, I always understood that this factor affected only one species of insect.

Bowers: I believe so. The effect appeared to them to be highly specific but they did not test it very widely. I have found, in fact, that juvabione is also active against a number of other insects (Bowers et al 1966).

Bell: Surely in the use of these substances as potential insecticides it does not matter whether they kill the juvenile or the adult form of the insect because one is considering vast populations of insects rather than individuals? If it kills off or reduces the next generation the substance can surely be considered effective?

Bowers: The issue does become relevant when an individual farmer is interested in preserving his crop for the current year.

Gressel: But is it not likely that the insect might normally go through about six generations per year in any case, and so removal of the first generation would still be effective?

Bowers: If the elimination of one generation would still allow the protection of the crop, then this would be acceptable. But nothing grows in isolation and although juvenile hormone may eliminate the insect from a few hectares of land, other similar insects can fly in from elsewhere and re-infest the crop.

Bell: But this is not an argument against the *method* but just against not using the substance widely enough.

Bowers: Right, except that the original damage would have to be endured, and that is not usually acceptable to the farmer.

Southwood: These substances may be more useful in the tropics than in temperate regions.

Elliott: We have to consider the possibility that most of these plant products may have no present function in the plant but may simply be metabolic dumps. Alkaloids and many other products seem to confer no particular advantages on the plant. The pyrethrins in *Tanacetum cinerariifolium* (the chrysanthemum), are present at a position in the plant, in the achenes, where they could not possibly exert any insecticidal activity. Growing in bright sunlight, as these plants do, if the pyrethrins were accessible to insects they would be so rapidly decomposed that no insecticidal activity could be seen.

Bradshaw: But the seed (the achene) is a very important part of any plant, and if the pyrethrins occur in the seed, this must be significant. After all, every plant, unless it is produced vegetatively, has to arise from a seed.

Southwood: Yes, and there are very many seed-feeding beetles known.

Elliott: Well I can only say that I am not aware of the pyrethrins having protected plants from insects that prey on them.

Bradshaw: But other insects might have eaten those seeds if the pyrethrins had not been there.

Elliott: Evolutionarily, perhaps that is so, but at the moment the pyrethrins do not appear to confer any advantage on the plant.

Sawicki: If the pyrethrins did prevent insect attack, the natural cultivars should have the largest amounts of pyrethrins. This is not so. Breeding programmes have increased the content of pyrethrins in the achenes from 0.5 to 2.5%.

Bell: When considering whether a secondary compound has an ecological role, one must bear in mind that a plant that devoted, say, 10% of its resources to making a secondary compound would not have survived long in competition with a form that used all its resources in the synthesis of primary metabolites unless the secondary compound conferred some selective advantage on the plant.

Clarke: But this view presupposes that there is no threshold of safety in the plant's metabolism that can tolerate 10% being wasted. We do know of situations in which 10% wastage is perfectly tolerable at some stage of the life-cycle.

Elliott: In fact only by very selective breeding can as much as 3% (dry weight) of pyrethrins be obtained. The general level is 1.5% dry weight or less, and the content in undried plant material is therefore much lower still.

Hartl: The present use of structures and products of metabolism need not have been why they evolved. What was once a rare secondary metabolite used for some other purpose may have turned out by chance to be an antibiotic or a fungicide, and then the plant would begin to produce it for that second reason. This would be what an animal breeder would call 'indirect' selection. Once the character has begun to develop one can select directly on that correlated response.

Bell: I agree entirely; my 10% argument is circumstantial and not really the best argument. One might have a perfectly useless compound which, because its gene happens to be in a certain position on the chromosome, might be carried along with a very important character, for a very long time. A better approach when looking for evidence of an ecological role is to identify what adaptations, if any, have been induced by the secondary compound in interrelated organisms.

Kuć: As I look around this room I see a tremendous diversity in shape and form, and I wonder how much of this diversity is really essential for survival!

Gressel: The critical experiment in the question of these pyrethrins is to look at different accessions of chrysanthemum and to see if there is a positive or a negative correlation between numbers of seeds and levels of pyrethrins.

Elliott: I am not aware of any information on that subject. Only a few varieties of the species *T. cinerariifolium* produce the insecticidal compounds. Closely related plants within the same family do not give any pyrethrins.

Gressel: Do the ones that don't produce it have more seed?

Elliott: I don't know.

Clarke: Presumably the direct test would be to put out seeds with and without pyrethrin and to watch whether insects eat them.

Sawicki: To our eternal shame *T. cinerariifolium* plants in Kenya are heavily treated with insecticides to protect them against various pests!

Bowers: Any chemical that is concentrated in a plant and has a biological action that could be appropriate to its protection must be considered as having played a possible part in the protection of the plant at some evolutionary stage. The tobacco plant, for example, is highly susceptible to attack from many insects if it does not contain nicotine or one of the other toxic alkaloids. It is hard to say that juvabione, which occurs principally in the heart wood of the balsam fir, *Abies balsamea*, is a protective chemical because of its remote siting within the tree. A boring insect would have to make contact with the juvabione before it would be affected, and the insect might not even survive such a long route anyway. Juvabione may not, therefore, be expressly a protective mechanism in current times against insects, but may have had this role earlier in evolution. On the other hand, juvabione can be shown to have significant fungicidal activity against a variety of fungi. So, although we can extract many toxic substances from plants and show that they have adverse effects on insects it does not mean that the substance is normally present purely to protect the plant. Some mechanisms of defence that are still present may no longer be used as they once were because the predatory insects, for example, may have evolved into another way of life. It is often impossible to demonstrate that a given mechanism was developed for plant protection. Yet one must attempt some cost accounting. There are enormous amounts of terpenes in pine trees and other conifers which seem to have little use. The allelopathy story of the soft chaparral is illustrative. These plants produce visible clouds of terpenes around themselves which inhibit the growth of competitive plants. There is no question that the plants protect themselves from competition with other plants. When weather conditions do not wash away the excess terpenes, these substances become autotoxic to the plants that produce them.

Clarke: Although the potential for ancient insecticides may be present in the genomes of plants, this is not a useful scientific explanation, in that it is too easy to 'invent' history.

Bowers: I believe that sesamin has also been reported in *T. cinerariifolium* (Doskotch & El-Feraly 1969).

Elliott: Yes, (+)-sesamin was isolated from an alcohol extract of these flowers (Doskotch & El-Feraly 1969) but only at a concentration (0.005% by weight) probably too low to influence any insecticidal activity shown by the pyrethrins. I am not aware that this isolation has been repeated.

Bowers: If sesamin were close to the pyrethrins, in the seeds, one would *have* to conclude that this was not accidental. Sesamin, after all has been demonstrated to be a synergist for the natural pyrethrin.

Clarke: In terms of cost-effectiveness, the optimal levels of these compounds will differ. The optimum from the point of view of protecting the plant in nature will not be the same as the optimum from the point of view of a producer who wants a lot of pyrethrins from the plants.

Bradshaw: In legumes (*Lotus* or *Trifolium* species) there is good evidence that the cyanogenic glucoside/enzyme system evolves only when the herbivore (e.g. the slug) is present (Jones 1973). The system acts as a defence in the presence of the herbivore, but has the disadvantage of conferring a susceptibility to frost damage. This system is therefore very carefully balanced for cost-effectiveness.

Graham-Bryce: If one is suggesting, as Dr Elliott has done, that many of these products may be metabolic dumping grounds, one ought to be able to show why, in biosynthetic terms, this is an advantage. There ought to be some related set of valuable processes within the plant, which create the need for this sort of dump.

Clarke: I agree.

Southwood: One also needs to explain why there should be so many different forms of dump, too.

Harborne: It is not only a question of making the compounds but of storing them as well. They will need to be channelled into an innocuous site within the plant—by packaging in some way in the vacuole or in special vessels—so that they do not interfere with the plant's metabolism.

Gressel: Yes. Insects or other animals would probably solve the problem simply by excreting these compounds, but plants cannot do that.

Elliott: Much manipulation is necessary to extract pyrethrins from the plant.

Harborne: With regard to the natural defence of *T. cinerariifolium* one would have to look at all the other types of secondary compounds. The pyrethrins are automatically considered to be the compounds that provide that plant with insecticidal protection but other compounds may be just as important.

REFERENCES

Bowers WS, Fales HM, Thompson MJ, Uebel EC 1966 Juvenile hormone: identification of an active compound from balsam fir. Science (Wash DC) 154:1020-1022

Doskotch RW, El-Feraly FS 1969 Isolation and characterization of (+)-sesamin and β-cyclopyrethrosin from pyrethrum flowers. Can J Chem 47:1139-1142

Jones DA 1973 Co-evolution and cyanogenesis. In: Heywood VH (ed) Taxonomy and ecology. Academic Press, London, p 213-242

Slama K 1979 Insect hormones and anti hormones in plants. In: Rosenthal GA, Janzen DH (eds) Herbivores: their interaction with secondary plant metabolites. Academic Press, New York and London, p 683-700

Slama K, Williams CM 1965 Juvenile hormone activity for the bug *Pyrrhocoris apterus*. Proc Natl Acad Sci USA 54:411-414

Insect–plant adaptations

T. R. E. SOUTHWOOD

Department of Zoology, University of Oxford, South Parks Road, Oxford OX1 3PS, UK

Abstract. The adaptation of insects to plants probably commenced in the early Permian period, though most current associations will be more recent. A major burst of adaptation must have followed the rise of the Angiosperms in the Cretaceous period, though some particular associations are as recent as this century. Living plants form a large proportion of the potential food in most habitats, though insects have had to overcome certain general hurdles to live and feed on them. Insects affect the reproduction and survival of plants, and thus the diversity of plant secondary chemicals may have evolved as a response. Where an insect species has a significant effect on a plant species that is its only host, coevolution may be envisaged. A spectacular example is provided by *Heliconius* butterflies and passion flower vines, studied by L. E. Gilbert and others. But such cases may be likened to 'vortices in the evolutionary stream': most plant species are influenced by a range of phytophagous insects so that selection will be for general defences—a situation termed diffuse coevolution. Evidence is presented on recent host–plant shifts to illustrate both the restrictions and the flexibility in current insect–plant associations.

1984 Origins and development of adaptation. Pitman Books, London (Ciba Foundation symposium 102), p 138-151

Of the five kingdoms into which living organisms are normally divided (Margulis & Schwartz 1982), the Plantae (green multicellular plants, from liverworts to flowering plants) date from the late Silurian period and so have much the shortest evolutionary history. Also, they are the only kingdom to have evolved on land, rather than in a marine or aquatic environment. The class Insecta also evolved on land around this period, but the adaptation to living and feeding on plants was a major step. This is evinced by the fact that of the 29 extant orders of insect only nine exploit the living tissues of higher plants for food (Southwood 1973). Fossil evidence for this adaptation is not conclusive until the Permian period, although there are indications that the process started in the Carboniferous period. The earliest insects were probably scavengers, as many insects and woodlice and millepedes are today, and in the Upper Carboniferous and Permian periods the spores and pollen, fallen from various primitive plants, may have formed a layer on the soil surface. Russian workers (Malyshev 1968, Rohdendorf & Raznitsin 1980) have postulated that the route to phytophagy for insects commenced by their scavenging on this pollen and spore layer on the ground, then by their ascending the plants and feeding on the strobuli or other reproductive organs, i.e. on the spores etc., at source. Such insects belonging to extinct orders like Dictyoneurida would live on the plants, but it seems likely that their beak-like mouth-parts pierced the cones and other fruiting structures of the plants (V. V. Zherikhim in Rohdendorf & Raznitsin 1980). True leaf-feeding probably did not occur until the Permian period (Smart & Hughes 1973, Rohdendorf & Raznitsin 1980). All the living orders that include phytophagous insects were present at the onset of the Cretaceous period when the explosive speciation of the flowering plants occurred.

By this period, then, insects had overcome the major 'hurdles' that occurred on the route from soil-surface scavenger to vegetation-dwelling phytophage. These hurdles are (Southwood 1973): (a) desiccation; (b) attachment; and (c) food quality.

Although the water content of living plant material is often relatively high, air movement often causes a marked deficit in saturation at a very small distance from the leaf surface. Most commonly, water loss from insects is reduced by modifications of their cuticle and tracheal system, whilst there is reduced sensitivity to the haemolymph changes associated with water loss (Willmer 1980). Plant-feeding insects, especially caterpillars, may drink free water, and many position themselves in folds or hollows in the leaf surface where restricted air movement raises the ambient humidity (Fennah 1963). Other insects have evolved to take this further by rolling the leaf or by joining adjacent leaves by silk; the silk alone may be formed into a web, whilst silk cases (as in the Coleophoridae) may be viewed as 'individually portable webs'. Insects that are leaf-miners and gall-formers have evolved life-styles in which they are not subject to desiccation, but if removed from their shelters they rapidly die from water loss (Willmer 1980).

Holding on to the aerial parts of plants, especially when these move in the wind, is impossible for insects with tarsal structures such as the pair of simple claws found in many primitive ground-dwelling groups. The pretarsal structures of many phytophagous insects aid in the gripping of plant surfaces; Miridae, for example, have additional sac-like structures between or under the claws, and adult Coleoptera have adhesive setae under the tarsi (Stark 1980). Caterpillars of both Lepidoptera and sawflies have pairs of sucker-like prolegs on the abdomen. That such structures have been developed independently, in two different orders, is evidence of the selective pressure for them. The silk of Lepidoptera and the secretions of whitefly larvae and scale insects also aid in overcoming the difficulties of adherence, a problem which, like water loss, is solved by leaf-mining or gall-forming.

The two hurdles of desiccation and attachment also face predators that live on plants, but there is a further hurdle to plant feeding, namely the composition of the plant material in relation to an insect's needs. Insects have a high protein content, but plants are predominantly carbohydrates, and thus nitrogen in protein is often a critical component of plant tissues for phytophages (McNeill & Southwood 1978, Mattson 1980, Newbery 1980). Indeed, as indicated already, the first insects that became adapted to plants did so with relatively protein-rich portions—pollen and spores—or with material already partially digested by microorganisms. The larvae of Diptera, e.g. the onion fly, *Hylemya antiqua* (Friend et al 1959) or the frit fly, *Oscinella frit* (Ryzhkova 1962), often depend on microorganisms in plant wounds, whilst other plant-feeding species in the Coleoptera, Lepidoptera and Hemiptera have internal symbiotic microorganisms (Crowson 1981).

Any group of insects that becomes adapted to plants in general has a wide range of potential foods for, in most terrestrial ecosystems, the biomass of plant material greatly exceeds that available to predators. But the attacks of insects affect the plants, thereby generating selective pressures on the plants to defend themselves against insect attack. The influence of the insect community on the shrub Scotch broom (*Sarothamnus scoparius*) was investigated by Waloff & Richards (1977) over the life of the plants (10 years). They found that the shrubs that were largely protected from insect attacks by regular pesticide application lived longer and produced more seeds than those that were unprotected and therefore were attacked by a wide range of insects, although none of these were ever sufficiently abundant to cause conspicuous damage. Recent work by Brown (1982) on plots of early successional plants has also indicated subtle effects: the protected plots had a greater percentage of the area covered by vegetation, of which grasses were a larger proportion, than did the natural plots. When this evidence is considered alongside that from the biological control of weeds (Holloway 1964), where insects may greatly modify the abundance and perhaps even the range of a plant, it seems safe to conclude that phytophagous insects are a significant factor in the dynamics of a plant species. In most plant communities interspecific competition is a potent force; anything that reduces the germination rate, leaf-area or growth rate will influence the success of particular species.

Although insect communities in general influence plant success and, as evinced by biological control work, particular insect phytophages may dramatically reduce the survival of their plant hosts, not every herbivorous insect has a measurable effect on its plant host (Hopkins & Whittaker 1980).

However, the balance of evidence is that most insects have some effect on plants and some have a considerable impact. A number of features of plants may be regarded as the plant's response to this pressure. These defences are:

(a) *chemical*: that is, plant secondary substances that are repellant, toxic or may inhibit digestion;

(b) *nutritional*: that is, a balance of amino acids that is unfavourable for insect nutrition, or low water or nitrogen levels in the plant;

(c) *physical*: that is, a tough cuticle and epidermal hairs, which will impede the ability of the insect to hold on, move and feed on the plant; certain epidermal outgrowths—glandular hairs (trichomes)—combine physical and chemical defences.

Most of these features may be associated with other functions in the plant; secondary plant substances are often the end-products of various metabolic pathways, and it has been suggested (Jermy 1976) that their role in defence is largely fortuitous. However, it seems unlikely that their great diversity would have been generated by the production of metabolic waste products alone. I believe that in general 'insects and plants must be viewed as two co-evolving, competing and often mutually dependent biochemical systems' (Southwood 1973), a concept highlighted by Ehrlich & Raven (1964) in their seminal paper on coevolution between butterflies and plants. The key feature of coevolution is *reciprocal adaptation*. Ehrlich & Raven (1964) proposed that initially insects fed on plants that gained some protection from their secondary substances; thus, plants with increasingly noxious chemicals are selected for by the pressures of insect herbivory. Certain insects would evolve a tolerance of, or even an attraction to or utilization of, the novel and noxious plant compound. Both the plant group and the insect group would diversify if free from competitive pressures.

This model is well illustrated by Berenbaum's (1978, 1981) studies on coumarins in plants and swallow-tail butterflies. Coumarins occur in various forms, but angular furanocoumarins are found in only 11 genera of Umbelliferae and in two of Leguminosae. Many more swallow-tail species feed on these particular plants than on the many more genera of Umbelliferae that have other types of coumarins.

An outstanding example of reciprocal adaptation in insects and plants is provided by *Heliconius* butterflies and two types of vines—*Passiflora* (Passifloraceae, passion vines) and *Anguria* and relatives (Curcubitaceae). The details of this coevolutionary web from the Neotropical forest are being elucidated by L. E. Gilbert and colleagues (Gilbert 1972, 1982, Brown 1981, Turner 1981). The adult butterflies are conspicuous and long-lived and they roost gregariously; both males and females require additional nitrogenous food in the adult state and this they obtain, unusually for adult Lepidoptera, from pollen. This is procured from the male inflorescences of the *Anguria* vines, which are scattered throughout the forest. The butterflies have unusually large eyes and an economical gliding flight which are regarded as adaptations to the habit of 'trap-lining'—i.e. flying a fairly regular track round part of the forest to visit food sources. A vine that has an almost continuous supply of male flowers is more likely to be included on the flight track of a *Heliconius* butterfly than one that is only an occasional pollen source and thus may be missed altogether. The pollination of the infrequent female flowers therefore demands an excessive number of male flowers to ensure regular visits from *Heliconius* butterflies, and ratios of over 30 male flowers to every female one on a vine are common. The butterflies' behaviour and the vines' flowering regime have thus coevolved. (The relationship between pollinators and flowers, which provides many striking and some bizarre examples of insect–plant adaptation, is not considered elsewhere in this review.)

The larvae of *Heliconius* feed on the young growths of *Passiflora* vines and can destroy the whole of a shoot. The irregular and intermittent growth of the passion vines may be another plant defence. Female butterflies on their regular track round the tropical forest may be seen to inspect vines. If new growth is detected and an egg is laid then, in general, the shoot is doomed. The behaviour and visual acuity of *Heliconius* females are adaptations to the growth habit and spatial distribution of the larval host plant. However, there are, apparently, further steps in this coevolution. The food available to larvae on a young shoot is limited; larvae are cannibalistic ('winner-takes-all' or contest competition); and females tend to avoid shoots that already have an egg laid on them (Williams & Gilbert 1981). Several species of *Passiflora* have on their leaves remarkable egg mimics that have evolved from four different plant structures: stipules, accessory buds, leaf nectaries and petioles (Gilbert 1982). These, like real eggs, discourage females from laying and so give the vine a measure of protection. The leaf-form of *Passiflora* species is very variable (Gilbert 1972, 1982); this may also be an adaptation, for the female *Heliconius* appear to use sight extensively in finding oviposition sites.

How general is this tight coevolution between insect and plant? It seems likely to be relatively uncommon, because few plants are as strongly influenced by a single insect herbivore as *Passiflora* vines are by *Heliconius*. More frequently it is the influence of the whole community of phytophagous insects that reduces the plants' fitness: adaptation of the plants may then be more general. This has been termed 'diffuse coevolution' (Janzen 1979, Fox 1981). Tight reciprocal coevolution (as in *Heliconius* and the vines) is probably relatively atypical, like 'vortices in the evolutionary stream' (Strong et al 1983).

The generality of much of the adaptation of insects to plants is evinced by the frequency with which new associations are formed when the human introduction of plants to new areas brings together hitherto new combinations of insects and plants. Although the human race has been engaged in this process for thousands of years, the major spread of many plants of 'economic importance' has been within recent history and we can therefore distinguish between pests that have moved with the crop (e.g. the cabbage butterfly, *Pieris rapae*, Jones 1977) and those members of the native fauna that have adapted to the new crop. Sugar cane is a good example of a plant moved by humans (Strong et al 1977) and cacao is another (Strong 1974). The accumulation of insect species new to the area reflects the scale on which the plant is grown, a phenomenon termed the species–area relationship (Diamond & May 1981):

$$S = cA^z$$

where S is the total number of species, A is the area, c is a proportionality constant that depends, amongst other factors, on the group of animals and the units of measurement of area, and z is the slope of the regression. The more widely a plant is grown, the greater its exposure to native insects and, as I argued over twenty years ago (Southwood 1961), the greater the probability that a pre-adapted insect will encounter the plant: a phenomenon, I suggested, analogous to the development of resistance to pesticides. However, as the pool of potential colonists becomes relatively exhausted, so the rate of accumulation of new species will fall off: the number of species may appear to reach an equilibrium (Lawton & Strong 1981, Southwood & Kennedy 1983).

Within the pool of potential colonists different species will have widely different probabilities of having some individuals that can adapt to the new host plant: they will have different predilections for the new plant (Southwood & Kennedy 1983). In a study of the insects that have colonized soybean (*Glycine max*), Turnipseed & Kogan (1976) have recognized three groups:

(a) polyphagous species;

(b) species with a narrow range of host plants (stenophagous) that are related to the introduced plant (for soybean, wild legumes); and

(c) other stenophagous species whose host-plant shift is less explicable; e.g. the longhorn beetle, *Dectes texanus*, normally lives in cockleburs (*Xanthium*), members of the Compositae, but has adapted to soybean.

It is generally supposed that the similarity of chemical and physical defences in closely allied plants is responsible for the host-plant shifts that fall into the second category. The insects have a high predilection for colonizing related plants (Southwood 1961, Southwood & Kennedy 1983). The adaptation of the Colorado beetle, *Leptinotarsa decemlineata*, taken from a wild Rocky Mountain *Solanum*, to the introduced potato (*Solanum tuberosum*) was one of the first recorded host-plant shifts that appeared to depend on the chemical similarity of the old and new hosts. Another example, but one in which the plants were not related, is the adaptation of the cabbage-white butterflies, *Pieris brassicae* and *Pieris rapae*, to garden nasturtium (*Tropaeolum majus*), a plant introduced from Peru and belonging to an entirely different family to the cabbage, yet sharing with the cabbage family

(Cruciferae) secondary plant substances—mustard oils. Likewise, the few non-umbelliferous hosts of the swallow-tail butterflies (*Papilio machaon* complex) in North America all contain furanocoumarins (Berenbaum 1981). But the mechanisms that have predisposed other insect species to shift their hosts to unrelated plants, as *Dectes texanus* has to soybean, are less easily identified. Sometimes it seems that spatial factors are important: the new host grows abundantly in the same habitats as the original host, the abundance of which is, perhaps, reduced. This situation would certainly give many opportunities for the selection of any pre-adapted individuals. The possible shift of the Lygaeid, *Kleidocerys resedae*, from birch to rhododendron may be such an example.

Some recent studies of the insects on the introduced shrubs *Buddleia*, in Britain, show the various components of the new fauna (Owen & Whiteway 1980, Southwood & Kennedy 1983) (Table 1). The majority of the insects are

TABLE 1 The normal host-plant associations of insects that have in Britain adapted to live on *Buddleia* species (from Southwood & Kennedy 1983)

Normal plant host	*Number of insect species*
Verbascum	5
Urtica	5
Hedera	2
Tilia	2
Stachys	1
Salix caprea	1?
Polyphagous insects	14

polyphagous and we can say that polyphagous species have a strong predilection to colonize introduced plants. But five of the insect species are normally found on *Verbascum*, a member of the Scrophulariaceae, which is placed in a different order of plants. It is, however, the view of some botanists that the Buddleioidea group in the Loganiaceae are misclassified and are really related to the Scrophulariaceae; thus these data may be thought to support that arrangement and to represent another example of coevolution in insect and plant groups (Ehrlich & Raven 1964, Eastop 1979, Hodkinson 1983). There is, however, another explanation: both *Verbascum* and *Buddleia* have hairy undersurfaces. It may be this physical feature that is the basis of the predilection of insects found on *Verbascum* for *Buddleia*. This hypothesis gains some support from the observation that another species comes to *Buddleia* from *Stachys* and, probably, another (the identification is uncertain) from *Salix caprea*. Both these plants have dense coverings of fine hairs on the leaves, whilst *Urtica* also has longer glandular hairs.

However, the similarity or differences between the original and the new host plant are by no means the only determinants of the insects' potential for adaptation. It seems, perhaps not surprisingly, that a taxonomic group of insects may have certain characters that pre-adapt them for particular groups of plants. Neither *Buddleia* nor any member of its plant family is native to Britain; yet a comparison of the insects now found on it there with those found on it in South Africa, where the genus is native, showed that most of the abundant insects were in the same genera or the same tribes (Southwood et al 1982, Southwood & Kennedy 1983).

That the apparent closeness of plants in taxonomic terms may not give a good indication of the ease with which insects from one may adapt to another is shown by the fauna of native and introduced *Quercus* in Britain. The faunas of the introduced, deciduous *Q. cerris* and of the evergreen *Q. ilex* are small compared with those on the native oaks, *Q. robur* and *Q. petraea* (C. E. J. Kennedy, G. R. W. Wint and T. R. E. Southwood, unpublished results). But the surfaces of the leaves are very different, those of the introduced species having a much greater density of trichomes. Also the phenology of each species of oak is characteristic: they differ in the timing of bud burst and of tannin development, and in the levels of defensive compounds (Feeny 1970, G. R. W. Wint, unpublished results). This system is the focus of my current research with C. E. J. Kennedy and G. R. W. Wint (supported by the Natural Environment Research Council).

These examples show that the process of insect adaptation to a new host plant is one that varies greatly in the extent of the modification required. For a process that is so significant in forestry, agriculture and horticulture we are still woefully ignorant of its main principles.

REFERENCES

Berenbaum MR 1978 Toxicity of a furanocoumarin to armyworms: a case of biosynthetic escape from insect herbivores. Science (Wash DC) 201:532-534

Berenbaum MR 1981 Patterns of furanocoumarin distribution and insect herbivory in the Umbelliferae: plant chemistry and community structure. Ecology 62:1254-1266

Brown KS 1981 The biology of *Heliconius* and related genera. Annu Rev Entomol 26:427-456

Brown VK 1982 The phytophagous insect community and its impact on early successional habitats. In: Proc 5th Int Symp Insect-Plant Relationships. Pudoc, Wageningen, p 205-213

Crowson RA 1981 The biology of the Coleoptera. Academic Press, London

Diamond JM, May RM 1981 Island biogeography and the design of natural reserves. In: May RM (ed) Theoretical ecology. Blackwell Scientific Publications, Oxford, p 228-252

Eastop VF 1979 Sternorrhyncha as angiosperm taxonomists. Symb Bot Ups 22:120-134

Ehrlich PR, Raven PH 1964 Butterflies and plants: a study in co-evolution. Evolution 18:586-608

Feeny P 1970 Seasonal changes in oak leaf tannins and nutrients as a cause of spring feeding by winter moth caterpillars. Ecology 51:565-581

Fennah RG 1963 Nutritional factors associated with seasonal population increase of cacao thrips *Selenothrips rubrocinctus* (Giard) (Thysanoptera) on cashew, *Anacardium occidentale*. Bull Entomol Res 53:681-713

Fox LR 1981 Defense and dynamics in plant–herbivore systems. Am Zool 21:853-864

Friend WG, Salkeld EH, Stevenson IL 1959 Acceleration of development of larvae of the onion maggot *Hylemya antiqua* (Merg.) by micro-organisms. Can J Zool 37:721-727

Gilbert LE 1972 Pollen feeding and reproductive biology of *Heliconius* butterflies. Proc Natl Acad Sci USA 69:1403-1407

Gilbert LE 1982 The coevolution of a butterfly and a vine. Sci Am 247(2):102-107

Hodkinson ID 1983 Co-evolution between Psyllids (Homoptera: Psylloidea) and rain forest trees: the first 120 million years. Proc Leeds Philos Lit Soc Sci Sect (Symp Br Ecol Soc June 1982), in press

Holloway JK 1964 Projects in biological control of weeds. In: DeBach P (ed) Biological control of insect pests and weeds. Chapman & Hall, London, p 650-670

Hopkins MJG, Whittaker JB 1980 Interactions between *Apion* species (Coleoptera: Curculionidae) and Polygonaceae. II. *Apion violaceum* Kirby and *Rumex obtusifolius* L. Ecol Entomol 5:241-247

Janzen DH 1979 New horizons in the biology of plant defences. In: Rosenthal GA, Janzen DH (eds) Herbivores: their interaction with secondary plant metabolites. Academic Press, New York, p 331-350

Jermy T 1976 Insect-host plant relationship—co-evolution or sequential evolution? Symp Biol Hung 16:109-113

Jones RE 1977 Movement patterns and egg distribution in cabbage butterflies. J Anim Ecol 46:195-212

Lawton J, Strong D 1981 Community patterns and competition in folivorous insects. Am Nat 118(3):317-338

Malyshev SI 1968 Genesis of the Hymenoptera and the phases of their evolution. Methuen, London

Margulis L, Schwartz KV 1982 Five kingdoms. W. H. Freeman & Co., San Francisco

Mattson WJ Jr 1980 Herbivory in relation to plant nitrogen content. Annu Rev Ecol Syst 11:119-161

McNeill S, Southwood TRE 1978 The role of nitrogen in the development of insect/plant relationships. In: Harborne JB (ed) Biochemical aspects of plant and animal co-evolution. Academic Press, London, p 77-98

Newbery D McC 1980 Interactions between the coccid *Kerya seychellarum* and its host species on Aldabra Atoll. Oecologia (Berl) 46:171-179

Owen DF, Whiteway WR 1980 *Buddleia davidii* in Britain: history and development of an associated fauna. Biol Conserv 17(2):149-155

Rohdendorf BB, Raznitsin AP 1980 (eds) The historical development of the class Insecta (in Russian). Tr Paleontol Inst Akad Nauk SSSR 175:1-268

Ryzhkova YV 1962 Phytopathogenic symbionts of *Oscinella frit* L. and *O. pussila* Mg. (Diptera, Chlorophidae) and their practical use (in Russian). Entomol Obozr 41:789-795 and Entomol Rev (Engl Transl Entomol Obozr) 41:490-494

Smart J, Hughes NF 1973 The insect and the plant: progressive palaeoecological integration. Symp R Entomol Soc Lond 6:143-155

Southwood TRE 1961 The number of species associated with various trees. J Anim Ecol 30:1-8

Southwood TRE 1973 The insect/plant relationship—an evolutionary perspective. Symp R Entomol Soc Lond 6:3-30

Southwood TRE, Kennedy CEJ 1983 Trees as islands. Oikos, in press
Southwood TRE, Moran VC, Kennedy CEJ 1982 The richness, abundance and biomass of the arthropod communities on trees. J Anim Ecol 51:635-649
Stark RW 1980 Recent trends in forest entomology. Annu Rev Entomol 10:303-324
Strong DR 1974 Rapid asymptotic species accumulation in phytophagous insect communities: the pests of cocoa. Science (Wash DC) 185:1064-1066
Strong DR, McCoy ED, Rey JR 1977 Time and the number of herbivore species: the pests of sugar cane. Ecology 58:167-175
Strong DR, Simberloff D, Abele LG (eds) 1983 Ecological communities: conceptual issues and the evidence. Princeton University Press, Princeton, NJ
Turner JRG 1981 Adaptation and evolution in *Heliconius*, a defense of Neo-Darwinism. Am Rev Ecol Syst 12:99-121
Turnipseed SG, Kogan M 1976 Soybean entomology. Annu Rev Entomol 21:247-282
Waloff N, Richards OW 1977 The effect of insect fauna on growth mortality and natality of broom, *Sarothamnus scoparius*. J Appl Ecol 14:787-798
Williams KS, Gilbert LE 1981 Insects as selective agents on plant vegetative morphology: egg mimicry reduces egg laying by butterflies. Science (Wash DC) 212:467-469
Willmer PG 1980 The effects of a fluctuating environment on the water relations of larval Lepidoptera. Ecol Entomol 5:271-292

DISCUSSION

Clarke: About 15 years ago I was having an argument about genetic drift and natural selection—an habitual argument among population geneticists—and I asked somebody who was espousing drift for some really good example of a neutral character. He cited leaf shape in plants. I then listed ways in which natural selection could affect the shapes of leaves. You have added to that list movement of the leaf, which causes insects to drop off. Some of the other factors are: the flow of water on the leaf surface; surface protection against insect herbivores; aerodynamics, mimicry, and diversity in order to confuse a 'predator'. The number of possible ways in which natural selection can affect this character seems to be very large.

Fowden: Wild and cultivated forms of *Solanum* species tend to exhibit different levels of resistance to a range of attacking organisms. Gibson (1976) compared the leaf structure and found that *Solanum berthaultii*, a well worked wild-type that has high resistance, had a more hairy leaf than many of the cultivated species which were almost hairless. They recognized that a number of these hairs carried lobes at the top, which readily ruptured to release a sticky material that literally stuck down insects, especially aphids, and so limited their ability to transfer viruses between plants. Recently Gibson & Pickett (1983) have recognized a second type of hair that carries a liquid globule, among whose components is β-farnesene, the biologically active alarm pheromone of aphids. This second type of hair may deter aphids from settling, but if they get sufficiently near to the leaf surface they become excited, and are likely to move around more and encounter the sticky hairs. Therefore a protective mechanism appears to exist in which the plant uses the biologically active pheromone of the aphid in the manner of an allomone.

Kuć: How important are protease inhibitors in the resistance of plants to insects? The elicitation of protease inhibitors in various plants can result from insect feeding or from mechanical injury. What are the practical applications of this?

Southwood: There is good evidence, as I mentioned, that the level of nitrogen in the plant is often very limiting for insects feeding on the plant. Protease inhibitors like the tannins have been studied by Paul Feeny (1970), working on oak trees. The reason why so many caterpillars are found on *Quercus robur* early in the spring is because of the little 'window' of tannin-free time when the leaf is first produced. I have always assumed that this was because later on the tannin is a problem for the plant's own metabolism and so not only does the leaf have to expand quickly but, at that time, insects come in. The caterpillars that feed later on in the year can actually cope with the protease inhibitors, but they take a long time to become adult. Thus they are more exposed to their own predators and parasites. Other oaks such as *Quercus ilex* have, not surprisingly, different defences and different patterns of their production. My colleague Dr G.R.W. Wint has been analysing these and has found that the peak of tannin production in the British oaks *Q. robur* and *Q. petraea* is not reproduced in the Mediterranean oaks, *Q. cerris* and *Q. ilex*; it is later and at a lower level (much lower in *Q. ilex*). If we were concerned with breeding oaks that were more resistant to winter moths, we would probably want to bring forward this peak of tannin production as far as possible without inhibiting the growth of the oak.

Kuć: The protease inhibitors in tomato are proteins. Are these substances generally accepted as being deterrents to insect predation?

Bell: Gatehouse et al (1979) have shown that resistance of the cow-pea (*Vigna unguiculata*) to the bruchid beetle (*Callosobruchus maculatus*) is controlled by the level of trypsin inhibitor in the seed. Trials are being done at the International Institute of Tropical Agriculture in Nigeria on strains which the Durham group have identified as containing particularly high levels of this trypsin inhibitor.

Clarke: Could you justify, Professor Southwood, the distinction you made between specific coevolution and diffuse coevolution? So far in our considerations of coevolution nobody has mentioned gene-for-gene specificity, although it was implied in Professor Kuć's paper. Is there really a distinction between the specific and the diffuse? No doubt a specific relationship could induce a compound that will work, fortuitously perhaps, with other organisms as well. If this resistance were overcome, then other defences could be put forward, which might be of a specific nature.

Southwood: The two categories are not really separate, but it is reasonable to suggest, as Fox (1981) and Janzen (1980) have done, that the evolved response will often be to a community of herbivores rather than to one particular species. Under these circumstances the response is likely to be more general—e.g. in the form of protease inhibitors (tannins, polyphenols) rather than a specific qualitative chemical, such as an alkaloid or a glucoside, to which some polyphagous species might be able to develop a resistance.

Bell: Most legume seeds containing high concentrations of canavanine are not attacked by bruchid beetles. The seeds of *Dioclea megacarpa* that contain as much as 13% dry weight of canavanine are, however, attacked by the larvae of *Caryedes brasiliensis*, which possesses an arginyl *t*-RNA synthetase that discriminates against canavanine and prevents the incorporation of canavanine into protein (Rosenthal et al 1976). The beetle is successful as a predator because it is biochemically adapted to circumvent the plant's chemical defence. If that chemical did not have a protective role then it would not have elicited the adaptation found in the insect. When trying to determine whether a plant secondary compound has an ecological role, one does not necessarily need to look at the plant itself but rather at the other organisms that interact with the plant, to determine the response, if any, which that compound has elicited in these organisms.

Kuć: We may be discussing matters at different levels. For example, in the interaction between potato and the pathogen *Phytophthera infestans*, the potato has steroid glycoalkaloids in the foliage and in the peel of the tuber, and can accumulate these substances there when those tissues are injured. But the pathogen *P. infestans* has learned to cope with the steroid glycoalkaloids which, though they may be important as pre-formed inhibitors and as part of a wound response, no longer afford significant protection to the potato against *P. infestans*. However, some varieties of potato are resistant to *P. infestans* and we must therefore consider other response mechanisms.

Wood: Professor Bell's example is particularly interesting because the biochemical mechanism of tolerance to canavanine was precisely the same in the insect as in the plant.

Bell: Yes, although this may not be the only way in which the compound can act. Nevertheless, one sees modification of the amino-acid activating enzyme in both cases.

Wood: Would there be any remote possibility here that gene transfer might operate in this relationship?

Bell: Do you mean that the insect may have acquired the DNA from the plant?

Wood: Yes. The insect is feeding on the plant. Now that we know about transferable genetic elements, we cannot ignore this speculative explanation.

Fowden: One assumes that the plant itself (*D. megacarpa*) is resistant to canavanine in a similar way, but this is not proven by critical experiments, although the discriminatory mechanism was first established in plants for several amino-acid activating enzymes and different analogue substrates.

Bell: That is a fair comment. Although one can show the incorporation of canavanine, the actual enzyme has not been isolated. The interesting possibility of doing hybridization experiments would show whether the DNA responsible for the mutation is identical in all the relevant organisms.

Bradshaw: Another aspect of this example is that if there is some transfer of DNA processes, as Dr Wood suggests, it must be very infrequent; otherwise, many other species would have done it. Indeed, why is it that only one type of beetle has beaten the canavanine defence system? This relates to what Professor Southwood was saying in his paper about relative exhaustion of the species pool, and also to what one finds in heavy-metal tolerance. I am mystified by the fact that only one species overcomes the threat and that although all the other species have had the opportunity to evolve the same thing they have never done it.

Graham-Bryce: A possible answer to Professor Bradshaw's question is simply that of probability: the transfer of the DNA processes is a relatively random occurrence in the genetic material.

Bradshaw: That does not actually answer the problem but merely indicates that it only happens once, which is what we observe, without understanding why. In insect–plant coevolution, there is good evidence that an enormous degree of insect speciation appears to start from a change in a single original individual in a single species which allows a particular plant defence system to be overcome (Erhlich & Raven 1964). In each case the original evolutionary adaptation must have occurred only at low frequency. But why should it have occurred in one particular species and nowhere else?

Clarke: One can imagine as many as, say, three mutations in succession sometimes being necessary for an advantage to be conferred. The sequential occurrence of three such mutations might be an extremely rare event (see also p 202).

Southwood: That is the idea behind my model of decreasing probability of selection. Because insects of similar taxa attack *Buddleia* both in South Africa and in the UK, this suggests that their genes require perhaps only one mutation for tolerance to develop in that case (Southwood & Kennedy 1983).

Hartl: Is there evidence that the polyphagous species that have moved over to the introduced plant are genetically different from similar populations elsewhere?

Southwood: That needs to be tested, and we don't yet know.

Bradshaw: Much more detailed genetic analysis is necessary so that we can indeed know whether three sequential mutations, for instance, are necessary for a given adaptation.

Datta: But how could one know? If the event is very rare one might not be able to see it happening again.

Bradshaw: One could at least look at the structure of the gene that has actually allowed an adaptation to occur.

Datta: Yes, but to analyse the whole process would, of course, be impossible.

Kuć: Similar questions are asked in the study of insects as in the study of fungi, bacteria and viruses. My interpretation of the evidence is that we are dealing with multiple mechanisms. When a successful pathogen is found it is because, out of this composite of multiple mechanisms, there is one limiting factor. It would be very rare to have all conditions perfect for susceptibility. The susceptible interaction can be considered as the extremely fine tuning of metabolism of both host and successful pathogen. In our work it is intriguing that the plant cannot differentiate between many bacterial, fungal or viral insults, and therefore the plant mobilizes many defence mechanisms to take care of most possibilities.

Southwood: Such mechanisms might, of course, be able to deal with insect attack as well.

REFERENCES

Ehrlich PR, Raven PH 1964 Butterflies and plants: a study in co-evolution. Evolution 18:586-608

Feeny P 1970 Seasonal changes in oak leaf tanning and nutrients as a cause of spring feeding by winter moth caterpillars. Ecology 51:565-581

Fox LR 1981 Defense and dynamics in plant–herbivore systems. Am Zool 21:853-864

Gatehouse AMR, Gatehouse JA, Dobie P, Kilminster AM, Boulter D 1979 Biochemical basis of insect resistance in *Vigna unguiculata*. J Sci Food Agric 30:948-959

Gibson RW 1976 Trapping of the spider mite *Tetranychus urticae* by glandular hairs on the wild potato *Solanum berthaultii*. Potato Res 19:179-182

Gibson RW, Pickett JA 1983 Wild potato repels aphids by release of aphid alarm pheromone. Nature (Lond) 302:608-609

Janzen DH 1980 When is it coevolution? Evolution 34:611-612

Rosenthal GA, Dahlman DL, Janzen DH 1976 A novel means of dealing with L-canavanine as a toxic metabolite. Science (Wash DC) 192:256-258

Southwood TRE, Kennedy CEJ 1983 Trees as islands. Oikos, in press

Adaptation of insects to insecticides

R. M. SAWICKI and I. DENHOLM

Department of Insecticides and Fungicides, Rothamsted Experimental Station, Harpenden, Hertfordshire AL5 2JQ, UK

Abstract. Insects have successfully adapted to most insecticides by becoming resistant to them. This adaptation, of recent origin, has evolved rapidly and independently in a large number of species and is of serious economic and medical importance. The origins of resistance are still obscure, but resistance is assumed to be pre-adaptive, arising through recurrent mutation of existing alleles. However, in at least one well-researched case it probably originated by gene duplication. Resistance can be monofactorial or multifactorial. When several independent mechanisms confer resistance to the same group of insecticides the order in which these mechanisms are selected may reflect their effectiveness in protecting insects from the toxic effects of the different compounds. Our concept of resistance is changing as insects continue to adapt to the insecticide-containing environment. Criteria for defining resistance in both fundamental and practical terms are re-examined here in the light of recent work.

1984 Origins and development of adaptation. Pitman Books, London (Ciba Foundation symposium 102), p 152-166

Over the last 40 years insects have had to contend with an environment containing hazards such as they had never previously encountered—extremely toxic synthetic insecticides used on a massive scale against all insect pests. There can now be almost no area of the globe free from insecticides. Yet as far as is known not a single species of insect pest has been eradicated. Indeed not only have all arthropod pests survived this holocaust, their threat has in many cases increased because they have rapidly adapted to this new environmental hazard by becoming resistant to insecticides. By 1980, at least 428 species had developed strains resistant to one or more pesticides in areas where chemical control has been extensively practised (Georghiou 1981) and this number is an underestimate because apparently it omits species in which resistance does not present serious control problems. Insecticide resistance thus represents a rapid and recent evolutionary phenomenon in a taxonomically diverse group of organisms, and its detailed examination is essential not only because it provides an insight into the genetic and biochemical nature of adaptations, and into their development by natural selection, but also because the serious economic implications of resistance demand that we find means to counter this form of adaptation.

Origins of resistance

Insecticide resistance is generally assumed to be a pre-adaptation conferred by novel alleles that arise only rarely in untreated populations by recurrent mutation. Insecticides are not normally considered mutagenic, and experiments to demonstrate that individuals can acquire heritable resistance as a consequence of insecticidal treatment have so far failed. The role of enzyme induction in resistance has not yet been established satisfactorily, but is likely to be only slight. Increased tolerance caused by environmental factors such as diet and changes in temperature are transient and outside the scope of this paper.

Mechanisms leading to the production of enzymes that cause resistance

Resistance-causing enzymes are thought to be selected by insecticides because they are effective in protecting the insect from poisoning. Such enzymes can originate in two ways: either through single point mutations that create aberrant structural genes that code enzymes with different properties through alterations in their amino-acid sequences; or through mutations that control the amounts of enzyme produced. This covers both regulatory mechanisms and gene amplification (Schimke 1980).

Qualitative changes in enzymes leading to resistance

Since the amino-acid sequences of enzymes involved in insecticide resistance have not been established, direct evidence for qualitative differences between allozymes that contribute to resistance (R enzymes) and those that do not (S enzymes) is not available. However, where enzymes of susceptible and resistant insects differ in substrate specificity, although their catalytic centre activities and molar quantities are the same (Smissaert et al 1975, Devonshire 1980), indirect evidence for qualitative differences is very strong.

The widespread use of organophosphorus insecticides and carbamates, which are inhibitors of acetylcholinesterase (EC 3.1.1.7), has selected in arthropods structural variants of this enzyme that are less sensitive to inhibition (reviewed by Devonshire 1980). These changes have been studied in greatest detail in the cattle tick *Boophilus microplus* (Nolan & Schnitzerling 1976), the spider mite *Tertranychus urticae* (Smissaert et al 1975) and the housefly *Musca domestica* (Devonshire 1975). In the cattle tick (Stone et al 1976) and the housefly (Oppenoorth 1982) there is evidence of at least two organophosphorus-insensitive mutant enzymes in different strains that differ in sensitivity to various inhibitors. In strains of species where this sensitivity has been measured, equal molar amounts of S and R enzyme are produced in the heterozygote and each type is independently expressed. The R enzyme has decreased sensitivity to the inhibitors and this has parallel effects on the catalytic centre activities of the enzyme with the substrate (Devonshire 1980). Although the acetylcholinesterases of resistant strains may thus have a much lower rate of inhibition by insecticides, their activities towards acetylcholine are often only slightly less than in susceptible strains, demonstrating that the alteration of the enzyme does not necessarily mean it is less able to function normally (Oppenoorth & Welling 1976). The differences in hydrolysis of acetylcholine are believed to be associated with restricted access to the catalytic centre. The different forms of acetylcholinesterase demonstrate a qualitative change in the enzyme that is selected for by acetylcholinesterase-inhibiting insecticides.

Quantitative changes in enzymes leading to resistance

The peach-potato aphid *Myzus persicae* provides the best documented case of resistance due to an increased production of enzyme. The enzyme responsible for resistance to organophosphorus insecticides, carbamates and, possibly, pyrethroids is the carboxylesterase E4 (EC 3.1.1.1) which has a broad substrate specificity (Devonshire 1977, Devonshire & Moores 1982).

Carboxylesterase E4 was isolated from other esterases in homogenates of susceptible and resistant aphids by gel filtration and ion-exchange chromatography. The purified enzyme, whether isolated from susceptible or resistant aphids, had identical catalytic centre activities towards both paraoxon and l-naphthylacetate, demonstrating that resistant aphids must produce a larger amount of the same enzyme.

The variants of *M. persicae* studied differ in their amount of E4 (Fig. 1) and in their resistance to insecticides (Sawicki et al 1980). Although assays of total esterase in these variants showed a progressive increase in enzyme activity with an increasing resistance, the quantitative relationships were affected also by hydrolysis of l-naphthylacetate by esterases other than E4, and were most significant in variants with little E4.

The rate of diethyl phosphate production by homogenates from each of the variants, incubated with [^{14}C]paraoxon at concentrations that give V_{max} for the enzyme, doubled between each of the seven variants; since catalytic

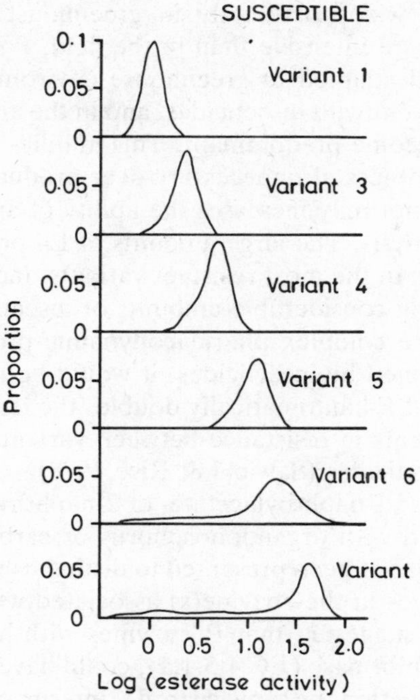

FIG. 1. Frequency distribution curves for total esterase activity for six of the seven variants of *Myzus persicae* (the peach-potato aphid) (after Sawicki et al 1980).

centre activity for paraoxon is the same in the different variants ($0.6\,h^{-1}$), and only E4 hydrolyses paraoxon, the rates gave a direct measure of the molar quantity of E4 in each variant. Thus, the amounts of E4 in the seven clones form a geometric series, doubling between each variant. This doubling is best explained by a succession of duplications of the structural gene for E4, with the susceptible aphids having one copy (2^0), variant 2 having two copies (2^1), and so on, with the most resistant clone having 2^6, i.e. 64 copies of the gene (Devonshire & Sawicki 1979).

In the two most resistant variants (variants 6 and 7 in Fig. 1) instability of E4 production leads to the presence of some individuals with less enzyme activity, and this is reflected in the strong asymmetry of distribution curves of the esterase activity for these variants (Sawicki et al 1980). This instability of enzyme production in the absence of selection is also a feature of bacterial, yeast and eukaryotic cell cultures that show excess enzyme production through gene duplication.

The increased production of E4 clearly confers a selective advantage on *M. persicae* in the presence of insecticides. In Great Britain, the two most resistant variants are widespread only in greenhouses where insecticidal application is much more intensive than in the field. Populations containing these variants are well adapted to greenhouse environments; variants with most E4 survive treatment with insecticides, and in the absence of insecticides those with least E4 become predominant. This nullifies the possible adverse effects on population fitness of unnecessary over-production of the enzyme.

Gene amplification not only increases the ability of aphids to detoxify the insecticide(s) by hydrolysis. The large amounts of E4 produced, up to about 3% of the total protein in the most resistant variants, increase the chances of survival by sequestering considerable amounts of insecticide (Devonshire & Moores 1982). Because complex pharmacodynamic processes are involved when insects are poisoned by insecticides, it would be naive to assume that doubling the amount of E4 automatically doubles the level of resistance, and in *M. persicae* increments in resistance between variants are not consistent, and differ betwen insecticides (Sawicki & Rice 1978).

In other insects where 1-naphthylacetate- or 2-naphthylacetate-hydrolysing enzymes are associated with organophosphorus or carbamate resistance no conclusive evidence has yet been presented to distinguish between qualitative and quantitative changes in the enzyme(s) associated with resistance.

Oppenoorth (1965) suggested that R enzymes with high activity, such as strong DDT-dehydrochlorinase (EC 4.5.1.1), could have arisen by a series of consecutive mutations within the same cistron if intermediate levels of activity had sufficient survival value to enable their spread upon selection. However, such enzymes have not as yet been sufficiently characterized genetically or biochemically to support this interesting suggestion.

Nature and value of cross-resistance in the adaptation of insects to insecticides

In most early studies the genetic and biochemical nature of cross-resistance to different insecticides was examined in insect populations at early stages of resistance, in which either the single mechanism present or the major mechanism detected had a readily recognizable resistance spectrum. These specific cross-resistance spectra were usually restricted to close chemical analogues of the insecticide and became designated according to the insecticidal group to which they conferred resistance, i.e. cyclodiene resistance, organophosphorus resistance, etc. However, it soon became apparent that resistance is often multifactoral; it is *multiple* when distinct resistance mechanisms each protect against different types of poison, or *multiplicate* when two or more co-existing mechanisms protect the insect against the same poison. When resistance is multiplicate the resistance mechanisms selected by the insecticide can have very different cross-resistance spectra even against insecticides of the same chemical group (the so-called sub-group resistance [Georghiou & Hawley 1971]), and some can even confer resistance to insecticides of chemically distinct groups. This multiplicity of resistance mechanisms in pest populations demonstrates that there are often several different genetic responses to a particular change in environment, and this has had serious repercussions on insect control (Sawicki 1975).

The resistance spectrum of the gene *kdr* (Farnham 1977) illustrates how cross-resistance that covers two chemically different groups can lead to the early ineffectiveness of insecticides of one group through selection of resistance by compounds of the other. The gene *kdr*, which determines knockdown resistance in houseflies and decreases the sensitivity of the nervous system to both DDT (dichlorodiphenyltrichloroethane) and pyrethroids, confers moderate to strong resistance to DDT and is the major resistance mechanism to pyrethroids. It sometimes co-exists with DDT-dehydrochlorinase, the major R enzyme that detoxifies DDT. In Danish populations of houseflies *kdr* is believed to have been responsible for much of the DDT resistance reported in the late 1940s, but in the USA, where *kdr* was apparently rare, DDT resistance is attributed to DDT-dehydrochlorinase. In Denmark pyrethroid resistance developed rapidly when synthetic pyrethroids were introduced for housefly control in the 1970s (Keiding 1976) but in the USA prolonged use of synthetic pyrethroids on animal farms in New York State (G. P. Georghiou, personal communication, 1981) failed to expose any pyrethroid resistance. However, in animal farms in Ontario (Harris et al 1982), and in a Californian strain selected for several years in the laboratory (De Vries & Georghiou 1981), the prolonged and intensive use of pyrethroids resulted in pyrethroid resistance that was largely caused by *kdr* (De Vries & Georghiou 1981) or by its more potent *super-kdr* allele (A. W. Farnham et al, unpublished results).

Thus where DDT-selected *kdr* is one of two major DDT resistance mechanisms, pyrethroid resistance is likely to manifest itself rapidly when these insecticides are introduced for pest control, as happened in Denmark. Indeed, even a lapse of 20 years between the withdrawal of DDT in Denmark and the introduction of synthetic pyrethroids failed to prevent the rapid reappearance of *kdr* in houseflies (Keiding 1976).

Detailed studies of the biochemistry and genetics of resistance to organophosphorus insecticides in Danish houseflies have shown why mechanisms that confer resistance to insecticides of the same group are often selected sequentially (Sawicki 1975).

From 1952, for almost 20 years, housefly control on Danish farms relied exclusively on organophosphorus insecticides supplemented by pyrethrum sprays. For the first 10 years the insecticides used most widely were diazinon and parathion, but from 1965 onwards dimethoate was introduced, following the discovery that diazinon-resistant and parathion-resistant houseflies were susceptible to this organophosphorus insecticide. Parathion and diazinon selected several genes that confer resistance to these and several other organophosphorus insecticides, but not to dimethoate. Widespread use of dimethoate led to the disappearance of diazinon-specific and parathion-specific resistance factors which were substituted by factors selected by dimethoate. These new factors conferred resistance not only to dimethoate but to most other organophosphorus insecticides including diazinon and parathion (Sawicki & Keiding 1981).

A major factor in resistance to dimethoate appears to have been the presence of acetylcholinesterase that was less readily inhibited by organophosphorus insecticides than the S enzyme (Devonshire 1975). When this mutant became common in Demark is uncertain, but it was not detected in any of the organophosphorus-resistant strains before dimethoate was used, and was even absent from strains intensively selected in the laboratory. Therefore unless mutation was fortuitously delayed for at least 10 years, the insensitive acetylcholinesterase was selected much more readily by dimethoate than by either parathion or diazinon.

It is interesting to speculate why the insensitive acetylcholinesterase was not selected by diazinon or parathion in Danish houseflies, even though it is the target enzyme for organophosphorus insecticides. In both phytophagous mites and ticks, which have relatively few organophosphorus-detoxifying enzymes, insensitive acetylcholinesterase is the major and most readily selected resistance mechanism. A reasonable explanation is that dimethoate selected the insensitive acetylcholinesterase because the biochemical characteristics of dimethoate were better suited than those of diazinon or parathion for selecting the mutant enzyme. The slight decrease in sensitivity of the mutant acetylcholinesterase would be of little importance against good inhibitors such as paraoxon, but could be more important in improving survival with dimethoxon, which is a poor inhibitor of the enzyme. Whereas only four times more paraoxon is needed to inhibit the resistant enzyme, 10 times more dimethoxon is required to obtain the same result (Table 1). The

TABLE 1 K_i values[a] of the organophosphorus-sensitive (AChES) and organophosphorus-insensitive (AChER) forms of acetylcholinesterase to dimethoxon and paraoxon in the housefly

Insecticide	K_i (mM^{-1} min^{-1})		*Ratio of sensitive: insensitive enzyme*
	AChES	*AChER*	
Paraoxon	578	139	4
Dimethoxon	20	2	10
K_i (paraoxon):K_i (dimethoxon)	29	70	—

[a] K_i is the bimolecular rate constant which measures the susceptibility of the enzyme to inhibition; the greater the K_i the faster the enzyme is inhibited and the more susceptible the insect to the insecticide (after Sawicki & Keiding 1981).

difference in amounts of the two poisons needed to inhibit the mutant enzyme is even more striking when the K_i values of dimethoxon and paraoxon are compared: 70 times more dimethoxon than paraoxon is needed to inhibit the less sensitive enzyme. Such strong concentrations of dimethoxon in the vicinity of the target are probably subject to rate-limiting effects which restrict the amounts of insecticide that reach the target site. Thus, K_i values influenced the rate of selection of the R enzyme by affecting the degree of phenotypic dominance of the heterozygote; the larger the difference in K_i between the S and R enzymes, the greater the phenotypic dominance in response to the insecticide. This may explain differences in the phenotypic expression of gene Opl, responsible for the organophosphorus-insensitive acetylcholinesterase in the mite *Tetranychus urticae* (Dittrich 1972). In this species the expression of gene Opl ranges from full dominance to parathion to full recessiveness to monocrotophos.

These two examples of multifactorial resistance emphasize the need to use insecticides in the way most likely to delay or prevent insects from adapting to them by becoming resistant.

Examination of criteria for defining insecticide resistance

Although resistance to insecticides is a very good example of an adaptation to a change in environment, it is first and foremost a practical problem in the field, and it is defined by criteria that do not always correspond to those used in laboratory studies. As a result there are misunderstandings about what constitutes resistance, and how to measure it.

The definition of resistance to insecticides is least controversial at the level of the individual insect; here it refers to the biochemical mechanism(s) or the phenotypic expression of one or more genes that improve, however slightly, the ability of the individual to survive and reproduce in the presence of insecticides.

At the population level, resistance can equally denote a *change in response* to an insecticide, through selection, or a *control failure* when the recommended dose no longer gives adequate control of the insects. Although the two are often interdependent, they are not synonymous because a change in response need not lead to loss of control. Fig. 2 illustrates this point: although field populations of houseflies on animal farms near Rothamsted were between five and 10 times less susceptible to permethrin, a synthetic pyrethroid, than was *S*, the standard susceptible laboratory strain shown, they were well controlled by the doses of pyrethroids recommended by the manufacturers (see Sawicki et al 1981). This disparity in tolerance was unimportant because it did not impair the efficacy of the recommended dose, which had been based on results of field trials. Hence, absolute differences in tolerance determined by comparing LD_{50} values of susceptible and field strains are largely irrelevant; what matters is if the tolerance of field populations has changed sufficiently to render the recommended dose ineffective. Loss of control through resistance is more likely to be detected from a comparison with the tolerance of field populations that are well controlled by the field dose than by reference to the response of susceptible strains.

This field dose is not immutable. It is usually calculated to give not only maximal effectiveness to the user but also to ensure adequate return to the

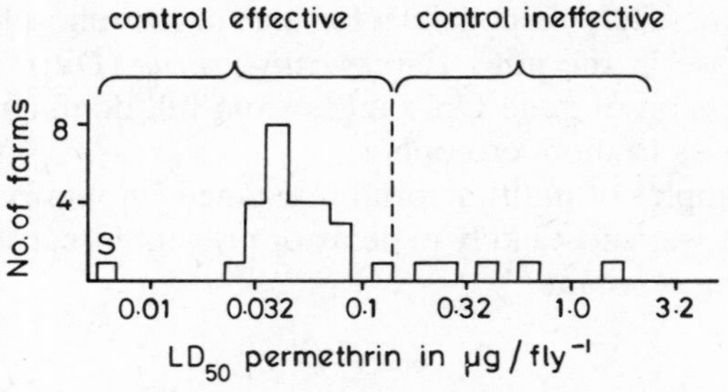

FIG. 2. Relationship between the tolerance of field populations of houseflies to permethrin, as revealed by laboratory tests, and the effectiveness of recommended doses of pyrethroids in the field. S shows, for comparison, results from a standard, susceptible laboratory strain (after Sawicki et al 1981).

producer. When resistance develops, control can sometimes be restored by increasing the dose or the frequency of application, but this requires safety clearance by regulatory agencies and must be commercially worthwhile. Thus, within limits, both resistance and its control are variables that depend on economic factors.

Resistance is also a dynamic phenomenon. International bodies such as the World Health Organization and the Food and Agriculture Organization of the United Nations ignore this when they continue to recommend, for detecting resistance, so-called normal populations that are defined as 'populations never subjected to insecticidal pressure and in which resistant individuals are rare' (World Health Organization 1981). The term 'normal' applied to susceptible strains is a misnomer and is more appropriate for field strains, many of which not only resist one or more groups of insecticides but also differ from susceptible strains in their tolerance even to compounds with which they have never been in contact.

As insects continue to adapt to insecticides in the environment, susceptible strains will be even less representative, and will come even more to resemble living fossils. Their use, particularly in detecting resistance and in screening for new insecticides, will thus become increasingly inappropriate. It is therefore now our turn to adapt to the changing environment.

REFERENCES

Devonshire AL 1975 Studies of the acetylcholinesterase from houseflies (*Musca domestica* L.) resistant and susceptible to organophosphorus insecticides. Biochem J 149:463-469

Devonshire AL 1977 The properties of a carboxylesterase from the peach-potato aphid *Myzus persicae* (Sulz.) and its role in conferring insecticide resistance. Biochem J 167:675-683

Devonshire AL 1980 Insecticide resistance caused by decreased sensitivity of acetylcholinesterase to inhibition. In: Insect neurobiology and pesticide action. Society of Chemical Industry, London, p 473-480

Devonshire AL, Moores GD 1982 A carboxylesterase with broad substrate specificity causes organophosphorus, carbamate and pyrethroid resistance in peach-potato aphids (*Myzus persicae*). Pestic Biochem Physiol 18:235-246

Devonshire AL, Sawicki RM 1979 Insecticide-resistant *Myzus persicae* as an example of evolution by gene duplication. Nature (Lond) 280:140-141

DeVries DH, Georghiou GP 1981 Decreased nerve sensitivity and decreased cuticular penetration as mechanisms of resistance to pyrethroids in a (1R)-trans-permethrin-selected strain of the housefly. Pestic Biochem Physiol 15:234-241

Dittrich V 1972 Phenotypic expression of gene OP^L for resistance in twospotted spider mites treated with various organophosphates. J Econ Entomol 65:1248-1255

Farnham AW 1977 Genetics of resistance of houseflies (*Musca domestica* L.) to pyrethroids. I. Knockdown resistance. Pestic Sci 28:631-636

Georghiou GP 1981 The occurrence of resistance to pesticides in arthropods. Food and Agriculture Organization of the United Nations, Rome

Georghiou GP, Hawley MK 1971 Insecticide resistance resulting from sequential selection of houseflies in the field by organophosphorus compounds. Bull WHO 45:43-51

Harris CR, Turnbull SA, Whistlecraft JW, Surgeoner GA 1982 Multiple resistance shown by field strains of housefly, *Musca domestica* (Diptera: Muscidae), to organochlorine, organophosphorus, carbamate and pyrethroid insecticides. Can Entomol 114:447-454

Keiding J 1976 Development of resistance to pyrethroids in field populations of Danish houseflies. Pestic Sci 7:283-291

Nolan J, Schnitzerling HJ 1976 Characterization of acetylcholinesterases of acaricide resistant and susceptible strains of the cattle tick *Boophilus microplus* (Can.) II. Pestic Biochem Physiol 6:142-147

Oppenoorth FJ 1965 Some cases of resistance caused by the alteration of enzymes. Proc 12th Intern Congress Entomol, London, p 240-242

Oppenoorth FJ 1982 Two different paraoxon-resistant acetylcholinesterase mutants in the housefly. Pestic Biochem Physiol 18:26-27

Oppenoorth FJ, Welling W 1976 Biochemistry and physiology of resistance. In: Wilkinson CF (ed) Insecticide biochemistry and physiology. Hegden, London, p 507-568

Sawicki RM 1975 Effects of sequential resistance on pesticide management. Proc 8th Br Insecticide and Fungicide Conf p 799-811

Sawicki RM, Keiding J 1981 Factors affecting the sequential acquisition by Danish houseflies (*Musca domestica* L.) of resistance to organophosphorus insecticides. Pestic Sci 12:587-591

Sawicki RM, Rice AD 1978 Response of susceptible and resistant peach-potato aphids, *Myzus persicae* (Sulz.) to insecticides in leaf-dip bioassays. Pestic Sci 9:513-516
Sawicki RM, Devonshire AL, Payne RW, Petzing S 1980 Stability of insecticide resistance in the peach-potato aphid, *Myzus persicae* Sulzer. Pestic Sci 11:33-42
Sawicki RM, Farnham AW, Denholm I, O'Dell K 1981 Housefly resistance to pyrethroids in the vicinity of Harpenden. Proc 1981 Br Crop Protect Conf: Pests and Diseases, p 609-616
Schimke RT 1980 Gene amplification and drug resistance. Sci Am 243(5):60-69
Smissaert HR, Abdul el Hamid FM, Overmeer WPJ 1975 Minimum acetylcholinesterase (AChE) fraction compatible with life derived by aid of a simple model explaining the degree of dominance of resistance to inhibitors in AChE "mutants". Biochem Pharmacol 24:1043-7
Stone BF, Nolan J, Schuntner CA 1976 Biochemical genetics of resistance to organophosphorus acaricides in three strains of the cattle tick, *Boophilus microplus*. Aust J Biol Sci 29:265-279
World Health Organization 1981 Instructions for determining the susceptibility for resistance of adult mosquitos to organochlorine, organophosphate and carbamate insecticides—diagnostic tests. WHO/VBC/81.806, World Health Organization

DISCUSSION

Clarke: I agree that 'normal' is only a relative term and that one should never regard organisms as static. What is considered as 'normal' will change with time. We should always consider 'resistant' and 'susceptible' as relative terms.

Georghiou: I am intrigued by the consequences of gene duplication in insecticide resistance especially as regards the useful life of new insecticides. Dr Sawicki and co-workers have described gene duplication in phosphate-resistant *Myzus persicae*, and we at Riverside, in collaboration with Dr Nicole Pasteur, have found what appears to be a similar phenomenon in phosphate-resistant *Culex quinquefasciatus* (Pasteur et al 1980). The types of dose–response curve published for many other cases of phosphate resistance suggest that gene duplication may not be as rare as we initially thought. If this is so, can it be assumed that as the number of gene duplicates increases, so does the probability of relevant mutations, so that subsequently introduced phosphate insecticides have a shorter useful life than earlier phosphates? This has happened with mosquito control in California and other countries.

Clarke: This must certainly be true in the statistical sense, but if many loci are serially duplicated this will increase the chances of getting more duplications or, indeed, of losing the duplicated gene itself because of the possible mispairing. This will produce the instabilities that Dr Sawicki has been talking about. Thus, one can more easily get increases in numbers of genes once the initial duplication has occurred. Presumably the more duplicates that are present, the more easy will it be to change the number of duplicates because they are all in series with one other and recombination may occur in the wrong place. All organisms seem to contain several duplicate esterases, and I don't know why. Unless all the esterase polymorphisms in the world came about recently as a consequence of the spraying of insecticides, we start from a position in which these enzymes are already rather peculiar in having a lot of duplicates, and a large amount of polymorphism.

Sawicki: Another peculiar thing about the esterases is that we don't know what they are for!

Gressel: In relation to the terminology of gene duplication here, most molecular biologists would prefer either a cytogenetic approach—i.e. showing that gene duplications are present—or an approach involving cloning of the gene and probing to prove duplication. I would not use the term 'gene duplication' unless you have shown gene duplication. It has been well demonstrated, for example, in animal cell cultures with the anti-tumour drug, methotrexate. There is more than one way for these doublings in esterase activities to arise. They could be by changes in a gene controlling the levels, and not by duplications. I am surprised this has not been done, because the E4 enzyme sounds easy to extract for such studies.

Sawicki: This problem is at present being tackled.

Clarke: Does *M. persicae* have chromosomes that one can easily see?

Sawicki: Yes, but it is complicated because some of the variants of *M. persicae* have no translocation and others do have it. Some people consider that translocation is involved in this amplification of genes. Current work on the molecular biology of this problem (A.L. Devonshire & L.M. Searle) should clarify what is going on, and whether all 64 copies are there to start with, some of them being switched on, or whether gene amplification is indeed occurring. We have demonstrated that there are quantitative and not qualitative differences in the pure enzyme and we are now attempting to clone the particular gene.

Hodgson: Are there parallel increases in other esterases apart from E4?

Sawicki: No.

Hodgson: So are you sure that you are not duplicating other genes?

Sawicki: We may be duplicating other genes, but certainly none of those responsible for paraoxon-hydrolysing enzymes.

Graham-Bryce: The exceptions to some of these general trends might be of interest, for example, in the chemical control of *Phorodon humuli* (damson hop aphid) which is being studied at East Malling. This aphid has not been studied in as much detail as *M. persicae* (the peach-potato aphid) but the history of resistance has followed the general pattern that Professor Georghiou described. Compounds such as demeton-*S*-methyl were used successfully for about 10 years before resistance rendered them ineffective. The effective life of subsequently introduced organophosphorus insecticides became progressively shorter until, with omethoate for example, it had become less than one year (Muir & Cranham 1979). Carbamate resistance also developed, so esterases are presumably involved in the mechanism. Mephospholan appears to be a notable exception to this general picture. Being a phosphoramidate, it differs from most of the other organophosphates used. This compound is used extensively now in English hop gardens, and is almost the last line of protection against *P. humuli*. It is a systemic insecticide applied as a soil drench and the uptake patterns suggest that it declines to sublethal concentrations in the course of the season. Therefore all the conditions of use imply strong selection pressure which would be expected to favour the development of resistance. The compound was introduced 10 years ago and has been used very heavily for several years. Yet so far there has been no evidence of development of significant resistance. The eventual explanation for this will be of considerable interest.

Bowers: The more genes present for an esterase, the more esterase can be produced. Insects are known to undergo extremes in their polyploidy. Could a large increase in polyploidy explain why more esterase is present in some of the more resistant insects?

Sawicki: I don't know. Cytological investigations have given no evidence for polyploidy.

Bowers: I can think of a parallel with the juvenile hormone. When this appears in the adult insect there is suddenly a tremendous increase in polyploidy in the fat body (Nair et al 1980), and this is necessary for the production of sufficient amounts of vitellogenin. Perhaps a signal induced by the insecticide could cause an enormous increase in the nuclear material and then an induction of large amounts of esterases.

Sawicki: We can't answer that one until we have more evidence, from the gene in *Escherichia coli*.

Clarke: It is perhaps surprising how rare gene amplification is, even when there is a large demand for a protein. As far as I know, the only duplications of genes clearly associated with a need for increasing the quantity of protein are those for histones and ribosomal proteins. The event seems to be exceptional—but then, perhaps, being subjected to insecticides is also exceptional.

Gressel: Is the instability that you mentioned possibly due to a lack of fitness in making this high percentage (3%) of esterase?

Sawicki: No. It is most certainly not due to a lack of fitness. Provided the aphids are under insecticide selection they reproduce very fast indeed, and they produce the most resistant variant. As soon as the insecticidal selection pressure is stopped, the population starts to produce less resistant or fully susceptible individuals. The interesting thing is that the jump back to susceptibility can occur in one generation. We have analysed the effect statistically, and demonstrated that the long tail in the E4 distribution curve consists of peaks that are in direct correlation with the other variants. So there can either be a stepwise loss of esterase activity or it can happen in one go. If it is a recombinational event that happens in one go from 64 to 1, it seems rather unlikely.

Gressel: That is why I wondered if competition with other variants and a lack of fitness were involved.

Sawicki: Probably when up to 16 copies of the gene are present the whole system is stable. When individuals with 16 copies are placed in colonies with individuals containing a single copy, the ensuing population contains only these two variants. Variants with intermediate levels of E4 (2,4,8,16 etc.) appear only when the unstable variants (32 and 64 copies of E4) are present.

Clarke: To what extent are you observing clones; do they go through meiosis?

Sawicki: Almost certainly not; these are single individuals and they go through viviparous reproduction.

Clarke: Is a meiosis not involved in that?

Sawicki: Probably not. There was supposed to be a process called endomeiosis which was put forward as an explanation for the viviparity in *M. persicae* (Cognetti 1961), but the hypothesis has not been verified at all by Blackman (1980).

Kuć: Is it possible, for example, that the large number of esterases in nature may have another, very different, enzymic activity?

Sawicki: They are most unlikely in insects to hydrolyse solely 1-naphthylacetate, and have many different functions.

Kuć: Take, as an example, a key enzyme in photosynthesis, ribulose 1,5-diphosphocarboxylase-oxygenase, an enzyme that catalyses two completely different reactions depending on whether there are high concentrations of CO_2 or O_2 in the environment.

Clarke: Insects containing a lot of esterases might use esters as a food source.

Sawicki: Sugar seems to be the most important food source of the aphids. The role of esterases on the whole in insect metabolism still seems to be a mystery.

Hodgson: Esterases, classically, are known as phase I detoxification enzymes which increase the water solubility of lipophilic compounds. I don't think it's necessary to invoke a so-called 'normal' substrate for a detoxification enzyme; their proper role *is* detoxification. I'm sure this is true for some of the esterases. Although some of them, such as acetylcholinesterase, will hydrolyse many esters, they have clear-cut normal functions. Some of the other esterases may be non-specific simply because nature has selected them as detoxification mechanisms.

Gressel: Is there a K_m difference between acetylcholinesterase and the non-specific esterases? If the non-specific ones had a much lower affinity this would suggest that they would have a high K_m.

Sawicki: Yes. For example, Jiang et al (1982) have recently found that in *Culex pipiens* the K_i for trichlorphon for a carboxylesterase is 30 times greater than that for the acetylcholinesterase; in other words, this esterase provides a very good buffer in protecting the acetylcholinesterase.

REFERENCES

Blackman RL 1980 Chromosomes and parthenogenesis in aphids. In: Blackman RL et al (eds) Insect cytogenetics. Blackwell Scientific, Oxford (Symp Royal Entomol Soc) vol 10:133-148

Cognetti G 1961 Citogenetica della partenogenesi degli Afidi. Arch Zool Ital 46:89-122

Jiang JL, Chen QY, Huang G, Zhang QZ 1982 On the properties of carboxylesterase in OP resistant and susceptible mosquitoes, *Culex pipiens palens*. Contrib Shanghai Inst Entomol 1:69-76

Muir RC, Cranham JE 1979 Resistance to pesticides in damson hop aphid and red spider mite on English hops. Proc Br Crop Protection Conf on Pests and Diseases 1:161-167

Nair KK, Unnithan GC, Wilson PR, Koohran CJ 1980 Phytochemistry of corpora allata, fat body and follicle cells of pre-precocene treated *Schistocerca gregaria*. Sci Papers of Inst Org Phys Chem, Wroclaw Tech Univ, no 22, p 389-404

Pasteur N, Georghiou GP, Ranasinghe LE 1980 Variations in the degree of homozygous resistance to organo-phosphorus insecticides in *Culex quinquefasciatus*. Proc Calif Mosq Control Assoc 48:69-73

Biochemical mechanisms of resistance to insecticides

ERNEST HODGSON and NAOKI MOTOYAMA

Interdepartmental Toxicology Program, School of Agriculture and Life Sciences, North Carolina State University, Raleigh, North Carolina 27650, USA

Abstract. The principal biochemical mechanisms of resistance to insecticides involve either modified, less sensitive cholinesterase, esterase action, glutathione *S*-transferase action or cytochrome *P*-450-dependent monooxygenation. Both quantitative and qualitative differences in cytochrome *P*-450 isozymes are under genetic control and both are related to resistance. Recent characterization studies involving ligand binding and multiplicity of isozymes in *Musca domestica* (the housefly) are discussed in relation to resistance. The recent demonstration that multiple isozymes of glutathione *S*-transferase exist in susceptible and resistant insects is of interest, and some re-examination of their role in the mechanism of resistance is required. Esterases are a heterogeneous group of enzymes whose role in resistance has often been suggested but seldom rigorously defined. Purification studies in the green rice leafhopper, *Nephotettix cincticeps*, have involved an enzyme with carboxylesterase, phosphotriesterase and pyrethroid esterase activities. A similar enzyme, but without pyrethroid esterase activity, is also found in the housefly. In resistance such enzymes may serve either to catalyse hydrolysis or as binding proteins. It has been suggested, from time to time, that regulator genes, enzyme induction and gene magnification all play a part in controlling biochemical mechanisms of resistance, although clearly defined evidence has not always been brought forward. These hypotheses are re-examined.

1984 Origins and development of adaptation. Pitman Books, London (Ciba Foundation symposium 102), p 167-189

Resistance to toxic agents is usually defined as the increased ability of a population to survive the effect of a toxic agent, this increase being brought about by genetic selection, either as a result of exposure to the agent in question or, in cross-resistance, by exposure to some other toxic agent. Tolerance, on the other hand, is the ability of previously unexposed individuals or populations to survive toxic effects better than an otherwise similar individual or population. While both phenomena are adaptations that may be inter-related the mechanisms are clearly different. The most recent summary

TABLE 1 Resistance mechanisms in insects[a]

Category	*Example or site*
A. BEHAVIOURAL	Resting behaviour Olfactory behaviour
B. PHYSIOLOGICAL	
1. Reduced penetration	Cuticular
2. Increased storage	Adipose tissue
3. Reduced nerve sensitivity	Impulse transmission and 'knockdown'
4. Increased excretion	Nicotine
C. BIOCHEMICAL	
1. Reduced target sensitivity	Cholinesterase
2. Receptor by-pass	HCN in scale insects
3. Increase in non-oxidative enzymes	Glutathione transferases and esterases
4. Increase in oxidative enzymes	Cytochrome *P*-450-dependent monooxygenase system

[a] Modified from Guthrie (1980).

of the mechanism of resistance to pesticide is the proceedings of a conference held in 1979 (Georghiou & Saito 1983).

While all resistance mechanisms (Table 1) have a biochemical or biophysical basis, either the mechanism or the intermediate steps between the mechanism and the expression of resistance may be unknown. Induction of xenobiotic-metabolizing enzymes is clearly of *adaptive* significance to the individual, but its relationship to *heritable* resistance is less clear.

This paper will be restricted to mechanisms that involve differences in identifiable enzymic components between susceptible and resistant strains of the same species, particularly the cytochrome *P*-450-dependent monooxygenase system (EC 1.14.14.1), glutathione *S*-transferases (EC 2.5.1.18) and various esterases. Since the genetics of insecticide resistance is covered separately at this symposium (Sawicki & Denholm, this volume), genetics will be discussed here only in relation to the control of enzymes and their isozymes.

Altered acetylcholinesterase

Since the discovery of an acetylcholinesterase (EC 3.1.1.7) insensitive to inhibition in the two-spotted spider mite (Smissaert 1964), the presence of an alteration in acetylcholinesterase as a mechanism of resistance to organophosphates and carbamates has been reported for various strains of mite, tick and insect (Oppenoorth & Welling 1976). Kinetic studies all suggest that the

TABLE 2 Comparison of kinetic constants for inhibition of normal and altered acetylcholinesterase (AChE)

Insect	*Inhibitor*	*Kinetic constant*	*Normal AChE*	*Altered AChE*
Housefly[a]				
	Tetrachlorvinphos	K_d (M)	4.8×10^{-8}	2.8×10^{-5}
		k_2 (min^{-1})	0.6	1.6
		k_i ($M^{-1}min^{-1}$)	1.2×10^{7}	6.0×10^{4}
Green rice leafhopper[b]				
	2-*sec*-Butylphenyl *N*-methylcarbamate	K_d	1.5×10^{-6}	5.0×10^{-4}
		k_2	1.3	0.9
		k_i	9.0×10^{5}	1.7×10^{3}
	2-*sec*-Butylphenyl *N*-propylcarbamate	K_d	1.2×10^{-4}	9.4×10^{-5}
		k_2	0.6	2.3
		k_i	5.1×10^{3}	2.5×10^{4}

[a] Tripathi & O'Brien (1973); [b] Yamamoto et al (1977).

major difference between the altered enzyme and normal acetylcholinesterase is in its affinity (K_d) to inhibitors (Table 2).

$$E + AX \underset{k_{-1}}{\overset{k_1}{\rightleftharpoons}} EAX \xrightarrow{k_2} EA + X \xrightarrow{k_3} E + A$$

$$(E + AX \xrightarrow{k_i} EA + X)$$

$$K_d = \frac{k_{-1}}{k_1}$$

E, enzyme; AX, inhibitor; EAX, enzyme–inhibitor complex; EA, acylated enzyme; X, leaving group, A, acyl group; K_d, affinity constant; k_i, bimolecular rate constant; k_1, k_{-1}, k_2, and k_3 are rate constants for the reactions indicated.

A comparison of inhibition by tetrachlorvinphos of altered and normal acetylcholinesterase from the brain of the housefly, *Musca domestica*, showed that the bimolecular rate constant (k_i) of the altered enzyme was 206-fold lower than that for the normal enzyme, although k_2 was 3-fold higher (Table 2). The principal difference was due to a 573-fold greater K_d of the altered acetylcholinesterase. Similar results have been reported for other organophosphate and carbamate inhibitors and also for acetylcholinesterase from *Nephotettix cincticeps*, the green rice leafhopper (Hama et al 1980). Substrate specificity studies showed that the altered enzyme has a higher Michaelis constant (K_m) for butyrylcholine than does the normal enzyme (O'Brien et al 1978) although there is no significant difference in their K_m values for the intrinsic substrate, acetylcholine. Thus, the change in affinity of acetylcholinesterase to inhibitors must involve a site that does not affect acetylcholine binding.

One of the most interesting recent findings about acetylcholinesterase is the relationship between the resistance levels to *N*-methylcarbamates and the anticholinesterase activity of *N*-propylcarbamates in the green rice leafhopper (Fig. 1) (Yamamoto et al 1977). Resistance to *N*-methylcarbamates showed a

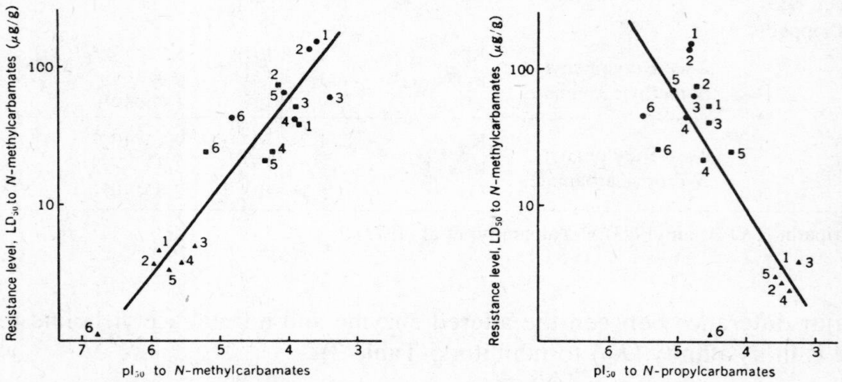

FIG. 1. Relationship between resistance level and anticholinesterase activity in the green rice leafhopper, *Nephotettix cincticeps*. 1 = 2-isopropylphenyl N-methylcarbamate; 2 = 2-*sec*-butylphenyl N-methylcarbamate; 3 = 3-methylphenyl N-methylcarbamate; 4 = 3,4-dimethylphenyl N-methylcarbamate; 5 = 3,5-dimethylphenyl N-methylcarbamate; 6 = carbaryl; ▲, susceptible; ■, resistant, RN-N; ●, highly resistant, RN-4; pI_{50}, negative logarithm of the inhibitor concentration at 50% inhibition. (From Yamamoto et al 1977.)

positive correlation with the anticholinesterase activity of *N*-methylcarbamates themselves, to which altered acetylcholinesterase is less sensitive than the normal enzyme. In contrast, the resistance level showed a negative correlation with the anticholinesterase activity of *N*-propylcarbamates, to which altered acetylcholinesterase is more sensitive than the normal enzyme. This is a classic example of the sensitivity of acetylcholinesterase as the main factor in resistance to carbamates in an insect, and it also raises the possibility of specific anti-resistance insecticides being developed.

Glutathione conjugation

Glutathione conjugation is an important detoxication route of many organophosphate insecticides, contributing to selective toxicity and to resistance.

At least three glutathione conjugation reactions are known in which organophosphate insecticides serve as substrates, conjugations with alkyl, aryl and phosphonate groups. A partially purified housefly enzyme catalysed both glutathione–alkyl and glutathione–aryl conjugation, the ratio of the two being determined by the structure of the organophosphate insecticide (Motoyama & Dauterman 1977a). Glutathione–phosphonate conjugation, on the other hand, was found only when the soluble fraction of the housefly was used as

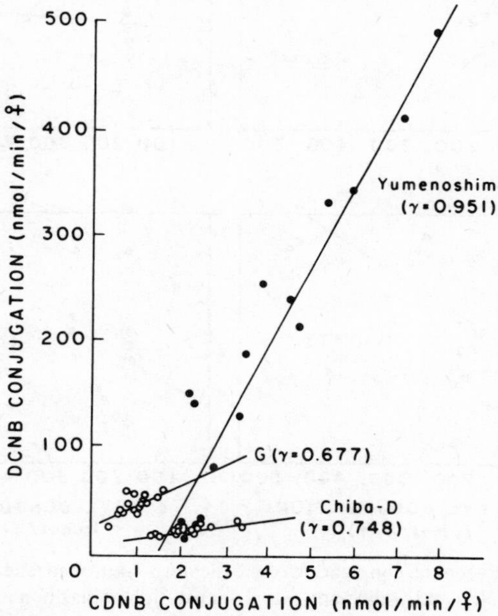

FIG. 2. Relationship between the conjugation of 1,2-dichloro-4-nitrobenzene (DCNB) and that of 1-chloro-2,4-dinitrobenzene (CDNB) in individuals of three strains of the housefly. ♀, enzyme from an individual female housefly; γ, correlation coefficient.

the enzyme source and when EPN (*O*-ethyl *O*-*p*-nitrophenyl phenylphosphonothioate) was the substrate (Hajjar et al 1980).

The involvement of glutathione *S*-transferase in resistance was first suggested by Lewis (1969). It may play a major role (Motoyama et al 1971), a minor role (Motoyama & Dauterman 1977b), or a partial role in addition to other mechanisms (Motoyama et al 1981).

In apparent contrast to mammalian liver, the housefly was originally believed to have only one glutathione *S*-transferase with a broad substrate specificity, but recent evidence indicates multiple forms of the enzyme. This evidence includes: (a) a strain of the housefly that differs from other strains in

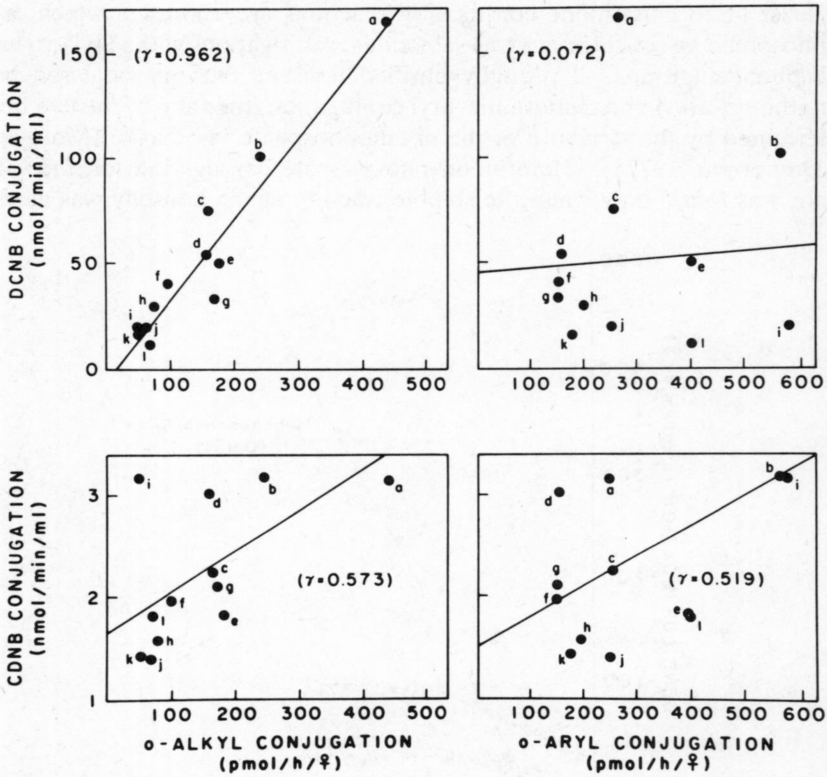

FIG. 3. Relationship between the conjugation of 1,2-dichloro-4-nitrobenzene (DCNB), 1-chloro-2,4-dinitrobenzene (CDNB) and *o*-alkyl and *o*-aryl conjugation of diazinon in different strains of the housefly. a–l are twelve different strains of the housefly.

the ratio of glutathione–alkyl to glutathione–aryl conjugation of parathion (Oppenoorth et al 1977); (b) isoelectrofocusing bands that have different relative activities towards methyl iodide, depending on the strain (Clark & Dauterman 1982); (c) a marked variation both in overall activity of the enzyme and in the ratio of DCNB (1,2-dichloro-4-nitrobenzene) conjugation to CDNB (1-chloro-2,4-dinitrobenzene) conjugation among strains and individuals (Fig. 2) (N. Motoyama et al, unpublished).

In this latter study, we also observed similar interstrain variation for glutathione–alkyl and glutathione–aryl conjugation of diazinon. To examine the relation between these activities and conjugation of DCNB and CDNB, the former two activities were plotted against the latter (Fig. 3). A close relationship ($r = 0.962$) was shown only between DCNB conjugation and glutathione–alkyl conjugation of diazinon, with no other clear-cut relationship being apparent. The simplest explanation of these results is provided by assuming that at least two forms of glutathione *S*-transferase exist in the housefly. One of the two forms would be active for DCNB and CDNB conjugation, for all the glutathione–alkyl conjugation of diazinon, as well as for part of the glutathione–aryl conjugation of diazinon. The other form would be predominantly active towards CDNB and also responsible for the remainder of the glutathione–aryl conjugation of diazinon. This second enzyme appears to be similar to glutathione *S*-transferase in the green rice leafhopper, which is active towards CDNB but not DCNB (N. Motoyama et al, unpublished).

Ester hydrolysis

Increased hydrolysis of organophosphate insecticides by esterases was one of the first mechanisms of resistance studied, the early studies leading to the 'mutant aliesterase hypothesis' of Oppenoorth & Van Asperan (1960).

According to this hypothesis an esterase, by mutation to a phosphotriesterase, carboxylesterase or other esterase, acquires the ability to hydrolyse insecticides. There is a concomitant decrease in activity of the altered enzyme towards aliesters such as methyl butyrate and naphthyl acetate. Although this appears to be the case in many examples of resistance, questions remain. For example, if the hypothesis is correct, one would expect to find pairs of esterases with contrasting specificities in resistant and susceptible strains of insect. However, chromatofocusing of various forms of esterase from the E_1 strain[a] of housefly, on which this hypothesis was originally developed, and from susceptible strains, shows no such clear-cut relationship (L. R. Kao et al, unpublished).

In the green rice leafhopper, five (E_1–E_5) out of the eight esterase bands that are separated on thin-layer agar gel electrophoresis are associated with

[a] There is no generally recognized convention for the nomenclature of insecticide-resistant strains of insect. Although often named according to the insecticide involved with original selection, such strains are commonly broadly cross-resistant. In other cases, strains are named according to localities or are given identifying numbers or letters that are of significance only to the original researcher. Housefly strains mentioned in this paper are as follows: CSMA, a susceptible strain originally maintained by the Chemical Specialties Manufacturers Association; Diazinon-R (originally Rutgers), resistant to diazinon and also broadly cross-resistant; E_1, broadly cross-resistant; Fc, broadly cross-resistant; *Fc,bwb,stw*, derived from Fc with stubby-winged and brown-body visible mutant marker genes; Orlando-R, DDT resistant with some cross-resistance; R-Baygon, resistant to propoxur and broadly cross-resistant; *R-Baygon,bwb,ocra*, derived from R-Baygon with brown-body and ocra-eye visible mutant marker genes.

resistance to organophosphate insecticides, including malathion (Ozaki 1969). Although E_2 and possibly E_1 and E_3 overlapped with malathion carboxylesterase activity (Miyato & Saito 1976), the role of these and the other esterase forms in cross-resistance to other organophosphate insecticides remained unclarified. Recently (N. Motoyama et al, unpublished), by chromatofocusing esterases from the organophosphate-resistant Kannonji strain, we resolved five peaks (I–V) which corresponded to E_1–E_5, respectively, on thin-layer agar gel electrophoresis (Fig. 4). The esterase peaks were all active

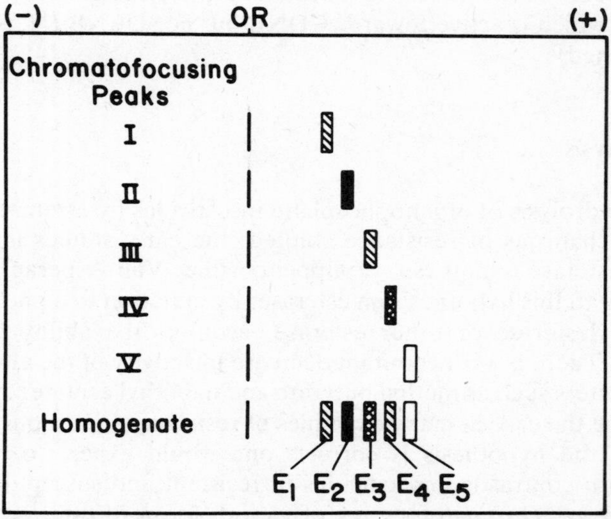

FIG. 4. Electrophoresis of esterase activity in whole homogenate and five electrofocusing peaks from the resistant Kannonji strain of the green rice leafhopper.

for the three non-insecticide substrates tested as well as for malathion, paraoxon (diethyl *p*-nitrophenyl phosphate) and fenvalerate, except for peak I, the overall activity of which was low (Table 3). These esterases are under independent genetic control, and different combinations of esterases are found in resistant strains, depending on the agent used for selection (Ozaki 1969). A strain with higher activity than a susceptible strain in both E_2 and E_3 is more resistant than a strain with higher activity in E_2 only.

Time-course studies showed that hydrolysis of malathion and fenvalerate continued to increase with time, while hydrolysis of paraoxon reached a plateau within 15 min. A considerable amount of *p*-nitrophenol was detected even at 0 °C and zero time, indicating that paraoxon hydrolysis is probably not due to phosphotriesterase action but to phosphorylation of the esterase. A similar result was previously reported as a mechanism of resistance to

TABLE 3 Substrate specificity of esterases separated on chromatofocusing[a]

		Activity					
		μmole min^{-1}ml^{-1}			nmole h^{-1}ml^{-1}		
Esterase peak	*Approximate pI*	p-*nitrophenyl acetate*[b]	*β-naphthyl acetate*[b]	*methylthiobutyrate*[b]	*malathion*[c]	*paraoxon*[c]	*fenvalerate*[c]
I	6.0	<0.01	0.06	0.01	ND	ND	0.08
II	5.7	1.65	1.43	0.43	0.63	0.16	0.26
III	5.4	0.54	0.52	0.15	0.25	0.06	0.26
IV	5.2	0.58	0.78	0.16	0.28	0.07	0.22
V	4.9	0.21	0.39	0.06	0.22	0.03	0.23

[a] N. Motoyama et al, unpublished data; [b] products were determined spectrophotometrically; [c] [^{14}C]-products were separated on thin-layer and chromatography and quantitated using a scintillation counter; ND, not detectable under the assay conditions used; pI, isoelectric point.

diazinon in insects (M. Hosokawa & N. Motoyama, unpublished paper, Annu Meet Jpn Soc Pestic Sci, 1982). In this work, an unknown factor in the soluble fraction of the resistant strain decreased the concentration of the active metabolite, diazoxon, significantly. Since the factor was a protein and the reaction occurred even at 0 °C and zero time, the mechanism appeared to be binding to protein, rather than an enzymic reaction. The difference in the amount of diazoxon removed between resistant and susceptible strains was more than enough to explain the differences between the strains in the LD_{50} of diazinon. These results suggest dual roles for esterases in resistance mechanisms. For malathion and fenvalerate, the esterase serves as a catalyst for hydrolysis, while for the other organophosphate insecticides, especially the oxygen analogues, the esterase serves as a binding protein, thereby protecting acetylcholinesterase from inhibition. This dual role may explain the cross-resistance observed between malathion and other organophosphate insecticides in many species.

Cytochrome *P*-450-dependent monooxygenase system

The observation, in 1960 (Eldefrawi et al), that carbaryl and sesamex, topically applied in combination, could reverse resistance to carbaryl in the housefly indicated that monooxygenase activity might be important, since methylenedioxyphenyl compounds had been shown to inhibit monooxygenation both *in vivo* and *in vitro*. Many later studies reinforced the idea that resistance could be due to monooxygenase reactions, and inhibition by synergists has often been used to demonstrate monooxygenase involvement. As pointed out by Wilkinson (1983), in any attempt to characterize the relative importance of monooxygenase enzymes in resistance, *in vivo* studies should be correlated with *in vitro* determination of enzyme activity.

Since the cytochrome *P*-450-dependent monooxygenase system is relatively non-specific, its high activity in many resistant strains is a probable cause of cross-resistance. Tsukamoto & Casida (1967) demonstrated that microsomes from carbamate-selected strains of the housefly showed an increased capacity for hydroxylation, *N*-dealkylation, *O*-dealkylation, epoxidation and oxidative desulphuration when compared to microsomes from susceptible strains. This broad specificity has been demonstrated many times since.

The presence of multiple forms of cytochrome *P*-450 is doubtless one of the reasons why all cross-resistance patterns in strains with high oxidase activity are not the same. In addition, resistant strains with such high oxidase activity may also possess high activity of other xenobiotic-metabolizing enzymes, such as glutathione *S*-transferases, and cross-resistance patterns may be influenced by the relative levels of oxidative and non-oxidative enzymes.

Increased cytochrome *P*-450 concentrations have been reported for a number of resistant housefly strains possessing high oxidase activities (see Hodgson 1984 for references) although there is not necessarily a correlation between high cytochrome *P*-450 and monooxygenase activity. For example, the two resistant housefly strains—*Fc, bwb, stw* and *R-Baygon, bwb, ocra*—have high monooxygenase activity, yet their cytochrome *P*-450 concentration does not exceed that seen in susceptible strains. No resistant strain is known to have more than a two-fold increase in cytochrome *P*-450 over the susceptible strains, whereas overall detoxication rates in resistant strains are frequently several times those in the susceptible strains. These results have led some investigators to suggest that cytochrome *P*-450 may not be the limiting factor in the monooxygenase activity of resistant strains.

Qualitatively different cytochrome *P*-450s were first described in the multiresistant Diazinon-R housefly strain by Perry & Buckner (1970) who reported that the wavelength for maximum absorption (λ_{max}) for the microsomal carbon monoxide (CO) difference spectrum of cytochrome *P*-450 was several nanometres lower than that in other strains. We also demonstrated this lowered λ_{max} for the CO spectrum, as well as other qualitative differences in cytochrome *P*-450s from Diazinon-R and the susceptible CSMA strain. The microsomal spectral characteristics generally differ between susceptible and resistant houseflies (Table 4). High levels of 'resistant' cytochrome *P*-450

TABLE 4 Microsomal spectral characteristics of typical susceptible and resistant housefly strains

Spectrum	*'Susceptible'*	*'Resistant'*
CO spectrum (λ max)	451–452	448–449
Relative *P*-450 level (CO spectrum)	100	150–200
Formation of type I difference spectrum	−	+
Type II *n*-octylamine spectrum	double trough	single trough
Relative magnitude of type III ethyl isocyanide 455 nm peak	100	60–70

have been detected in the Fc strain and Dimethoate-R strain, although the 'susceptible' type was detected in another Fc strain as well as in R-Baygon and in the DDT-resistant Orlando-R strain of houseflies (see Hodgson 1984 for references).

The significance of these qualitatively different cytochrome *P*-450s is unclear. No specific enzymic reaction or resistance characteristic has been correlated with the 'resistant' cytochrome. However, high oxidase activity can always be correlated with the presence of the type I binding spectrum of cytochrome *P*-450, whereas other 'resistant' cytochrome characteristics are not necessarily related to high oxidase activity. These binding spectra, with a peak at about 385 nm and a trough at about 420 nm, are caused by ligand binding to a lipophilic site adjacent to the haem moiety of the cytochrome and are believed to manifest substrate binding to the oxidized cytochrome. Type I spectra can be readily demonstrated by standard techniques, using microsomes from a number of resistant strains of the housefly, and they are identical to those of mammalian liver microsomes (Fig. 5). Microsomes from the abdomens of Fc and *Fc,bwb,stw* housefly strains, which are high in

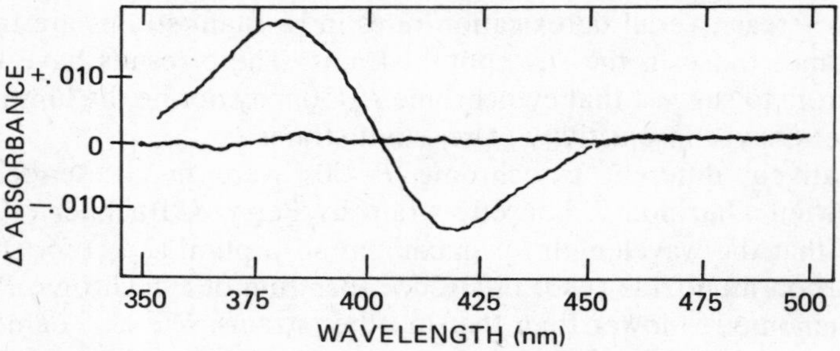

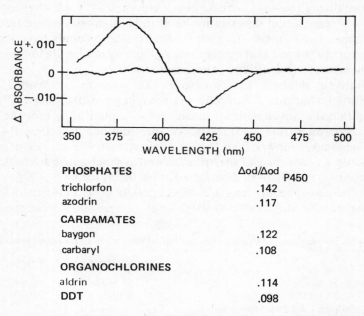

	Δod/Δod P450
PHOSPHATES	
trichlorfon	.142
azodrin	.117
CARBAMATES	
baygon	.122
carbaryl	.108
ORGANOCHLORINES	
aldrin	.114
DDT	.098

FIG. 5. Type I optical difference spectrum obtained using abdominal microsomes from the Fc resistant strain of housefly. (From Hodgson et al 1974)

cytochrome *P*-450-dependent monooxygenase activity, form binding spectra with type I substrates but lack the other resistant characteristics. Genetic studies suggest that, in strains exhibiting significant type I binding, increased oxidase activity segregates on the same chromosome that confers type I binding regardless of the cytochrome *P*-450 titre (see Hodgson 1984 for references).

Two or more semi-dominant genes for high oxidase activity are known to occur in the housefly, and the resistance attributable to them can usually be blocked by synergists and confer cross-resistance. These genes are associated with chromosomes II and V, although resistance involving high oxidase activity is much more commonly associated with chromosome II than with chromosome V. Both genes can occur in the same strain. Genetic studies of other mechanisms of resistance have also been done, particularly in the housefly, and genes have been located which are involved in all the various mechanisms listed: penetration, altered cholinesterase, resistance to paralysis by DDT (knockdown resistance) and non-oxidative enzymes (Plapp 1976).

We also studied the genetics of cytochrome *P*-450 variants in the Diazinon-R housefly strain, of previously unknown genetic constitution, and in the Fc strain, both of which contain genes for high-oxidase activity on chromosome V. Resistant strains were crossed with a susceptible strain that carries visible, recessive markers on chromosomes II, III and V. To eliminate crossing over in the F_1 generation, F_1 males were back-crossed to females of the marked susceptible strain. The eight phenotypic progenies resulting from the back-cross each contained specific combinations of resistant and susceptible chromosomes. These substrains were analysed for qualitative characteristics of cytochrome *P*-450. The characteristics seen in the Diazinon-R strain are all inherited as semidominants on chromosome II. In the Fc strains, however, type I binding segregates with chromosome V, and chromosome II contains gene(s) that have quantitative effects on the expression of cytochrome *P*-450.

In further studies of crossing over in the Diazinon-R strain, we demonstrated that, of the cytochrome *P*-450 parameters tested, only type I binding is linked always to resistance. This strongly suggests that this characteristic is related to some crucial enzymic activity in the resistance mechanism. Moreover, clearly more than one gene is involved in the inheritance of cytochrome *P*-450 in the housefly.

Detailed study of the type I binding spectrum is thus essential to an understanding of the role of cytochrome *P*-450 in resistance. The apparent lack of type I binding in microsomes from susceptible houseflies was first demonstrated in our laboratory with benzphetamine as a ligand and, subsequently, with a large number of pesticidal and non-pesticidal chemicals, all of which act as type I ligands with microsomes from mammalian liver and those from susceptible houseflies. Subsequent studies, however, revealed a significant type I spectrum with the insecticide synergist, sulfoxide, and a small type I spectrum with other methylenedioxyphenyl compounds. Furthermore, benzphetamine gave a type I spectrum with some microsomal subfractions from susceptible houseflies. These results suggest that although a type I binding site(s) exists in the cytochrome *P*-450s of susceptible houseflies, it is qualitatively different, probably being more remote from the haem, than the type I binding site in the cytochrome *P*-450 of housefly strains such as Fc and Diazinon-R. Moldenke & Terriere (1981) reported the presence of type I binding in susceptible houseflies, apparently unaware of our own work in which its presence is reported and discussed.

Several workers have failed to observe normal type I spectra with oxidized cytochrome *P*-450, even from resistant houseflies. The reasons for this are entirely methodological since we have re-examined all these strains and have demonstrated normal type I spectra in them for several different ligands.

Different approaches to the study of the multiplicity of forms of cytochrome *P*-450 in the housefly are summarized in Table 5. While all these

TABLE 5 Demonstration of multiple forms of cytochrome *P*-450 in insects

Type of study	*Principal references*
A. Purification	Capdevila et al 1975 Schonbrod et al 1975 Agosin 1976 Capdevila & Agosin 1977
B. Induction	Perry et al 1971 Capdevila et al 1973a,b
C. Strain differences	Perry & Bucknor 1970 Philpot & Hodgson 1972
D. Genetic studies	Tate et al 1973, 1974 Plapp et al 1976
E. Off-balance spectra	Kulkarni & Hodgson 1980
F. Tryptic digestion	Kulkarni & Hodgson 1980
G. Differential centrifugation	Kulkarni & Hodgson 1980
H. SDS–PAGE (electrophoresis)	Stanton et al 1978 Terriere & Yu 1979

methods have drawbacks, and some are seriously flawed, the weight of evidence is overwhelming that several isozymes of cytochrome *P*-450 exist. Since spectral characteristics appear to be under genetic control and also related to resistance, these isozymes are probably involved in resistance and the apparent inheritance of spectral characteristics probably reflects the dominant spectral characteristic(s) of the inherited isozyme(s).

Discussion

The 'normal' function of detoxication enzymes is often discussed, and it appears reasonable that it is primarily the metabolism of exogenous chemicals. It is doubtful whether animal life, at least on land, would be possible without the ability of animals to render lipophilic compounds water-soluble even when they are of low toxicity. Resistance can thus be seen as an adaptive mechanism by which organisms can select from the range of biochemical detoxication reactions, to enable the population to survive in an otherwise toxic environment.

The biochemical mechanisms of resistance to insecticides are considerably more complex than previously believed, largely due to the presence of isozymes of the detoxication enzymes. Further progress in understanding the biochemistry of resistance mechanisms depends on the purification of these isozymes from susceptible and resistant insects and on determination of the genetic events occurring during their selection. This is particularly true for the isozymes of cytochrome *P*-450, which are particularly recalcitrant to purification.

Plapp (1976 *et seq.*) proposed that regulatory gene(s) were involved in resistance because the genes for many enzymes that take part in resistance are located on the same chromosome and because, in resistant insects, several such enzymes often appear to be selected for simultaneously. The fact that enzyme inducers, such as phenobarbitone, may induce more than one such enzyme was also held to support this idea. This mechanism would allow the inheritance as a unit of the enzymes involved in resistance. While this hypothesis has proved useful as a stimulus to further investigation it leaves enough discrepancies unexplained to preclude its unqualified acceptance.

Problems include the following: isozymes of cytochrome *P*-450 are not inherited as a group, since different strains have different spectral characteristics; cytochrome *P*-450 and glutathione transferase, the genes for which are both on chromosome II in the housefly, are sometimes inherited together and sometimes not; the relationship between induction and heritable resistance is still not clear. Indeed, in mammals, it is now known that the induced cytochrome *P*-450 isozymes are quantitatively unimportant in the uninduced animal. Thus, heritable resistance may be due to the 'constitutive cytochrome *P*-450s' while induction involves a separate set of inducible isozymes. Further work in biochemistry and biochemical genetics is essential for a resolution of these problems.

REFERENCES

Agosin M 1976 Insect cytochrome *P*-450. Mol Cell Biochem 2:33-44

Capdevila J, Agosin M 1977 Multiple forms of housefly cytochrome *P*-450. In: Ullrich V (ed) Microsomes and drug oxidations. Pergamon Press, Oxford

Capdevila J, Perry AS, Morello A, Agosin M 1973a Some spectral properties of cytochrome *P*-450 from microsomes isolated from control, phenobarbital and naphthalene-treated houseflies. Biochim Biophys Acta 314:93-103

Capdevila J, Morello A, Perry AS, Agosin M 1973b Effect of phenobarbital and naphthalene on some of the components of the electron transport system and the hydroxylating activity of housefly microsomes. Biochemistry 12:1445-1451

Capdevila J, Ahmad N, Agosin M 1975 Soluble cytochrome *P*-450 from housefly microsomes. Partial purification and characterization of two hemoprotein forms. J Biol Chem 250:1048-1060

Clark WC, Dauterman WC 1982 The characterization by affinity chromatography of glutathione *S*-transferases from different strains of the housefly. Pestic Biochem Physiol 17:307-314

Eldefrawi ME, Miskus R, Sutcher V 1969 Methylenedioxyphenyl derivatives as synergists for carbamate insecticides on susceptible, DDT-, and parathion-resistant houseflies. J Econ Entomol 53:231-234

Georghiou GP, Saito T (eds) 1983 Pest resistance to pesticides. Plenum Press, New York

Guthrie FE 1980 Resistance and tolerance to toxicants. In: Hodgson E, Guthrie FE (eds) Introduction to biochemical toxicology. Elsevier/North-Holland, New York, p 357-375

Hajjar NP, Nomier AA, Hodgson E, Dauterman WC 1980 S-(O-ethyl phenylphosphonothionyl) glutathione; evidence for its formation in the *in vitro* metabolism of EPN in houseflies. Drug Chem Toxicol 3:421-433

Hama H, Iwata T, Miyata, T, Saito T 1980 Some properties of acetylcholine esterases partially purified from susceptible and resistant green rice leafhoppers, *Nephotettix cincticeps* Uhler. Appl Entomol Zool 15:249-261

Hodgson E 1984 Microsomal monooxygenase. In: Kerkut GA, Gilbert LI (eds) Comprehensive insect physiology, biochemistry and pharmacology. Pergamon Press, London, in press

Hodgson E, Tate LG, Kulkarni AP, Plapp FW 1974 Microsomal cytochrome *P*-450: characterization and possible role in insecticide resistance in *Musca domestica*. J. Agric Food Chem 22:360-366

Kulkarni AP, Hodgson E 1980 Multiplicity of cytochrome *P*-450 in microsomal membranes from the housefly, *Musca domestica*. Biochim Biophys Acta 632:573-588

Lewis JB 1969 Detoxication of diazinon by subcellular fraction of diazinon-resistant and susceptible houseflies. Nature (Lond) 224:917-918

Miyata T, Saito T 1976 Mechanism of malathion resistance in the green rice leafhopper, *Nephotettix cincticeps* Uhler. J Pestic Sci (Nihon Noyaku Gakkaishi) 1:23-29

Moldenke AF, Terriere LC 1981 Cytochrome *P*-450 in insects. 3: Increase in substrate binding by microsomes from phenobarbital induced houseflies. Pestic Biochem Physiol 16:222-230

Motoyama N, Dauterman WC 1977a Purification and properties of housefly glutathione *S*-transferase. Insect Biochem 7:361-369

Motoyama N, Dauterman WC 1977b Genetic studies on glutathione-dependent reactions in resistant strains of the housefly, *Musca domestica* L. Pestic Biochem Physiol 7:433-450

Motoyama N, Rock GC, Dauterman WC 1971 Studies on the mechanism of azinphosmethyl resistance in the predaceous mite, *Neuseiulus fallacis*. Pestic Biochem Physiol 1:205-215

Motoyama N, Hayaoka T, Dauterman WC 1981 Multiple factors for organophosphorus resistance in the housefly, *Musca domestica* L. J Pestic Sci (Nihon Noyaku Gakkaishi) 5:393-402

O'Brien RD, Tripathi RK, Howell LL 1978 Substrate preferences of wild and mutant housefly acetylcholinesterases and a comparison with the bovine erythrocyte enzyme. Biochim Biophys Acta 526:129-134

Oppenoorth FJ, Van Asperan K 1960 Allelic genes in the housefly producing modified enzymes that cause organophosphate resistance. Science (Wash DC) 132:298-299

Oppenoorth FJ, Welling HR 1976 Biochemistry and physiology of resistance. In: Wilkinson CF (ed) Insecticide biochemistry and physiology. Plenum Press, New York, p 507-551

Oppenoorth FJ, Smissaert HR, Welling N, Van der Pas LJT, Hitman KT 1977 Insensitive acetylcholinesterase, high glutathione S-transferase and hydrolytic activity as resistant factors in a tetrachlorvinphos-resistant strain of housefly. Pestic Biochem Physiol 7:34-47

Ozaki K 1969 The resistance to organophosphorus insecticides of the green rice leafhopper, *Nephotettix cincticeps* Uhler, and the small brown plant hopper, *Laodelphox striatellus* Fallen. Rev Plant Prot Res 2:1-15

Perry AS, Bucknor AJ 1970 Studies on microsomal cytochrome *P*-450 in resistant and susceptible houseflies. Life Sci 9:335-350

Perry AS, Dale WE, Bucknor AJ 1971 Induction and repression of microsomal mixed-function oxidases and cytochrome *P*-450 in resistant and susceptible houseflies. Pestic Biochem Physiol 1:131-142

Philpot RM, Hodgson E 1972 Differences in the cytochrome *P*-450s from resistant and susceptible houseflies. Chem-Biol Interact 4:399-408

Plapp FW 1976 Biochemical genetics of insecticide resistance. Annu Rev Entomol 21:179-197

Plapp FW, Tate LG, Hodgson E 1976 Biochemical genetics of oxidative resistance to diazinon in the housefly. Pestic Biochem Physiol 6:175-182

Schonbrod RD, Terriere LC 1975 The solubilization and separation of two forms of microsomal cytochrome *P*-450 from the housefly, *Musca domestica* L. Biochem Biophys Res Commun 64:829-835

Smissaert HK 1964 Cholinesterase inhibition in spider mites susceptible and resistant to organophosphates. Science (Wash DC) 143:129-131

Stanton RH, Plapp FW, White RA, Agosin M 1978 Induction of multiple cytochrome *P*-450 species in housefly microsomes—SDS-gel electrophoresis studies. Comp Biochem Physiol 61:297-305

Tate LG, Plapp FW, Hodgson E 1973 Cytochrome *P*-450 difference spectra of microsomes from several insecticide-resistant and -susceptible strains of the housefly *Musca domestica* L. Chem-Biol Interact 6:237-247

Tate LG, Plapp FW, Hodgson E 1974 Genetics of cytochrome *P*-450 in two insecticide-resistant strains of the housefly *Musca domestica* L. Biochem Genet 11:49–63

Terriere LC, Yu SJ 1979 Cytochrome *P*-450 in insects. 2: Multiple forms in the flesh fly (*Sarcophaga bullata*, Parker) and the blow fly (*Phormia regina*, Meigen). Pestic Biochem Physiol 12:249-256

Tripathi RK, O'Brien RD 1973 Insensitivity of acetylcholinesterase as a factor in resistance of houseflies to the organophosphate, Rabon. Pestic Biochem Physiol 3:495-503

Tsukamoto M, Casida JE 1967 Metabolism of methylcarbamate insecticides by the $NADPH_2$-requiring system from houseflies. Nature (Lond) 213:49-51

Wilkinson CF 1983 Role of mixed function oxidases in insecticide resistance. In: Georghiou GP, Saito T (eds) Pest resistance to pesticides. Plenum Press, New York, p 207-228

Yamamoto I, Kyomura N, Takahashi Y 1977 Aryl N-propylcarbamate, a potent inhibitor of acetylcholinesterase from the resistant green rice leafhopper, *Nephotettix cincticeps*. J. Pestic Sci (Nihon Noyaku Gakkaishi) 2:463-466

DISCUSSION

Wood: On the question of genes that give resistance to different insecticides being linked, I have a few comments. In *Drosophila*, Tsukamoto & Ogaki (1954) and Kikkawa (1964b) found a single locus (*RI*) that gave a much wider spectrum of cross-resistance than had been found in other insects. They may have been dealing with a cluster of linked genes, although irradiation experiments suggested a single mutation site (Kikkawa 1964a).

We have been investigating resistance in the Mediterranean fruit fly, *Ceratitis capitata* (Busch-Petersen & Wood 1983). In two populations we looked at the resistance to dieldrin and found a very heterogeneous response to it. Using mass selection followed by single family sib selection, we finally produced a homozygous dieldrin-resistant strain. When we tested this strain against other insecticides it showed an expected cross-resistance to the cyclodienes aldrin and endrin, and cross-tolerance to HCH (lindane). But it also showed cross-resistance to permethrin and very high cross-resistance to malathion, which was quite unexpected. We thought that either we had a single gene like the *RI* gene in *Drosophila*, which has a very wide spectrum of cross-resistance, or we had a cluster of linked genes. Therefore we have now been doing a series of back-crosses to susceptible, with selection in successive generations: malathion selection in one direction and dieldrin selection in the other. We seem to be separating these two kinds of resistance, but it is taking some time, suggesting that these genes may, indeed, be closely linked. There seems to be such a powerful linkage equilibrium that when we select with one insecticide we produce resistance to a wide variety of others.

Georghiou: Published work so far has shown that resistance to dieldrin is not metabolic but is due to insensitivity of the target, whereas resistance to malathion (with the exception of the insensitive cholinesterase type) is due to metabolic processes, involving carboxylesterases etc. (see review by Tsukamoto 1983). So I would doubt that we are dealing with a single pleiotropic gene. Close linkage might be a better hypothesis.

Sawicki: It is uncanny that in the housefly most of the major resistance genes to major classes of insecticides are on chromosome II. We do not believe that a 'supergene' is present because the resistance genes seem to be scattered along the chromosome. There is a whole series of different mechanisms—the glutathione transferases, esterases and mixed function oxidases. I don't know whether this is a random arrangement or whether something in the structure of this particular chromosome makes it unique.

Clarke: In various organisms, non-random arrangements of genes with respect to chromosomes have been observed. In the chicken such arrangements are statistically non-random, but the reason for this is unknown.

Wood: In *Drosophila*, McCartney et al (1977) tested whether the mutants affecting resistance tended to be associated on a particular chromosome. They did the same for wing shape and eye colour and for 42 other properties. Statistical analysis revealed that there was very little association between the location of genes that affected any individual property.

Hartl: Can anybody explain why the resistances that are found in nature seem to be simple Mendelian resistances? In contrast, early laboratory work done in the 1950s by Crow (1957) and others in *Drosophila* seemed to generate multifactorial resistances to insecticides.

Sawicki: I would disagree about this. For example our study of pyrethroid resistance in field populations of housefly has revealed at least seven different factors. This is not the only case of polyfactorial resistances in the field. I have not yet met polygenic resistance, but the polyfactorial type certainly exists.

Hartl: Could you clarify your use of those terms?

Sawicki: Polyfactorial is when one can isolate individual factors; polygenic is in the sense used by Mather where one has a whole series of genes controlling a quantitative character.

Gressel: When there have been slight increases in tolerance to herbicides, and not a total resistance, the explanation is usually polygenic with low broad-sense hereditability. Total resistances always seem to be monogenic.

Georgopoulos: This is probably true also for the resistance of plants to diseases: the field resistance is often polygenic, while in the greenhouse one usually selects for major gene resistance.

Hartl: Are there any generalizations about when one can expect to see simple Mendelian resistances and when one can expect to see multifactorial or polygenic resistances? Or is it just a matter of chance?

Georghiou: Studies involving insecticides and genetic analysis with chromosomal markers have, so far, led to the conclusion that each mechanism is due to a single gene. The sum total of resistance may be due to more than one gene, but each one is responsible for a single factor that segregates in a Mendelian fashion (see review by Tsukamoto 1983).

Hodgson: The Fc strain of housefly is a little odd but it does have genes on both chromosome II and chromosome V that affect the cytochrome *P*-450 system. This may turn out to be linked to two different forms of *P*-450, one coded for by a gene on chromosome II and the other by a gene on chromosome V. So, what Dr Georghiou has said is generally correct, but it is not a necessary condition.

Hartl: To consider the number of genes that influence a resistance trait is a little misleading because what we have is a distribution of effects; some loci are more important than others. In quantitative terms the effect of a gene on the variance in a population is a function of the gene frequency. We really need to know what number of genes accounts for what fraction of the variance in resistance. If one gene accounts for 80% of the variance we can probably implicate a simple Mendelian trait; if two genes account for 80% of the variance then we could say that there are two genes, and so on. I want to know what accounts for *most* of the important genetic variation and not what accounts for all the variation.

Gressel: But surely we cannot generalize about this? If the resistance is due to a modified enzyme that does not bind the pesticide, then this is a single gene effect that is responsible for more than 80% of the variance. If resistance is due to a detoxification by a multiplicity of enzymes, or by a single enzyme, or by multiple doses, then the variation will change.

Hartl: But all these effects depend on gene frequency. As the gene frequency approaches 50% in the population the proportion of the variance attributable to that gene increases. The number of genes in that sense is not a biological question but a statistical one.

Bradshaw: The corollary of that is that as the gene reaches a frequency of 100% then its effect (i.e. the variance) returns to zero. We must always bear in mind what we select to look at. Thirty or forty years ago most characters in plants were thought to be due to a single major gene, because people were choosing to pick out for study genes whose effects could be seen distinctly. So when discussing any case one should be aware of how it was analysed; it could have been analysed essentially to identify major segregations, when a differently and more properly conducted analytical experiment would have shown that only 40% of the resistance was due to a gene with major effects and the rest was due to genes with minor effects.

Sawicki: This is relevant to the esterase assays on *Myzus persicae* mutants that I mentioned in my paper. With the mutant that has two copies of the gene, the resistance factor is less than 1.5: it is virtually within the limits of experimental error in the bioassay. This is why I consider the term *tolerance* inapplicable; anything that leads to survival is resistance, in my mind. A more precise technique for detecting a difference in response (and the biochemical technique in this case is much more precise than the bioassay) will reveal this difference, which could not be identified otherwise.

Clarke: It is almost inevitable that if we look sufficiently accurately at an enzyme we shall come up with a simple Mendelian result, unless we measure its quantity, or variables related to its quantity.

Wood: Even in the field one can come up with a fairly simple result, as Dr Sawicki's earlier example of the housefly illustrated. The two genes for DDT resistance—*kdr* and *Deh*—tend to exclude each other. Houseflies in the United States tend to have *Deh* and a very low frequency of *kdr* (or none at all). In Canada *kdr* is found at high frequency.

Sawicki: But there is a very high DDTase in the Canadian ones as well.

Wood: Yes, and this also happens in Japan. I am not suggesting that the two genes are always mutually exclusive. But in Italy, where *kdr* was first identified by Milani (1954), it can no longer easily be found, and *Deh* has been selected. Nature tends to simplify the situation. Although, as Dr Sawicki points out, seven gene loci are involved in pyrethroid resistance, I would be very surprised if the *kdr* locus did not exert the major effect (in Danish and English populations).

Clarke: We also greatly simplify the environment by treating it with vast amounts of insecticide. When one first observes resistance it is almost bound to be, statistically, a single gene effect. The response depends on the rate at which useful variation appears.

Bradshaw: This raises considerations of rates of evolution. A high level of selection will automatically favour genes with major effects.

Georghiou: The question of distribution of the *kdr* gene is interesting. Published work often states that Italian houseflies possess *kdr*, as do Canadian flies, while American flies show little evidence for it. I cannot see what is so unique about the US–Canadian border! I suspect that the intensity of selection pressure and ecological aspects decide whether *kdr* is selected or not. In the Canadian environment the population of houseflies on farms may be more circumscribed, and inward migration may be smaller and thus selection pressure high, favouring a more decisive mechanism such as *kdr*. In southern climates, where flies breed freely beyond the area in which they are economically controlled, selection pressure is probably lower, so there is a higher chance of survival of individuals with less efficient resistance mechanisms.

Clarke: There may be another factor involved here. When a gene is first being selected, say, for insecticide resistance, and if the selection is then taken away, the resistance may carry with it considerable disadvantages that cause it to disappear again from the population. If, on the other hand, one selects through to fixation, the population is then exposed to the resistance gene in large numbers for the first time. Other genes that are compatible with it, or that interact favourably with it, will be selected. The genetic background will thus be selected to match the particular gene that one has selected. The result would be to improve the general environment for that gene relative to its alleles, and to other genes at other loci. This could have two effects. First, the resistance gene might exclude other such genes. Second, when one took away the selection, the resistance gene would not then be eliminated by its non-resistant allele(s), particularly if there were a lot of genetic interaction. The population would be sitting on a new adaptive peak, with a valley between it and the previous state that can no longer be crossed. The state present before selection for the resistance could not be attained again. This argument depends on the existence of much interaction between genes. Whichever resistance gene appears first may 'win' in that it encourages the accumulation of 'modifiers' that favour it.

Sawicki: We have now found major differences in the genetic make-up of housefly populations, and are beginning to wonder about the implications of this phenomenon.

We have recently studied an esterase which we suspect to be involved in resistance to pyrethroids and to some organophosphorus insecticides. Not one of 17 Danish strains that we examined, collected from 1947 to 1982, had this esterase. Yet every one of over 40 British strains that we have examined has it, and some appear to be homozygous for it. Elsewhere in Europe some strains are heterozygous for this esterase, while others lack it. This makes us think that housefly populations can be quite distinct.

Another point concerns sex determinants in houseflies. In Europe, Franco et al (1982) have demonstrated the existence of three types of housefly population along a latitudinal line. Populations of Northern Europe were of the standard type (XX females and XY males); those of Central and Southern Italy were autosomal (XX females and males); and in the large intermediate zone the populations were mixed and had several karyotypes in both sexes.

In British populations most of the sex determination is autosomal. One of the X chromosomes acts as a male determinant. There is also an additional male determinant on autosome III, and an autosomal female sex-determinant.

Our results (I. Denholm, M.G. Franco, P.G. Rubini) do not suggest that selective pressure with insecticides has contributed towards the evolution of populations with different sex determinants. They indicate, however, that there are micro-evolutionary processes in housefly populations even within neighbouring populations in the United Kingdom. We find these genetic differences within housefly populations very interesting and puzzling.

Clarke: In the presence of two different adaptive peaks and a lot of migration, the population may not be displaced from its peak. Even if mating is random, when selection is strong enough it can divide a population into two types (Thoday 1972). It is a matter of the balance between selection and gene flow. For example, between snail populations that are only 5 metres apart one can see significant genetic differences.

REFERENCES

Busch-Petersen E, Wood RJ 1983 Dieldrin resistance as a prospective candidate for genetic sexing in the Mediterranean fruitfly *Ceratitis capitata*. In: Fruitflies of economic importance. EEC, Brussels, in press

Crow JF 1957 Genetics of insect resistance to chemicals. Annu Rev Entomol 2:227-246

Franco MG, Rubini PG, Vecchi M 1982 Sex-determinants and their distribution in various populations of *Musca domestica* L. of Western Europe. Genet Res 40:279-293

Kikkawa H 1964a Genetic studies on the resistance to parathion in *Drosophila melanogaster*. II. Induction of a resistance gene from its susceptible allele. Botyu-Kagaku 29:37-42

Kikkawa H 1964b The genetic study on the resistance to Sevin in *Drosophila melanogaster*. Botyu-Kagaku 29:42-46

McCartney PR, Renwick JH, Munday MR 1977 The pattern of loci on *Drosophila* chromosomes. Heredity 38:37-45

Milani R 1954 Comportamento mendeliano della resistenza alla azione abbattente del DDT e correlazione fra abbattimento e mortalità in *Musca domestica*. Riv Parassitol 15:513-542

Thoday JM 1972 Disruptive selection. Proc R Soc Lond B Biol Sci 182:109-143

Tsukamoto M 1983 Methods of genetic analysis of insecticide resistance. In: Georghiou GP, Saito T (eds) Pest resistance to pesticides. Plenum Press, New York, p 71-98

Tsukamoto M, Ogaki M 1954 Gene analysis of resistance to DDT and BHC in *Drosophila melanogaster*. Botyu-Kagaku 19:25-32

Adaptation of fungi to fungitoxic compounds

S. G. GEORGOPOULOS

Laboratory of Plant Pathology, Athens College of Agricultural Sciences, Votanikos, Athens 301, Greece

Abstract. Some fungitoxic chemicals often fail to protect crops because the target fungi develop resistance. Depending on its mechanism of action, the fungicide may lose its effectiveness completely after a limited number of applications, it may show a gradually less satisfactory performance or it may retain its effectiveness unchanged despite extensive use. In all cases investigated, resistance to agricultural fungicides has been shown to result from mutations of chromosomal genes. Biochemical mechanisms of adaptation have been recognized mainly as alterations of the cellular component with which the fungicide interacts, as reduced uptake or increased efflux, and as decreased conversion into a more toxic compound. However, most studies on biochemical genetics have utilized laboratory mutants and it is not always certain that similar mechanisms are responsible for adaptation in nature.

1984 Origins and development of adaptation. Pitman Books, London (Ciba Foundation symposium 102), p 190-203

Chemicals are often used for the control of fungi whose activities are undesirable to human beings. In agriculture, fungicides together with gene-controlled disease resistance of cultivated plants constitute the main means for protecting crops against fungal diseases. For reasons which are beyond the scope of this paper, the biological control of fungal pathogens, where it is possible at all, has failed to acquire the importance attached to the biological control of insect pests. At present, antifungal chemicals are often the only reliable and economically acceptable way to control plant diseases. The fungicides currently available for controlling fungi that are pathogenic to higher plants have a great variety of chemical structures. Many different mechanisms of fungicidal action have become known. Some of the agricultural fungicides interfere with fungal respiration, in a specific or a non-specific way, or with oxidative phosphorylation. Others are inhibitors of biosynthetic processes, such as the synthesis of nucleic acids, proteins, lipids or chitin. Sometimes, antifungal activity is due to interference with formation or orientation of microtubules, or to a more general disruption of cellular structure. Furthermore, a group of compounds effective against plant disease has recently been recognized to inhibit the biosynthesis of melanin which appears to be important for host penetration in certain cases (Woloshuk & Sisler 1982).

Obviously, large-scale and repeated application of a chemical that prevents or severely curtails survival or multiplication in a target fungus gives plenty of opportunity for the selection of fungal forms that have adapted, more or less, to solve this problem. The fact that many fungi pathogenic to plants are haploid in their parasitic phase removes many of the complexities of adaptation encountered by other eukaryotes. In such fungi, resistant mutants are subject to immediate scrutiny by selection because they are not shielded by dominance. Despite this, fungi have not adapted successfully to all agricultural fungicides.

Changes at the population level

The use of fungitoxic compounds in agriculture has permitted different patterns of response to be recognized in fungal populations. Although other factors may also be important, the results so far indicate that the response that should be anticipated depends mainly on the mechanism of action of the chemical. Thus, agricultural fungicides can be divided into three main groups, depending on the success, if any, that fungal populations have had in adapting to them.

'High-risk' fungicides

With fungicides of the first group, which might be called 'high-risk' fungicides, one gets the impression of an 'all-or-none' concept of adaptation (Stern 1970): the high initial effectiveness is followed, after a few applications, by a complete and sudden breakdown of the disease-controlling ability of the fungicide. This happens only with compounds whose mechanism of action permits highly successful adaptive changes, leading to: (a) a high degree of resistance, so that the strains that are selected live and reproduce normally in the presence of the fungicide at practically applicable rates; and (b) a high relative fitness of the selected strains during any periods between applications when little or no selective chemical is around.

Because of the high degree of resistance possible, the frequency distribution of the target fungus in the field consists of two or more distinct, non-overlapping subpopulations around widely differing means. A distinction

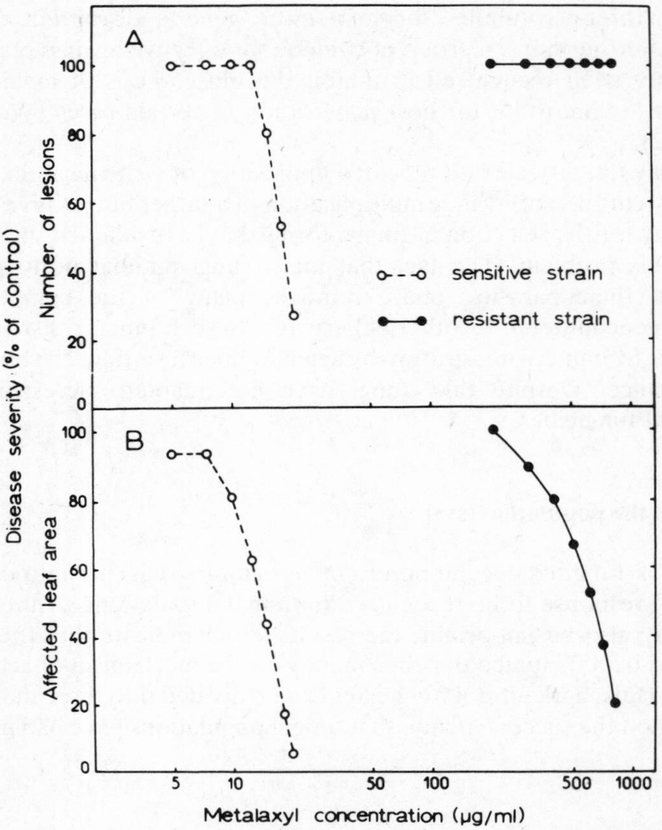

FIG. 1. Differences in metalaxyl sensitivity between two strains of *Pseudoperonospora cubensis*. Cucumber leaves were placed inverted on aqueous solutions of metalaxyl, at concentrations ranging from 5 to 800 μg/ml, beyond which phytotoxicity was observed. Then, on the lower surface of each leaf, 20 inoculations were made with droplets of a suspension of sporangia of the respective strain (approximately 100 sporangia per droplet). The results were recorded as number of lesions (A) and as total diseased area (B) after 10 days' incubation at 22 °C. (A. C. Grigoriu & S. G. Georgopoulos, unpublished results.)

between the two subpopulations can very easily be made in the laboratory, as shown in Fig. 1, by differences in sensitivity to metalaxyl, an acylalanine that is believed to act as an inhibitor of RNA synthesis in fungi. This distinction can also be made in the field: with frequent application of such a chemical the resistant subpopulation predominates and the fungicide becomes useless in that particular situation.

The high relative fitness of the selected strains does not permit strong selection against the resistant subpopulation in the periods between the applications of fungicide. This leads to a very fast increase in the proportion of resistant strains and a sudden loss of fungicidal effectiveness, which is particularly striking with fast diseases (that have a high infection rate) and with treatments that do not allow for many escapes (Skylakakis 1982).

A specific example of this comes from our work on the control of *Cercospora beticola* of sugar beet in Greece (Georgopoulos 1982b). Soon after its introduction, benomyl, a benzimidazole derivative that prevents the polymerization of tubulin, was shown by many field experiments to be superior to other fungicides available at the time. Commercial application of benomyl in Greece gave excellent results from 1970 until midsummer 1972, when this compound suddenly lost the ability to control leaf spot. With the exception of fields that were sprayed with non-benzimidazole fungicides in July 1972, the destruction of sugar-beet foliage was complete by the middle of August. In experimental plots, plants sprayed with benomyl could not be distinguished from those sprayed with water. As shown in Fig. 2, benomyl-resistant strains of the fungus on artificial substrates remain unaffected by

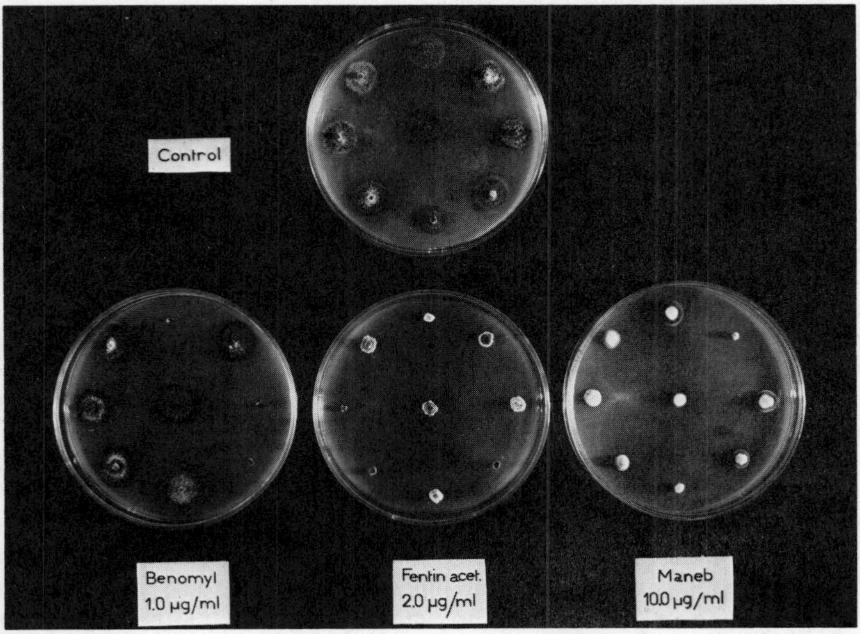

FIG. 2. Response of nine isolates of *Cercospora beticola* to three fungicides used for its control in northern Greece. The isolates were selected to represent the range of sensitivity of the population of *C. beticola* in the area.

concentrations of the fungicide that are completely inhibitory to the wild-type. In a field experiment conducted in 1973, we found that plots having only 3% of leaf spot caused by resistant strains at the end of June had nearly 92% at the beginning of August, as a result of two benomyl applications. This, of course, explained the sudden loss of effectiveness. The high degree of resistance and the high relative fitness permit the proportion of resistant strains to become very high quickly and to remain high many years after the use of benzimidazoles has been discontinued in the area (Table 1).

TABLE 1 Mean frequencies of fungicide-resistant isolates of *Cercospora beticola* in northern Greece

Year	*Percentage of isolates resistant to:*	
	Benzimidazoles	*fentin*[c]
1973[a]	97.3	—
1974[a]	92.6	—
1977[a]	—	83
1982[b]	91.0	65

[a] Data from references cited by Georgopoulos (1982b). [b] Unpublished results. [c] Isolates moderately resistant to fentin have also been included.

'Moderate-risk' fungicides

The 'all-or-none', very restricted definition of adaptation does not apply to fungicides of the second group, which are of 'moderate risk' and for which only small and slow decreases in effectiveness have been observed. The fungal population does appear to have adapted somewhat to the chemical after many applications, but seldom are treatments found to be without any effect at all, thus giving the impression of a 'relative adaptation' (Stern 1970). These 'moderate-risk' fungicides can be divided into two subgroups. Highly successful adaptation of fungi to fungicides in the first subgroup is impossible because of a considerable difference in fitness between sensitive and resistant fungal subpopulations which are distinct and well separated. A good example is provided by the response of *Pyricularia oryzae*, the fungus that causes rice blast, to the antibiotic kasugamycin, which inhibits protein synthesis. A decline in effectiveness of this fungicide in some areas of Japan has been associated with the presence of highly resistant strains. In mixed inoculations, however, the resistant strains of the fungus were shown to be inferior to the sensitive strains in their ability to infect rice plants in the absence of kasugamycin (Ito & Yamaguchi 1979). This seems to account for the slow decline in effectiveness of kasugamycin during its use, and for the fast decline in the proportion of resistant isolates after the antibiotic was withdrawn

(Uesugi 1982). With fungicides of this subgroup it is only in 'closed systems', such as glasshouses or fruit packing houses, that we may observe a complete loss of effectiveness.

With chemicals of the second subgroup no clear-cut differences in sensitivity between field isolates can be found, even after many years of intensive use. The sensitivity of the whole population seems to have only one, rather wide and unimodal, frequency distribution. The mechanism of toxicity of such fungicides, in contrast with benzimidazoles, acylalanines or kasugamycin, does not permit the use of laboratory and/or field treatments that are highly effective against one fraction of the population and completely without effect on another fraction. With exposure, however, directional selection may operate to shift the whole distribution in the direction of increased resistance with the result being a 'relative adaptation'.

This last pattern of response may be clarified by reference to the organotin fungicides, fentin acetate (triphenyltin acetate) and fentin hydroxide (triphenyltin hydroxide) (Georgopoulos 1982b), which are believed to inhibit oxidative phosphorylation, like oligomycin. These fungicides were found preferable to other protectants available at the beginning of the 1960s for the control of *C. beticola* in Greece and were used exclusively for almost 10 years, with 5–8 applications per year. The control of the disease was more satisfactory in some years than in others but if there was any progressive decrease in effectiveness, it was not sufficiently great to alert the non-specialist. Fentins were used to some extent between 1970 and 1972, and returned to almost exclusive use again after the failure of benzimidazoles, until 1977. During the latter years of this period, poor performance of the fentins was observed in locations where the population of *C. beticola* contained a higher proportion of forms that were more resistant to the fungicide. Thus, the mean colony diameter of *C. beticola* after 6 days' incubation on medium containing the fungicide was 6.0 mm for isolates from an area in which fentins were used for 12 years, as compared with 4.0 mm for isolates from an area never sprayed with fentin (Giannopolitis 1978). The distribution of resistance within the population, however, remained unimodal in both cases. That any great variation in fentin sensitivity between isolates of *C. beticola* cannot be found is also shown in Fig. 2. It is only with considerable difficulty that strains may be classified as sensitive, moderately resistant and resistant. There is no concentration of the fungicide at which the most sensitive isolates will not grow at all, while the most resistant isolates will remain unaffected. The difference between fentin and benomyl is very striking (Fig. 2). This difference, together with the somewhat reduced ability of the least sensitive members of the fungal population to compete on unsprayed sugar beets explains: (a) why no adaptation to fentin has been reported from countries other than Greece; (b) why fentins are still used to a considerable extent in Greece; (although the results are not as satisfactory as they were originally, there has not been a complete loss of effectiveness like that seen with the benzimidazoles) and (c) why isolates of *C. beticola* that are classified as fentin-sensitive continue to form a considerable proportion of the population of the fungus (Table 1), indicating no strong selection against them.

A somewhat different pattern has been recognized with ethirimol, a fungicide that apparently interferes with adenine metabolism. Use of this fungicide on barley results in a reduced mean sensitivity of the population of *Erysiphe gramins* f.sp. *hordei* from treated as compared to untreated fields. A five-year survey, however, showed a reduction in the frequency of both the most sensitive and the most resistant forms of the pathogen, with forms of intermediate sensitivity becoming predominant (Brent 1982).

'Low-risk' fungicides

Fungi that are pathogenic to plants have had no success at all in adapting to fungicides of the third group, which might be called 'low-risk' fungicides, the effectiveness of which has never yet been reported to decline. This is particularly true of many protectant fungicides of the multi-site inhibitor type, e.g. the dithiocarbamates and the phthalimides, which have been in widespread use for decades. It is important to note that no fungal strain with any considerable resistance to such fungicides has so far been isolated from nature or induced in the laboratory. In its sensitivity to such a chemical the population of a target fungus apparently consists of one unimodal frequency distribution. This is illustrated by the example of *C. beticola* and the dithiocarbamate maneb (Fig. 2), which has been used since 1977 in Greece with no decline in effectiveness. Apparently, if directional selection causes any shift in the direction of increased resistance, the limits set by the type of toxic mechanism make this shift insignificant relative to the rates of application of these fungicides in practice. It so happened that practically all the fungicides available until the middle of the 1960s belonged to this third group, and it is therefore not surprising that adaptation to fungicides was of little importance 15 years ago (see Georgopoulos 1977).

Genetic and biochemical mechanisms

Although fungi have been shown capable of phenotypic adaptation to toxic compounds in several cases, such adaptation has never been shown to be responsible for fungicide failures in agriculture. Similarly, no association of such failures with extra-chromosomal genetic elements has ever been reported. Chromosomal genes for resistance to agricultural fungicides have been identified in many cases and, as a rule, reduced sensitivity to a given chemical may result from mutation of any one of a number of different genes. With regard to biochemical changes, modification of the cellular component with which the fungicide interacts appears to be the most common mechanism of resistance. However, there are also well documented cases of resistance due to a decrease in uptake or to an increase in efflux of the fungicide, as well as to a reduced conversion of the fungicide into a more toxic compound.

In some instances we have a rather lopsided picture of mechanisms of adaptation to fungicides, with only limited (if any) information on the genetic control or the biochemical changes involved. The best exception to this is the *ben*-A gene for resistance to the benzimidazoles in *Aspergillus nidulans*. This gene is now known as the structural gene for β-tubulin, one of the subunits of the tubulin molecule. Mutations of this gene affect the electrophoretic properties of β-tubulin and, at the same time, affect the ability of the protein to bind the fungicide. Binding prevents polymerization of tubulin and is, therefore, inversely correlated to resistance. One mutation at the *ben*-A locus gives resistance to all benzimidazoles, while an allelic mutation leads to a resistance only to thiabendazole and an extra-sensitivity to carbendazim and carbendazim generators, such as benomyl (see review by Davidse 1982). While differential binding of the fungicide seems to be the basis for resistance to benzimidazoles in a few more fungal species, other mechanisms cannot be excluded, even in *A. nidulans* which, in addition to the β-tubulin gene, has two other genes responsible for low resistance to carbendazim. The mechanisms responsible for the great number of failures of benzimidazole fungicides in practice have seldom been investigated.

As a second example, I shall examine the mutations responsible for resistance to carboxin and to other carboxamides in *Ustilago maydis*. These fungicides are potent and highly specific inhibitors of the succinate dehydrogenase complex in mitochondria. Work in our laboratory (Ziogas & Georgopoulos 1980) has shown that wild-type strains of *U. maydis* possess the genetic information for the synthesis of two alternative, cyanide-insensitive mitochondrial electron-transport systems, in addition to the normal cytochrome pathway. Of the two alternative systems, one is formed constitutively and is insensitive to derivatives of hydroxamic acid, and the other is induced by inhibitors of mitochondrial translation and is resistant to hydroxamates.

Most of *U. maydis* mutants selected on a medium that contains carboxin carry a mutation at the locus *ants*. These mutants are only slightly resistant to carboxin and, unlike the wild-type, they are unable to grow in media that contain small concentrations of antimycin A. The data in Table 2 indicate that

TABLE 2 Sensitivity to cyanide (1.0 mM) of NADH oxidation by mitochondria from wild-type and *ants* mutant cells of *Ustilago maydis* grown for 20 h in a glucose medium with and without chloramphenicol (10.0 mM)

Treatment	*Respiration rate (nmoles O_2 min^{-1}mg protein^{-1}*	
	Wild-type[a]	ants *mutant*[b]
(1) *Mitochondria from untreated cells*		
Control	321	360
Cyanide	146	3
(2) *Mitochondria from chloramphenicol-treated cells*		
Control	381	380
Cyanide	339	290

[a] From Ziogas & Georgopoulos (1980). [b] B. N. Ziogas & S. G. Georgopoulos, unpublished results.

ants mutant cells differ from the wild-type cells in their requirement for growth in the presence of chloramphenicol in order to develop cyanide-insensitive mitochondrial electron transport. In addition, this alternative respiration is mediated only by one pathway and not by two as in the chloramphenicol-treated wild-type cells. This pathway is considerably resistant to hydroxamates and resembles the inducible alternative system of the wild-type in its other characteristics (B. N. Ziogas & S. G. Georgopoulos, unpublished work). Thus, chloramphenicol does not remove the block imposed by the *ants* mutation which, itself, prevents formation or function of the constitutive alternative system. The fact that strains carrying this mutation are selected for by carboxin, a selective inhibitor of succinate dehydrogenase, supports the view that the constitutive system is located in the area of this enzyme. This agrees with the current view on the alternative system of higher plant mitochondria, which is considered to be part of the ubiquinone cycle (Rich & Moore 1976).

A different class of *U. maydis* mutants that are selected for by carboxin carry a mutation at the *oxr*-1 locus. These mutants form all three electron-transport pathways found in the wild-type, but the succinate dehydrogenase of their mitochondria is resistant (*oxr*-1A) or highly resistant (*oxr*-1B) to carboxin (Georgopoulos 1982a). Mitochondrial preparations from such mutants can be recognized also in the absence of inhibitor because of the higher temperature sensitivity of their succinate dehydrogenase complex (Georgopoulos et al 1975). Finally, carboxin analogues are now known that can recognize the *oxr*-1A or *oxr*-1B enzyme complex, for which they are much better inhibitors than for the wild-type complex.

The evidence described indicates that both the *ants* and the *oxr*-1 mutations, which decrease sensitivity to carboxin in *U. maydis*, modify the site of action of the fungicide itself. Since the two genes differ in their action on cyanide-insensitive respiration, they probably modify different components of the succinate dehydrogenase complex. The nature of these components, however, remains to be determined.

The genetic and biochemical mechanisms of adaptation to other agricultural fungicides are discussed in previous reviews (Dekker 1976, Georgopoulos 1977, 1982a). Unfortunately, laboratory-induced mutants of non-pathogenic species, such as *A. nidulans*, or of pathogenic species, such as *U. maydis*, have usually been used, and it is not known whether similar mechanisms are responsible for adaptation in nature. Most pathogenic fungi are not easily subjected to genetic analysis and the study of field isolates therefore presents difficulties. For example, in *C. beticola*, neither the genetic control nor the biochemical mechanism of resistance to benomyl or to fentin has been studied. For obligately parasitic fungi there is no information available about the genetics of fungicide resistance. This is rather curious, of course, since methods of genetic analysis have been worked out for several species of rusts and powdery mildews.

REFERENCES

Brent KJ 1982 Case study 4. Powdery mildews of barley and cucumber. In: Dekker J, Georgopoulos SG (eds) Fungicide resistance in crop protection. Centre Agric Publishing & Documentation, Wageningen, p 219-230

Davidse LC 1982 Benzimidazole compounds: selectivity and resistance. In: Dekker J, Georgopoulos SG (eds) Fungicide resistance in crop protection. Centre Agric Publishing & Documentation, Wageningen, p 60-70

Dekker J 1976 Acquired resistance to fungicides. Annu Rev Phytopathol 14:405-428

Georgopoulos SG 1977 Development of fungal resistance to fungicides. In: Siegel MR, Sisler HD (eds) Antifungal compounds, vol 2. Marcel Dekker, New York, p 439-495

Georgopoulos SG 1982a Genetical and biochemical background of fungicide resistance. In: Dekker J, Georgopoulos SG (eds) Fungicide resistance in crop protection. Centre Agric Publishing & Documentation, Wageningen, p 46-52

Georgopoulos SG 1982b Case study 1. *Cercospora beticola* of sugarbeets. In: Dekker J, Georgopoulos SG (eds) Fungicide resistance in crop protection. Centre Agric Publishing & Documentation, Wageningen, p 187-194

Georgopoulos SG, Chrysayi M, White GA 1975 Carboxin resistance in the haploid, the heterozygous diploid, and the plant parasitic dicaryotic phase of *Ustilago maydis*. Pestic Biochem Physiol 5:543-551

Giannopolitis CN 1978 Occurrence of strains of *Cercospora beticola* resistant to triphenyltin fungicides in Greece. Plant Dis Rep 57:321-324

Ito I, Yamaguchi T 1979 Competition between sensitive and resistant strains of *Pyricularia oryzae* Cav. against kasugamycin. Ann Phytopathol Soc Jpn 45:40-46

Rich PR, Moore AL 1976 The involvement of the protonmotive ubiquinone cycle in the respiratory chain of higher plants and its relation to the branchpoint of the alternative pathway. FEBS (Fed Eur Biochem Soc) Lett 65:339-344

Skylakakis G 1982 The development and use of models describing outbreaks of fungicide resistance. Crop Prot 1:249-262

Stern JT Jr 1970 The meaning of 'adaptation' and its relation to the phenomenon of natural selection. Evol Biol 4:39-66

Uesugi Y 1982 Case study 3. *Pyricularia oryzae* on rice. In: Dekker J, Georgopoulos SG (eds) Fungicide resistance in crop protection. Centre Agric Publishing & Documentation, Wageningen, p 207-218

Woloshuk CP, Sisler HD 1982 Tricyclazole, pyroquilon, tetrachlorophthalide, PCBA, coumarin and related compounds inhibit melanization and epidermal penetration by *Pyricularia oryzae*. J Pestic Sci (Nihon Noyaku Gakkaishi) 7:161-166

Ziogas BN, Georgopoulos SG 1980 Chloramphenicol-induction of a second cyanide- and azide-insensitive mitochondrial pathway in *Ustilago maydis*. Biochim Biophys Acta 592:223-234

DISCUSSION

Gressel: In the laboratory, the triazine-resistant biotypes are more sensitive to another group of photosystem II-inhibiting herbicides (Arntzen et al 1982), which isn't surprising. They are also more susceptible to thiocarbamates (Laval-Martin et al 1983), and the reason for this is unknown. We are likely to see cases that are just the opposite of the huge cross-resistances that have been described for insecticides. Does cross-resistance occur with fungicides?

Georgopoulos: Yes, among fungicides that act in the same way, at least. But we have a larger group—the ergosterol biosynthesis inhibitors—which are chemically very different and fall into 5–6 different classes, and we do see cross-resistance among those, with one mutation. We never see one mutation giving resistance to quite unrelated toxicants, as seen in the *kdr* mutation of insecticide resistance.

Davies: In your experiments that required the isolation of mitochondria from chloramphenicol-treated fungi, since chloramphenicol blocks the synthesis of mitochondrial proteins, I wouldn't expect there to be very good yields.

Georgopoulos: No; chloramphenicol is not effective enough in *Ustilago maydis* to prevent the formation of mitochondria. It definitely affects mitochondrial protein synthesis as shown by the reduced activity of the cytochrome pathway of mitochondria from cells grown in chloramphenicol-containing medium. Most aerobic fungi do grow in the presence of chloramphenicol or streptomycin and they do contain mitochondria.

Bradshaw: It is fascinating that whether or not evolution of resistance occurs does not seem to depend on the group of fungus but on the type of fungicide applied. Why do you think the low-risk fungicides have such a low risk?

Georgopoulos: At the time when no systemic fungicides that acted specifically on one process were available, many people who wanted to study modes of action of the multi-site fungicides tried very hard to obtain mutants for use in such studies. Even with different methods, no success was achieved. When the first systemic fungicides became available, the very first attempts to produce mutants were successful. So it seems that the appropriate variability is not available for the development of resistance to the low-risk fungicides. We do not yet know from any biochemical genetic studies why this should be so.

Wood: Perhaps the low-risk fungicides are not persistent in the environment?

Georgopoulos: Other people have suggested this, and it may sometimes be influential, but there are occasions when lack of persistence is not a problem because the fungicide is sprayed at frequent intervals, e.g. every week.

Wood: The problem with persistence is that one sees a gradual decline in effectiveness of the pesticide, and unless one re-sprays within a certain time, there will be a period when the rate of kill is reduced. In this case is there always a high level of mortality or does it decline?

Georgopoulos: One often has to spray much more frequently than would allow this period to develop because new leaves are produced that are not protected.

Graham-Bryce: Maneb and related compounds do not disappear suddenly; like all pesticides in the environment they decay according to broadly exponential kinetics, each at a different rate.

Hutchinson: Do you know what the particular mechanism is for maneb activity?

Georgopoulos: Maneb forms a complex with SH-containing compounds, and so it inhibits many different enzymes.

Hutchinson: So this is a fairly non-specific reaction?

Graham-Bryce: Yes, but I feel that one cannot conclude that multi-site activity or broad-spectrum activity is automatically associated with a relative inability to develop resistance.

Georgopoulos: I agree.

Datta: In my paper (p 204-218) I shall be considering resistance genes acquired from an extraneous source into an organism. Some of what has been said here about plants slightly suggests that something similar has happened with organisms that are higher than bacteria. For example, the genes that Professor Bradshaw described as being present or not present might have some extraneous source.

Bradshaw: There is no doubt that throughout evolution there have been millions of cases where adaptive solutions could have been expected, because there was ample opportunity, and yet they never occurred. Extinct species often demonstrate such evolutionary failures.

Clarke: It needs care to be scientific about this. It is sometimes easy to explain why something happened but always much more difficult to explain why something did not happen.

Dittrich: Mites are very adaptable animals, and there is hardly anything to which they do not develop resistance. However, one compound used widely in the USA and elsewhere on roses under glass is Pentac® (dienochlor), a cyclopentadiene which is highly chlorinated, and mites still seem to be unadapted to it according to McEnroe & Lakocy (1969). It has been used in greenhouses for about 21 years now, and no resistance has become apparent as far as I know.

Graham-Bryce: Perhaps mephospholan, which I discussed earlier in relation to aphid control, is a similar example (p 163). Resistance may eventually come, but we need to know why, for some compounds, it takes a particularly long time.

Datta: The disease gonorrhoea, caused by the gonococcus, was treated with penicillin from soon after 1940, but it was not until 1976 that the bacteria acquired the ability to destroy penicillin. I had believed that this could not happen, but eventually it did, by acquisition of a plasmid or plasmid genes from another bacterium.

Georgopoulos: So even in that case the gene was not present in the original bacteria.

Hartl: Perhaps the apparently insurmountable toxins are those that can act in several ways, so that the organism has more than a single metabolic problem to solve—perhaps more than, say, three such problems (see also p 150). The likelihood of the three necessary mutations occurring simultaneously would then be very small indeed. Perhaps the problem in adaptation to heavy metals, for instance, is not just the heavy metals themselves but also the very special characteristics of mine sites—high exposure, poorly aerated soil, and unusual drainage characteristics. The organisms would need, therefore, to adapt to many other things as well as to the heavy metals.

Bradshaw: I agree that mine sites are peculiar in several ways. On the other hand, different mine sites have different combinations of characteristics. But we have studied the plants in the laboratory as well, where the only problem for adaptation is the heavy metal itself, and we obtain similar results to those in the field.

Hutchinson: There are certainly complex nutritional and drainage problems in mine waste and mine tailings sites, but the chances of the grass *Deschampsia cespitosa* adapting in the way that it has done would, theoretically, have been considered to be very low by most people (see Hutchinson, this volume). It seems to have suddenly evolved tolerances to four metals that it had to handle, plus another four that it did not have to handle. It has also developed tolerance to SO_2 and to extreme acidity. This is surely a highly unlikely combination, but it happened. It makes me think that ultimately a maneb-resistant fungus may indeed arise.

Hartl: If insect populations are treated with mixtures of insecticides, do the insects acquire resistance to the mixture just as easily as they do to a single compound?

Georghiou: This would depend on the choice of compounds. Since genes for resistance in unselected populations are believed to be rare, the probability of two unrelated genes being present in the same individual must be very small indeed. Under these conditions, an insect that may survive one of the chemicals in a mixture would be killed by the other, and thus resistance would be delayed. Perhaps for similar reasons, multifunctional fungicides have been slower in inducing resistance than the single-gene, single-target compounds. Unfortunately very few studies have been done on this (Georghiou et al 1983).

Kuć: Are there any low-risk fungicides that are systemic?

Georgopoulos: Not that I know of; there may be some of moderate risk. A systemic compound, in order to enter and move within the plant has to be not very reactive with the plant constituents, or it might kill the plant. The SH-binding agents, for example could not be systemic in the plant.

Gressel: Also, plant-cell suspensions treated *in vitro* with maneb will die. The cuticle normally keeps the compound out.

Clarke: It might be appropriate to remind ourselves that natural selection is primarily a mechanism that generates improbability, so if we wait long enough the improbable may eventually happen!

REFERENCES

Arntzen CJ, Pfister K, Steinback KE 1982 The mechanism of chloroplast triazine resistance—alternations in the site of herbicide action. In: LeBaron HM, Gressel J (eds) Herbicide resistance in plants. Wiley Interscience, New York, p 185-213

Georghiou GP, Lagunes A, Baker J 1983 Effect of insecticide mixtures and rotations on the evolution of resistance. Pergamon Press, London (Proc 5th Int Congr Pestic Chem, Kyoto, Japan), in press

Laval-Martin D, Grizeau D, Calvayrac R 1983 Characterization of diuron-resistant *Euglena*: greater tolerance for various phenylurea herbicides and increased sensitivity of thylakoids to ethyl-*S*-dipropyl thiocarbamate. Plant Sci Lett 29:155-167

McEnroe WD, Lakocy A 1969 The development of Pentac® resistance in an outcrossing swarm of the two-spotted spider mite. J Econ Entomol 62:283-286

Bacterial resistance to antibiotics

NAOMI DATTA

Department of Bacteriology, Royal Postgraduate Medical School, Hammersmith Hospital, Du Cane Road, London W12 0HS, UK

Abstract. Effective antibacterial drugs have been available for nearly 50 years. After the introduction of each new such drug, whether chemically synthesized or a naturally occurring antibiotic, bacterial resistance to it has emerged. The genetic mechanisms by which bacteria have acquired resistance were quite unexpected; a new evolutionary pathway has been revealed. Although some antibiotic resistance has resulted from mutational changes in structural proteins—targets for the drugs' action—most has resulted from the acquisition of new, ready-made genes from an external source—that is, from another bacterium. Vectors of the resistance genes are plasmids—heritable DNA molecules that are transmissible between bacterial cells. Plasmids without antibiotic-resistance genes are common in all kinds of bacteria. Resistance plasmids have resulted from the insertion of new DNA sequences into previously existing plasmids. Thus, the spread of antibiotic resistance is at three levels: bacteria between people or animals; plasmids between bacteria; and transposable genes between plasmids.

1984 Origins and development of adaptation. Pitman Books, London (Ciba Foundation symposium 102), p 204-218

Sulphonamides were first used in medicine in the 1930s, penicillin in 1940 and, in the following decades, many more effective antibacterial drugs were discovered and developed. Totally new ones have not been found recently but modifications of older ones continue to be produced. As each new drug has been introduced, resistant strains of bacteria previously sensitive to it have appeared. Many strains of bacteria, especially in hospitals, have acquired resistance to so many antibiotics that hardly any effective agent is left with which to treat some infections; this means that the newest drugs have to be used with the probability that new resistances to them will emerge.

Multiple drug resistance in bacteria was already a clinical problem 25 years ago. It had then been proved experimentally that in pure bacterial cultures, streptomycin-resistant mutants were present at frequencies of about 1 in 10^9; it was assumed that the multiple resistance of clinical isolates had resulted from sequential mutations. Penicillin resistance in *Staphylococcus aureus*, however, was known to be mediated by the production of penicillinase, a character that could not be shown to be acquired by mutation.

In 1959, Japanese workers showed that multiple resistance could be transmitted from one bacterium to another. In outbreaks of dysentery caused by *Shigella flexneri* some isolates were resistant to four unrelated drugs, streptomycin, chloramphenicol, tetracycline and sulphonamides. The epidemiological evidence did not suggest sequential mutations, and *Escherichia coli* that were resistant to the same four drugs were sometimes excreted by the patients with dysentery. From mixtures of resistant *S. flexneri* and sensitive *E. coli*, or *vice versa*, it was possible to isolate clones of the initially sensitive organism that had acquired the multiple resistance pattern. The nature of the transmissible agent, called the R (resistance) factor, was unknown but analogies with the F (fertility) factor of *E. coli* K12 were recognized (Watanabe 1963). Both F and the R factors first identified are conjugative plasmids—DNA molecules that carry genetic information for their own replication in bacterial host cells and also for their transmissibility from cell to cell by direct contact (bacterial conjugation).

Since the discovery of the F factor and the first R factors, enormous advances have been made in research on plasmids, much of it consequent upon their importance in biotechnology. I shall discuss their natural history in relation to antibiotic resistance. Although plasmids are responsible for most of the antibiotic resistance found in pathogenic bacteria the more conventional evolutionary mechanism—selection of mutants—is also important.

Mutation to drug resistance

One-step mutations give resistance to some drugs, the most important being streptomycin, rifampicin, nalidixic acid. Not surprisingly, this type of resistance is seen in clinical isolates. Multi-step mutations probably gave rise to some clinically important resistance to penicillin and to cephalosporin. For example, strains of *Neisseria gonorrhoeae* isolated before 1976 were sometimes resistant to penicillin at levels that interfered with therapy. They withstood concentrations as high as 1 mg/l in test cultures. Since 1976, *N. gonorrhoeae* have appeared, which produce penicillinase determined by plasmids and which are resistant to much higher concentrations of penicillin.

Trimethoprim is a synthetic drug that inhibits bacterial dihydrofolate reductase and is very effective against a wide variety of bacteria. Resistance to trimethoprim, though often plasmid-determined, is also caused by various kinds of mutation. Exposure of pure cultures to trimethoprim allows the isolation of thymine-requiring mutants, and such selection has been reported during trimethoprim treatment. Other mutations affect bacterial dihydrofolate reductase, making it less susceptible to trimethoprim. In rare cases the bacterium overproduces its normal enzyme.

Resistance to antibacterial drugs is a serious problem in tuberculosis and leprosy, both caused by slow-growing *Mycobacteria* species. Although plasmids occur in *Mycobacteria*, the clinically important resistances are not plasmid-determined but result from mutations. The distinction is important since the emergence of resistant mutants can be prevented by using two or more unrelated drugs in combination, but such treatment may actually encourage the spread of plasmids that determine multiple resistance.

Plasmids

Plasmid DNA is easily demonstrated in bacterial lysates after agarose-gel electrophoresis and has been identified in all bacterial genera examined. The smallest plasmids, e.g. p15 in *E. coli* 15, have about 1.5 kilobases (kb) which is enough to encode only the two or three genes that determine their own replication in a growing host culture. Other plasmids are 100 times larger, possessing about 10% of total cell DNA. In most plasmids it is not possible to account for all the DNA as known genes. Every plasmid has essential genes whose organization is often complex. Many plasmids also determine bacterial conjugation and their own transfer between cells. Conjugative plasmids determine the production of protein outgrowths from the cell, called *conjugative pili*, which are required for conjugation but whose exact role is not known. At conjugation, one strand of the double-stranded plasmid molecule is cut and transferred longitudinally to the recipient cell where it is recircularized. Complementary strands of DNA are synthesized in both donor and recipient so that the transfer event also results in replication of the plasmid.

Conjugative plasmids bring about not only the transfer of their own DNA from donor to recipient but also the transfer of other DNA molecules of the donor, including its chromosome. The transfer of chromosome, with resulting chromosomal recombination, is usually at a frequency much lower (of the order of 1 in 10^4) than that of transfer of the plasmid. But many non-conjugative plasmids are mobilized and transferred by conjugative ones extremely efficiently, so that their frequency of transfer may be higher than that of the mobilizing plasmid. Such easily mobilizable plasmids are usually small molecules, 6 to 18 kb, whereas conjugative plasmids are larger, at least 30 kb. The operon of plasmid F which determines conjugation itself occupies 30 kb. Other plasmids have simpler mechanisms but none is as fully analysed as that of F (Willetts & Skurray 1980).

Replication, maintenance and transfer may be considered as true functions

TABLE 1 Antibacterial drugs to which plasmids determine bacterial resistance

Penicillins and cephalosporins	
Erythromycin, lincomycin and streptogramin B	
Streptomycin	Tetracyclines
Neomycin	Chloramphenicol
Kanamycin	Fusidic acid
Gentamicin	Sulphonamides
Tobramycin	Trimethoprim
Amikacin	

of plasmids. In addition, plasmids carry a great variety of genes that modify the host phenotype and determine the production of bacteriocins, toxins and adhesins, the enzyme pathways for degradation of toxic organic compounds, and even plant oncogenesis. More genes are constantly being dicovered. Antibiotic resistances are only some of the important bacterial characters that are plasmid-determined (Broda 1979).

Most of the antibiotic resistance found in bacteria that cause disease of humans and animals is plasmid-determined. Table 1 shows the antibacterial drugs to which plasmids confer resistance. It includes most of those in use in human and veterinary medicine. The pharmaceutical industry is still a jump ahead in having developed a variety of new semisynthetic penicillins and cephalosporins that are effective against bacteria whose plasmids determine resistance to naturally occurring cephalosporins as well as to every other drug listed here. In the last 25 years plasmids have appeared with new resistance genes, and resistance plasmids have appeared in more and more bacterial genera (Table 2).

Plasmids confer resistance by the production of a variety of enzymes whose structural and often regulatory genes are encoded in the plasmid DNA. The action of some is shown in Table 3 (se also Davies, this volume).

TABLE 2 Genera in which R plasmids have been found[a]

Enterobacteriaceae, i.e. *Escherichia, Salmonella, Shigella, Proteus, Providencia, Klebsiella, Serratia* etc.	
Pseudomonas	
Acinetobacter	*Staphylococci*
Vibrio	*Streptococci*
Yersinia	*Bacillus*
Pasteurella	*Clostridium*
Campylobacter	*Corynebacterium*
Haemophilus	
Neisseria	
Bacteroides	

[a] Not all the plasmid-borne resistances listed in Table 1 are found in every genus listed here.

TABLE 3 Mechanisms of R plasmid resistance

Drug	*Plasmid product*	*Effect*
Penicillins Cephalosporins	β-lactamases	Detoxification
Tetracyclines	Inducible proteins	Active exclusion of drug from cell
Chloramphenicol	Acetyltransferase	Detoxification
Erythromycin	Methylase	Modification of target, ribosomal RNA
Streptomycin Kanamycin Gentamicin	Various drug-modifying enzymes	Detoxification and interference with drug transport
Sulphonamides	Dihydropteroate synthetase	Provision of insensitive target enzymes
Trimethoprim	Dihydrofolate reductase	

See Davies & Smith 1978, and Davies, this volume.

Classification of plasmids

Each bacterial genus in nature possesses a range of plasmids. Many plasmids can replicate in a range of different bacteria and identification of the host range should be the first step in plasmid classification. Plasmids of gram-negative bacteria are not transferable to gram-positive hosts, or *vice versa*. Plasmids of gram-negative bacteria vary in their host ranges, those with the widest host ranges being referred to as 'promiscuous'. In the gram-positive genera, too, some plasmids are promiscuous, being transferable from *Streptococcus* species to *Staphylococcus* and *Bacillus* species (Engel et al 1980). The method of transfer is often by conjugation, but some plasmids depend on the presence of bacteriophage for their dissemination (Lacey 1980).

Within any bacterial genus, plasmids are classifiable by *incompatibility*. Closely related plasmids, including two variants of the same plasmid, do not replicate stably together in a growing cell culture. As bacterial numbers increase, clones are formed which carry one or other plasmid, but not both. Compatible plasmids are as stable in combination as either is alone. Testing a large number of plasmids, each against all, for stable co-existence, allows them to be arranged in incompatibility (Inc) groups. When plasmids from wild-type bacteria are so arranged, the results suggest that true evolutionary groups are identified, since two plasmids within a group have much DNA homology, whereas two plasmids of different groups nearly always have little. Often, incompatible plasmids exert *surface exclusion* on one another; that is, a resident plasmid in a recipient strain reduces the frequency of transfer of a plasmid of the same group, but does not affect that of other plasmids. Another characteristic common to plasmids within an Inc group is the structure of their pili. Conjugative pili vary morphologically and serologically and act as receptors for different phages (Bradley 1980).

Incompatibility can be explained by the fact that plasmids of the same group have a common system of replication, under negative control, and each plasmid of the group is subject to repression by another (Uhlin & Nordström 1975). I and my colleagues have worked with conjugative drug-resistance plasmids of gram-negative bacteria, transferred to *E. coli* K12, and have identified some 20 Inc groups. (We designate them by capital letters, IncF, IncI etc., dating from the time when only two kinds were known—the F-like ones, and those related to plasmids that determine colicin I.) Plasmids in other bacteria (e.g. *Pseudomonas* or *Staphylococcus* species) have also been classified by incompatibility. Having classified our plasmids, we then asked whether the resistance genes of each group had evolved separately. The answer was no; there was little correlation between the resistance genes, identified by their products, and the Inc group of the plasmid (Hedges et al 1974).

Transposons

The identification of the same enzyme, the TEM penicillinase, determined by otherwise unrelated plasmids, was circumstantial evidence of genetic exchange between DNA molecules. It was easy to show such exchange in the laboratory. Plasmid RP4 determines TEM penicillinase, but plasmid R64 does not; RP4 and R64 are compatible, and unrelated. If both plasmids are present in *E. coli* and selection is made for the conjugative transfer of ampicillin resistance (mediated by TEM penicillinase) to another strain, then the R64 plasmid, with newly acquired ampicillin resistance and usually without the RP4 plasmid, appears in transconjugants. No loss of ampicillin resistance by RP4 is observed (Datta et al 1971, Bennett et al 1977). Transposition of the TEM penicillinase gene from RP4 to the bacterial chromosome can be shown; if RP4 is eliminated, either by the introduction of an incompatible plasmid or by 'curing', ampicillin resistance is retained in the chromosome of some clones. From the chromosome it can 'jump' to further unrelated plasmids that are subsequently introduced. Hedges & Jacob (1974) showed that sequential transposition of the TEM penicillinase gene from plasmid to plasmid or from chromosome to plasmid was always accompanied by an increase in molecular mass of the plasmid, equivalent to about 5 kb—that is, more than can be accounted for by the TEM penicillinase gene alone. They called the moveable DNA sequence a *transposon* and pointed out its analogy with the insertion sequences (IS) that had already been described by Starlinger & Saedler 1972.

TABLE 4 Some drug-resistance transposons

Transposon	*Length* (kb)	*Resistance determinants*
Tn*1,2,3*	5.0	Ap
Tn*5*	5.7	Km
Tn*7*	13.5	TpSmSp
Tn*9*	2.5	Cm
Tn*10*	9.3	Tc
Tn*732*	10.7	Gm
Tn*1699*	9.3	GmApKm

See Kleckner (1981). Abbreviations: Ap = ampicillin, Km = kanamycin, Tp = trimethoprim, Sm = streptomycin, Sp = spectinomycin, Cm = chloramphenicol, Tc = tetracycline, Gm = gentamicin.

Many plasmid genes for resistance to antibiotics are transposable (Table 4) and it seems probable that all resistance plasmids have evolved by the insertion of new genes, transposed from elsewhere, into existing plasmids.

The origins of the resistance genes themselves are unknown; they may be chromosomal genes of other organisms (Hedges & Jacob 1977, and see Davies, this volume). Any gene, it may be supposed, can become transposable by linkage into the necessary sequence. The transposition function is encoded in transposons and ISs; it involves the replication of the transposing sequence so that, when transposition is completed, the donor molecule is unaltered whilst the recipient has acquired the new sequence. The entire sequences of some transposons and ISs are known. Although proteins required for transposition and for its regulation have been identified, the exact nature of the event itself is not understood (Kleckner 1981).

Transposition can overcome the limited spread of resistance genes imposed by the limited host range of their vector plasmids. Plasmid DNA can be introduced, by conjugation or other means, into a cell in which that plasmid cannot replicate; if the plasmid DNA includes a transposon, this transposon can be rescued by jumping to a resident replicon, either the chromosome or another plasmid. This can be seen in laboratory experiments (Barth & Datta 1977) and probably explains the acquisition of drug resistance by *Haemophilus* species (Laufs et al 1981) and *Acinetobacter* species (Devaud et al 1982).

The world-wide spread of resistance genes

The TEM penicillinase that is carried by transposons 1 and 3 (Tn*1* and Tn*3*) determines resistance to benzyl penicillin, ampicillin and carbenicillin. It was first identified in plasmids of *E. coli* and *Salmonella* species (Datta & Kontomichalou 1965) and soon afterwards in plasmids of *Pseudomonas* *aeruginosa* (Sykes & Richmond 1970). Within a decade it had appeared in plasmids of many incompatibility groups in bacteria isolated in all continents. In 1975 ampicillin-resistant strains of *Haemophilus influenzae* type b were first isolated, the resistance being conferred by production of TEM penicillinase. The strains came from children with meningitis, and similar strains appeared in widely separated geographical areas, with apparently no epidemiological connection. In 1976 the first gonococci with this same penicillinase appeared and rapidly spread. In *H. influenzae* and *N. gonorrhoeae* no other plasmid-determined penicillinase or other β-lactamase has been found, although plasmids of other gram-negative bacteria possess a variety of these enzymes. The spread of this transposable penicillinase is a particularly good example of adaptation of bacteria to the medical and veterinary use of a vast 'tonnage' of penicillins, especially ampicillin. It is only one example of the world-wide spread of antibiotic resistances carried by plasmids.

The spread of resistance genes must depend on several factors, of which one is the selection pressure imposed by antibiotic use. Among others is the efficiency (frequency) of the transposition process itself. The TEM transposon is highly efficient, and so is Tn7 which determines a trimethoprim-resistant dihydrofolate reductase. But gentamicin resistance, important in hospital medicine, is transposable at much lower frequencies; although gentamicin resistance has spread among bacteria in hospitals, it is determined by a narrower range of plasmids than is the production of β-lactamase or the Tn7 dihydrofolate reductase (Datta et al 1981).

The plasmid vectors of resistance genes are as widely disseminated as their bacterial hosts. In early studies, three plasmids of the same Inc group (Table 5) were shown to be almost identical in their entire molecules (about 100 kb), with the exception of the sequences that determine antibiotic resistance (Sharp et al 1973). Several 'epidemic' plasmids have been recognized that, over a period of years, have spread in a variety of bacterial hosts and over wide geographical areas. In the course of their spread they have often

TABLE 5 Closely related R plasmids

Plasmid[a]	*Resistances*	*Origin*	
		Bacterial	*Geographical*
R1	ApSmCmKmSu	*Salmonella paratyphi B*	Spain
R6	TcSmCmKmSu	*S. typhimurium*	West Germany
R100	TcSmCmSuHg	*Shigella flexneri*	Japan

[a] These three plasmids, group IncFII, of different origins, have different patterns of resistance genes but are otherwise almost identical in their DNA. Abbreviations as for Table 4, and Hg = mercuric chloride.

acquired new resistance genes, and sometimes undergone deletions or other DNA rearrangements. A good example is plasmids of IncM, clinically important because they determine gentamicin resistance. Labigne-Roussel et al (1982) have demonstrated four steps by which these particular 'epidemic' plasmids probably evolved from an ancestor, identified in Paris in 1972, to the current form, which is 29 kb larger, has four additional resistances, and is present to this day in hospitals of the London area.

Plasmids in 'pre-antibiotic' bacteria

Plasmids without genes for antibiotic resistance were common in bacteria *before* the medical use of antibiotics. The F factor of *E. coli* K12 (Hayes 1953) is one example, and others have been found in old cultures of *N. gonorrhoeae* (Roberts et al 1978). But the main evidence for this statement is our own work (Hughes & Datta 1983, and unpublished) on enterobacteria collected over the years 1917 to 1954 by the late E.G.D. Murray and sealed in glass tubes until 1980. From 433 of these isolates, 84 (19%) transferred conjugative plasmids to *E. coli* K12. None of them determined antibiotic resistance, one determined tellurite resistance and 36 determined the production of bacteriocins. The method used to indicate transfer was the mobilization of small, non-conjugative resistance plasmids introduced by us into the mating mixtures. In *E. coli* K12 we have labelled the conjugative plasmids with resistance transposons (converting them to R plasmids) and have classified them by compatibility-testing with present-day R plasmids. Of the 84, 63 fall into one or other of our Inc groups, which is consistent with the current view that antibiotic resistance plasmids have evolved by the insertion of new genes into existing plasmids. Of course, once resistance genes are acquired, the vector plasmids are at a selective advantage in environments where their hosts are exposed to antibiotics.

Conclusions

The spread of antibiotic resistance occurs at three levels, all of which can be recognized in short-term studies in a single hospital (Datta et al 1980): resistant bacteria spread between patients, R plasmids between bacteria and transposons between plasmids. Although the body of relevant knowledge accumulated in the last 20 years has not, unfortunately, been of much help in prevention, it is full of interest and has provided valuable 'spin-offs' in all fields of biology.

REFERENCES

Barth PT, Datta N 1977 Two naturally occurring transposons indistinguishable from *Tn*7. J Gen Microbiol 102:129-134

Bennett PM, Grinsted J, Richmond MH 1977 Transposition of *Tn*A does not generate deletions. Mol Gen Genet 154:205-211

Broda P 1979 Plasmids. W H Freeman & Co, Oxford

Bradley DE 1980 Morphological and serological relationships of conjugative pili. Plasmid 4:155-169

Datta N, Hedges RW, Shaw EJ, Sykes RB, Richmond MH 1971 Properties of an R factor from *Pseudomonas aeruginosa*. J Bacteriol 108:1244-1249

Datta N, Kontomichalou P 1965 Penicillinase synthesis controlled by R factors in Enterobacteriaceae. Nature (Lond) 208:239-241

Datta N, Dacey S, Hughes V et al 1980 Distribution of genes for trimethoprim and gentamicin resistance in bacteria and their plasmids in a general hospital. J Gen Microbiol 118:495-508

Datta N, Nugent M, Richards H 1981 Transposons encoding trimethoprim or gentamicin resistance in medically important bacteria. Cold Spring Harbor Symp Quant Biol 45:45-51

Davies J, Smith DI 1978 Plasmid-determined resistance to antimicrobial agents. Annu Rev Microbiol 32:469-518

Devaud M, Kayser FH, Bächi B 1982 Transposon-mediated multiple antibiotic resistance in *Acinetobacter* strains. Antimicrob Agents Chemother 22:323-329

Engel HWB, Soedirman N, Rost JA, van Leeuven WJ, van Embden JDA 1980 Transferability of macrolide, lincomycin and streptomycin resistance between group A, B and D streptococci, *Streptococcus pneumoniae* and *Staphylococcus aureus*. J Bacteriol 142:407-413

Hayes W 1953 Observations on a transmissible agent determining sexual differentiation in *Bacterium coli*. J Gen Microbiol 8:72-88

Hedges RW, Datta N, Kontomichalou P, Smith JT 1974 Molecular specificities of R factor determined β-lactamases: correlation with plasmid compatibilities. J Bacteriol 117:56-62

Hedges RW, Jacob AE 1974 Transposition of ampicillin resistance from RP4 to other replicons. Mol Gen Genet 132:31-40

Hedges RW, Jacob AE 1977 *In vivo* translocation of genes of *Pseudomonas aeruginosa* onto a promiscuously transmissible plasmid. FEMS (Fed Eur Microbiol Soc) Microbiol Lett 2:15-19

Hughes VM, Datta N 1983 Conjugative plasmids in bacteria of the 'pre-antibiotic' era. Nature (Lond) 302:725-726

Kleckner N 1981 Transposable elements in prokaryotes. Annu Rev Genet 15:341-404

Labigne-Roussel A, Witchitz, J, Courvalin R 1982 Evolution of disseminated Inc7-M plasmids encoding gentamicin resistance. Plasmid 8:215-231

Lacey RW 1980 Evidence for two mechanisms of plasmid transfer in mixed culture of *Staphylococcus aureus*. J Gen Microbiol 119:423-435

Laufs R, Riess F-C, Jahn G, Fock R, Kaulfers P-M 1981 Origin of *Haemophilus influenzae* R factors. J Bacteriol 147:563-568

Roberts M, Elwell L, Falkow S 1978 Introduction to the mechanisms of genetic exchange in the gonococcus: plasmids and conjugation in *Neisseria gonorrhoeae*. In: Brooks GF et al (eds) Immunobiology of *Neisseria gonorrhoeae*. Amer Soc Microbiol, Washington DC, p 38-43

Sharp PA, Cohen SN, Davidson N 1973 Electron microscope heteroduplex studies of sequence relations among plasmids of *Escherichia coli*. II: Structure of drug resistance (R) factors and F factors. J Mol Biol 75:235-255

Starlinger P, Saedler H 1972 Insertion mutations in micro-organisms. Biochimie 54:177-185

Sykes RB, Richmond MH 1970 Intergeneric transfer of a β-lactamase gene between *Ps. aeruginosa* and *E. coli*. Nature (Lond) 226:952-954

Uhlin BE, Nordström K 1975 Plasmid incompatibility and control of replication: copy mutants of the R-factor R1 in *Escherichia coli* K12. J Bacteriol 124:641-649

Watanabe T 1963 Infective heredity of multiple drug resistance in bacteria. Bacteriol Rev 27:87-115

Willetts N, Skurray R 1980 The conjugation system of F-like plasmids. Annu Rev Gen 14:41-76

DISCUSSION

Clarke: How do the incompatibility groups recognize each other? How are plasmids of the same incompatibility group excluded from the bacteria?

Datta: They are not excluded, but each plasmid has control over its own replication; it possesses genes that allow its replication, and repressors that limit replication. The repressor that controls plasmid A will also control the replication of a closely related plasmid, A′, but will have no effect on an unrelated one. This has been shown by analysis of plasmid mutants altered in their copy number: if the repressor is less effective, more copies and less incompatibility result (Nordström 1983).

Hartl: In the Murray collection of bacteria that you mentioned, can you recognize in the plasmids any determinants besides those for plasmid replication?

Datta: Their conjugative pili were of the same sort that we have today, and some of the plasmids determined colicinogeny. However, the only resistance determinant that we found among the 84 plasmids was to potassium tellurite (K_2Te0_3), in one case (Hughes & Datta 1983). Present-day plasmids commonly confer resistance to mercuric chloride, the reason for which is not obvious (Summers & Silver 1978), so we had expected to find this, but did not.

Sawicki: What is the possible explanation for there being no antibiotic resistance in the transposons of these 1917 bacteria?

Datta: I imagine that it was just that the bacteria had not been selected for resistance. The explosive spread of resistance genes has been in response to the human use of antibiotics.

Sawicki: Was resistance likely to have been present at all?

Datta: Yes, but rarely. Although it is likely to have been *somewhere*, it was not necessarily present at all in *Salmonella* or *Shigella* or other bacteria that live in humans. It might have come from some totally different source.

Bradshaw: Isn't it rather risky to claim that it was there already? If one takes the argument back too far it means that every gene that we see in the world today must have been present at time zero.

Datta: A resistance gene might have evolved from an enzyme with a different function. There has been a lot of argument about why bacteria have penicillinase at all: is it to protect them against penicillin which is, after all, a natural product? Smith & Marples (1964) described hedgehogs, in a remote part of New Zealand, which had ringworm. The ringworm fungus was producing a kind of penicillin, and the lesions were superinfected with penicillinase-producing *Staphylococcus aureus*, which probably possessed penicillinase plasmids like those found in human infection.

Bell: Can any part of a plasmid be transferred from one plasmid to another? Is it a totally random transmission?

Datta: No; it has to be connected into a specialized and specific transposon which 'jumps', replicating and leaving a copy of itself behind, where it started, and making another copy for the new position. The transposition requires genes and also requires terminal inverted repeats, which must be some sort of recognition site. Probably any DNA could become part of a transposon.

Graham-Bryce: You mentioned that there have been no major new developments in antibiotic structures for 10 years except for ringing the changes of substituents in the basic chemical. But these substituent modifications may overcome bacterial resistance to established compounds. Bacterial resistance thus appears to be relatively compound-specific, and can be overcome by analogue synthesis. If so, it would seem to differ from insecticide and fungicide resistance. Although there are examples, such as the one mentioned by Professor Hodgson (p 167-189) of *N*-propylcarbamates and *N*-methylcarbamates showing a negative cross-resistance, what one might call a *class* resistance seems to be more common. Is this a genuine difference and, if it is, why should it occur?

Datta: With different penicillins there may or may not be cross-resistance. Earlier in the symposium it has been mentioned that if the mode of action of an insecticide, say, is the same as that of another then one can get cross-resistance, even though the two molecules are quite different. I know of only one similar example in antibiotic resistance, and that involves macrolide antibiotics such as erythromycin, whose effect is on bacterial ribosomes. The plasmid genes that confer resistance to these antibiotics do so by producing a methylase that acts on ribosomal RNA, modifying it so that the target becomes insusceptible to the antibiotic; the methylated ribosomes are also insusceptible to certain other antibiotics that act at the same site, even though they are chemically different from the macrolides. Lincomycin and streptogramin are affected by that cross-resistance.

Davies: Target-site effects, whether by mutation or by some form of modification are specific for the molecules that act at that target, whereas other types of change, such as in cell-walls, can give rise to a more general type of resistance that would include a variety of different compounds provided they did not all have the same uptake mechanism (cross-resistance). One can distinguish whether a particular process is a group mechanism or a specific mechanism on the basis of the mode of action of the antibiotic, the way it gets into the cell, and the biochemical mechanism of resistance.

Graham-Bryce: I understood from Professor Datta's paper that even where target-site modification occurred, resistance was relatively compound-specific in bacteria.

Davies: That is not necessarily true, as Professor Datta's example illustrated; it depends on the target and the inhibitor.

Hutchinson: Is transmission of genetic information between different types of organisms relevant here, for example, between bacteria and yeast cells or between bacteria and algae?

Datta: Not a great deal is known, but I am sure that this transmission can occur. Restricting the discussion to bacteria, I can say that transposons can overcome the limitation put on the spread of a gene by the host range of the vector plasmid. This vector plasmid may be able to enter a bacterium in which it cannot replicate. Nevertheless, if it carries a transposon into that bacterium, then the transposon will have a chance to transpose and to be rescued while the rest of the plasmid DNA is lost. One can demonstrate this in the laboratory. *Haemophilus influenzae,* which causes meningitis in children, has only recently become resistant to antibiotics, by means of plasmids. It seems that plasmids which cannot replicate in *Haemophilus* have entered and stayed there long enough to let the transposon be established, leaving the resistance in the new host (Laufs et al 1981). Similar processes may take place in other creatures and, indeed, DNA from the Ti (tumour-inducing) plasmid of *Agrobacterium tumefasciens* can insert into the chromosomes of plants (Barton et al 1983).

Kuć: I have a naive question. If a single cell of *Escherichia coli* is grown in a nutrient medium, the cell will divide to form a colony of cells. Would the initial pattern of, say, esterase enzymes continue to be identical in each bacterial cell? And how would one identify this? Would a different esterase pattern be evident if the cell morphology, growth pattern, or efficiency of division were interfered with? How much unknown difference might there be between cells that we think are identical?

Datta: This is terribly unorthodox! Each cell has its complement of DNA, with the capacity for producing all the necessary proteins and controls upon those productions. In that sense every cell in the culture is similar, unless it mutates.

Kuć: It strikes me that there are many ways in which the 'deck' can be shuffled.

Datta: I disagree. The same morphology and the same genetic information is present, so what can be shuffled?

Hartl: If colonies of bacteria are studied by electrophoresis, the enzymes behave in a perfectly conventional, sensible way—as if, instead of grinding up a bacterial colony, one had ground up a fly or a cow liver.

Datta: But, of course, bacteria do not even have different tissues, so the pattern is even more uniform.

Hartl: Yes. So one does not see a mixture of electrophoretic mobilities because of any tremendous mutation rates in the bacterial colony; the genes are just as stable as any other kinds of genes.

Kuć: If different plasmids are regulated differently is it not possible that we will see slightly different expression of the same plasmid in different cells?

Datta: We do not have many possibilities because they are all encoded strictly in the relatively stable DNA.

Sawicki: In *Myzus persicae* cloned from a single individual there is a fair amount of variation within the progenies, even though these progenies are genetically identical. We find variations not only in behaviour, but in the types of setae and other morphological features from individuals bred from only one individual (Blackman 1979). This puzzles me.

Datta: Isn't that differentiation, which we don't understand anyway?

Hartl: Are these differences in the aphid population heritable?

Sawicki: No. If one starts with an individual aphid that is bred for *n* generations on chrysanthemums, for example, one will obtain a whole series of progeny, some of which will establish themselves on chrysanthemums very readily, while others will not take to them at all.

Clarke: This reminds me of an example in *Drosophila* that was analysed genetically. Selection for bristle number was exercised on an inbred line and the line responded (Frankham et al 1978). This turned out to be due to the ribosomal genes—bobbed alleles—which can change in number because they are serially duplicated, and consequently recombinational events can change the numbers of these genes up or down. A response to selection was therefore obtained in what was originally an inbred line. Perhaps a similar phenomenon could explain Dr Sawicki's results.

Hartl: It is now known that, in *Drosophila*, under certain conditions of nuclear genome and cytoplasm, certain classes of transposable elements called *P* factors become mobilized and transpose at a much higher frequency than normal. So if one had hybrid dysgenesis of this sort in an inbred line, or a genetically homogeneous line, one would have a flowering of mutations in each generation, in spite of the inbreeding and because of the increased mutation caused by the transposing of these elements.

Davies: Is there any example of a plasmid-determined antibiotic resistance in which, for the same antibiotic, both detoxification and target-site modification mechanisms have been found?

Datta: No; it is either one or the other, as far as I know.

Georghiou: In insects there are examples where both types of resistance—

site-insensitivity and enhanced metabolism—exist. Whether enhanced metabolism eventually becomes unimportant would depend, I presume, on the extent of protection that is conferred by the 'insensitivity' factor.

Datta: When these resistances are observed in bacteria, the bacteria do not show any other differences. Their pathogenicity and serology are unchanged, so they are not 'unfit' in the sense of having disadvantageous metabolic alterations.

REFERENCES

Barton KA, Binns AN, Matzke AJM, Chilton M-D 1983 Regeneration of intact tobacco plants containing full length copies of genetically engineered T-DNA, and transmission of T-DNA to R1 progeny. Cell 32:1033-1043

Blackman RL 1979 Stability and variation in aphid clonal lines. Biol J Linn Soc 11:259-278

Frankham R, Briscoe DA, Nurthen RK 1978 Unequal crossing over at the rRNA locus as a source of quantitative genetic variation. Nature (Lond) 272:80-81

Hughes VM, Datta N 1983 Conjugative plasmids in bacteria of the 'pre-antibiotic' era. Nature (Lond) 302:725-726

Laufs R, Riess F-C, Jahn G, Fock R, Kaulfers P-M 1981 Origin of *Haemophilus influenzae* R factors. J Bacteriol 147:563-568

Nordström K 1983 Review: control of plasmid replication. Plasmid 9:1-7

Smith JMB, Marples MJ 1964 A natural reservoir of penicillin-resistant strains of *Streptococcus aureus*. Nature (Lond) 201:844

Summers AO, Silver S 1978 Microbial transformations of metals. Annu Rev Microbiol 32:637-672

Evolutionary relationships among genes for antibiotic resistance

JULIAN DAVIES and GARY GRAY

Biogen S. A., 3 Route de Troinex, P.O. Box 1211, Geneva 24, Switzerland

Abstract. The genes that determine resistance to antibiotics are commonly found encoded by extrachromosomal elements in bacteria. These were described first in Enterobacteriaceae and subsequently in a variety of other genera; their spread is associated with the increased use of antibiotics in human and animal medicine. Antibiotic-resistance genes that determine the production of enzymes which modify (detoxify) the antibiotics have been detected in antibiotic-producing organisms. It has been suggested that the producing strains provided the source of antibiotic-resistance genes that were then 'picked-up' by recombination. Recent studies of the nucleotide sequence of certain antibiotic-resistance genes indicate regions of strong homology in the encoded proteins. The implications of these similarities are discussed.

1984 Origins and development of adaptation. Pitman Books, London (Ciba Foundation symposium 102), p 219-232

Studies of several ubiquitous biological functions have provided good bases for the prediction of protein evolution. The cytochromes are perhaps the prime example of protein phylogeny (Dickerson 1972); analysis of the sequences of several other groups of proteins has also provided information on evolutionary relationships. Among microbes, antibiotic-resistance mechanisms are widely distributed and are found in all microbial genera that have been shown to harbour resistance plasmids. In addition, other bacterial species are known to possess chromosomal determinants for these resistance mechanisms. In terms of biochemical function, at least, resistances to penicillins and tetracyclines are the most widely distributed. It is generally assumed that the dissemination of these resistance characters is the result of facile intergeneric transfer of resistance plasmids as a result of the strong selective pressure provided by antibiotic use for human and animal purposes.

Studies of plasmid evolution at the molecular level have demonstrated that the capacity of plasmids to undergo intergeneric transfer by conjugation, transduction, transformation or cell fusion, allied with mechanisms of

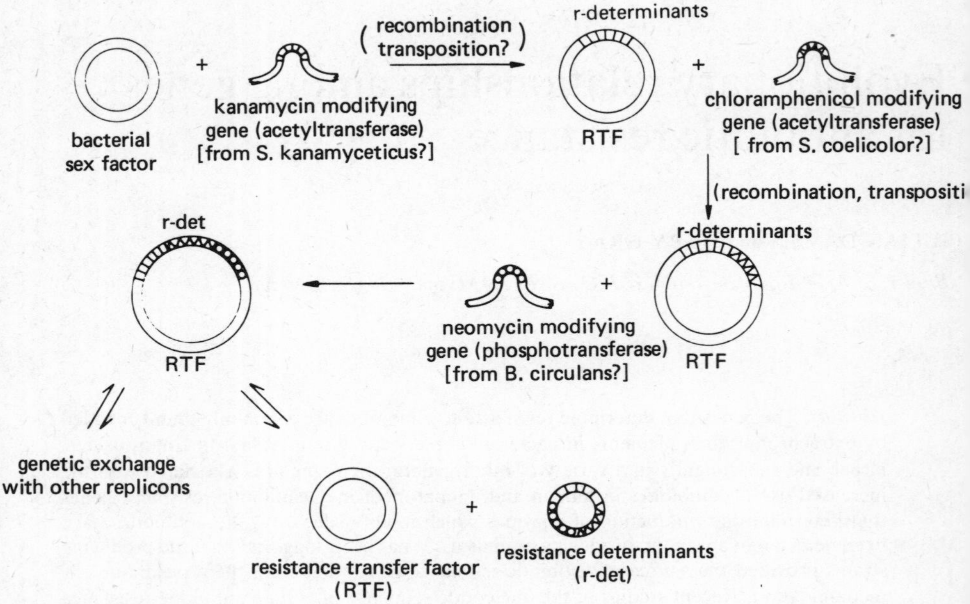

FIG. 1. The 'pathway' of evolution of resistance plasmids (modified from Watanabe 1971). *S. kanamyceticus* and *S. coelicolor* are *Streptomyces* spp; *B. circulans* is *Bacillus* spp.

non-homologous recombination promoted by insertion (IS) elements, provides satisfactory explanations for plasmid formation, rearrangement and dissemination (Kopecko et al 1976). Thus, the original proposals of Watanabe (1971) for mechanisms of plasmid evolution can be considered, in principle, as an established concept (Fig. 1). On the other hand, the origin of the various antibiotic resistance determinants remains a mystery. Although several possible origins have been proposed there is, in principle, no compelling reason why one should be considered in favour of the others. However, the fact that antibiotic-producing organisms must have mechanisms of self-protection leads us to favour the notion that antibiotic-producing organisms are a logical source of resistance mechanisms (Benveniste & Davies 1973). The resistance mechanisms must have been present long before the strong pressure for selection of resistant organisms was present owing to recent antibiotic use. The antibiotic-producing microorganisms may have been the primary source of genes for resistance enzymes but, since such bacteria exist in nature in close proximity with other antibiotic-sensitive bacteria, interspecific gene exchange probably occurs, as a result of localized selection pressure, and the recent donors of resistance determinants are unknown

(Davies & Kagan 1977). Koch (1981) has also considered the problem of resistance-determinant origin and has alternative proposals.

The notion that antibiotic-resistance mechanisms may have evolved as a result of substrate similarities (e.g. the aminoglycoside antibiotics may not be the *true* substrates of the aminoglycoside-modifying enzymes), together with the implication that the modifying enzymes may have other metabolic functions, cannot be eliminated. However, this seems unlikely since limited studies have shown that the aminoglycoside-modifying enzymes do not have other, non-antibiotic-related substrates. Nonetheless, the virtually ubiquitous distribution of β-lactamase (e.g. penicillinase) activities among microbes implies a key function for these enzymes in normal metabolic processes; such a function would presumably be a step in cell-wall matrix formation.

Aminoglycoside-resistance mechanisms

The aminoglycosides constitute a group of pseudosaccharide antibiotics of which the majority contain 2-deoxystreptamine as the basic structural unit (Table 1). The group includes the kanamycin and gentamicin antibiotics

TABLE 1 The aminoglycoside aminocyclitol antibiotics used in clinical practice

Streptomycin	Neomycin	Kanamycin A,B	Amikacin
Dihydrostreptomycin	Paromomycin	Tobramycin	Netilmicin
	Lividomycin	Gentamicin	
Spectinomycin	Ribostamycin	Sisomicin	Apramycin

commonly used for the treatment of serious infections caused by gram-negative bacteria. The first report of bacterial resistance to aminoglycoside antibiotics that was mediated by resistance-plasmid-encoded modifying enzymes was in 1952 (see Okamoto & Suzuki 1965); since that time a number of different modifications of aminoglycosides has been characterized (Fig. 2). These include the *O*-phosphorylation and *O*-nucleotidylylation of hydroxyl groups and the *N*-acetylation of amino substituents (Davies & Smith 1978). The aminoglycosides are unusual in the sense that often three, four or more different enzymes are capable of modifying each drug (Table 2). This contrasts with the β-lactam antibiotics (e.g. penicillin) and chloramphenicol, which possess a single type of enzymic modification. In addition, each aminoglycoside-modifying enzyme may exist in several isozymic forms that can be differentiated on the basis of substrate range; therefore, 10 or more enzymes may be capable of modifying each aminoglycoside. The isozymes presumably vary at the active (or binding) site of the enzyme since the

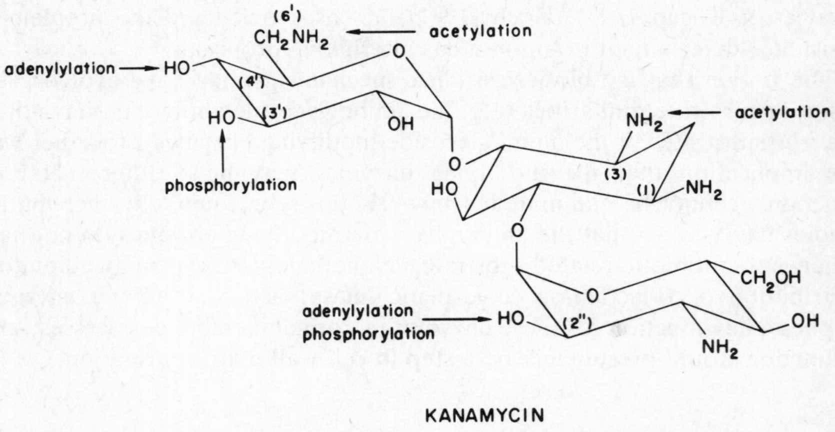

FIG. 2. The structure of kanamycin A, indicating enzymic modifications in resistant organisms.

substrate ranges can be altered by variations in the K_m for the aminoglycoside. Quite small differences in K_m (5–10 fold) alone are sufficient to explain the resistance spectra of organisms that harbour the different isozymes (Table 3).

The biochemical mechanism of resistance of microorganisms to aminoglycosides has now been established (Bryan & Van Den Elzen 1977, Davies & Kagan 1977, Shannon et al 1982). The aminoglycoside-modifying enzymes

TABLE 2 Aminoglycoside-modifying enzymes of resistant bacteria

Modification	*Enzyme*	*Typical substrates*[a]
Acetylation	AAC(2′)	Gentamicin, tobramycin
	AAC(6′)	Tobramycin, kanamycin, amikacin, neomycin, gentamicin[b]
	AAC(3)	Gentamicin, tobramycin, kanamycin
Nucleotidylylation	ANT(4′)	Amikacin, tobramycin, kanamycin
	ANT(2″)	Gentamicin, tobramycin, kanamycin
	ANT(3″)	Streptomycin, spectinomycin
	ANT(6)	Streptomycin
Phosphorylation	APH(3′)	Kanamycin, neomycin
	APH(3″)	Streptomycin
	APH(2″)	Gentamicin
	APH(5″)	Ribostamycin

AAC, aminoglycoside acetyltransferase; ANT, aminoglycoside nucleotidyltransferase; APH, aminoglycoside phosphotransferase. [a] Not all substrates are listed; each enzyme exists in a variety of forms with different substrate ranges. [b] Different, naturally occurring derivatives of gentamicin are gentamicin C_1, gentamicin C_{1a} and gentamicin C_2; gentamicin C_{1a} is a substrate for AAC(6′) but gentamicin C_1 is not.

TABLE 3 (a) Aminoglycoside 3′-*O*-phosphotransferases, their substrates and their K_m values

Isozyme type	*Substrate K_m values (μM)*				
	Neo	*Kan*	*But*	*Liv*	*ATP*
APH3′-II (Tn*5*)	3.9	2.3	10.7	n.a.	24.0
APH3′-III (*S. aureus*)	1.3	1.6	1.6	2.6	18.0
APH3′ (*B. circulans*)	3.2	3.3	1.6	n.a.	15.2

(b) Aminoglycoside 3-*N*-acetyltransferases, their substrates and their K_m values

Isozyme type	*Substrate K_m values (μM)*				
	Gen	*Tob*	*Neo*	*Apr*	*For*
AAC(3)-I	0.7	84.0	—	—	0.9
AAC(3)-II	1.1	5.1	—	—	—
AAC(3)-III	3.7	4.1	3.6	—	—
AAC(3)-IV	2.1	3.2	4.2	2.7	—

n.a., not available; —, not measurable because affinity too low. *Neo*, neomycin; *Kan*, kanamycin; *But*, butirosin; *Liv*, lividomycin; *Gen*, gentamicin C_{1a}; *Tob*, tobramycin; *Apr*, apramycin; *For*, fortimicin. For other abbreviations, see Table 2.

are intracellular and do not inactivate the drug in the culture medium. Since the transport of aminoglycosides into the cell across a membrane proton gradient is inefficient, it suffices for enzymic inactivation to occur inside the cell at a rate sufficient to counteract the uptake. Thus, although the aminoglycoside-modifying enzymes use important energy substrates of the cell (ATP and acetyl-coenzyme A), this use is limited and presumably does not cause a drain on cell resources. This is especially true of *O*-adenylylation, since the adenylic acid residue is irreversibly lost and not available for regeneration (as it is after *O*-phosphorylation).

Comparisons of gene and protein sequence

The foregoing discussion serves to illustrate that the aminoglycoside-modifying enzymes of bacteria represent a special and interesting group of functions. A large number of enzymes catalyse a limited number of ATP or acetyl-CoA-based modifications on a small number of structurally related substrates. Presumably the enzymes are related with respect to the binding sites for the two substrates (antibiotic and co-factor), and variation should occur in the different isozymes particularly when the K_m for two of the substrates differs by an order of magnitude. Nearly identical sequences in the different enzyme proteins may identify the position of the binding sites. The aminoglycoside-modifying enzymes have been isolated from numerous microbial genera. When resistance (R) factors are involved, it is difficult to say that the first isolate studied is the native source, since the organism will almost certainly have been selected out of a mixed population and may represent the result of numerous natural intergeneric and interspecific transfers. Thus, in considering evolutionary relationships and origins among the aminoglycoside-modifying enzymes, we are not able to identify true origins of plasmid-encoded enzymes. The aminoglycoside-modifying enzymes of antibiotic-producing organisms (in those cases examined) are encoded by (permanent) chromosomal genes (Komatsu et al 1981). This is circumstantial (and admittedly weak) evidence for the notion that the producing organisms may be the source of resistance genes for R-plasmids. The argument is weak since there is every reason to believe that the chromosomal or extrachromosomal locations of genes may be pure chance. An essential gene may have a favoured location on the chromosome; there is little doubt that in antibiotic-producing organisms, an accompanying antibiotic-resistance mechanism is essential to survival and production.

In the consideration of the aminoglycoside-modifying enzymes as a family, relationships have been examined by three different methods: (1) nucleic-acid sequence homology by hybridization; (2) protein sequence homology by antibody cross-reaction; and (3) complete primary nucleic-acid sequence comparisons. Of these, only the latter has provided any useful information since nucleic acid hybridization and immunological cross-reaction were limited to a few closely related sequences. For example, the four phosphotransferases to be described did not show evidence of similarities by methods (1) and (2).

The complete nucleotide sequences of four independently isolated aminoglycoside 3′-*O*-phosphotransferase genes (APH(3′) genes) have been determined (Thompson & Gray 1983) and the proposed amino acid sequences for the APH(3′) genes from these sources are compared in Fig. 3. The protein sequences include the gram-positive *Staphylococcus aureus* APH(3′)-III, those from the resistance genes of the gram-negative transposons Tn*5* and Tn*903* and that from the neomycin-producing *Streptomyces fradiae*. Three isozymic forms of APH(3′) with related but distinct substrate ranges are represented by this group of sequences. All four enzymes modify kanamycin and neomycin and are subgrouped according to their reactivity with butirosin, lividomycin and amikacin (Table 3). The availability of the four APH(3′) sequences permits the comparison of protein sequences of the three isozymic forms of APH(3′) and an estimation of their relationship with the putative ancestral protein form, the aminoglycoside-producing *S. fradiae*. As can readily be seen from an inspection of Fig. 3, the four amino acid sequences exhibit extensive regions of homology, with the carboxyl terminal segments

```
1.  Streptomyces fradiae    M - - - - - - - - - - - - - D D S I L R R K I Y R R E W R A V N E G D S
2.  Staphylococcus aureus   M - - - - - - - - - V K M R I S P E L K K L I E K Y R C V K D T E G M S
3.  Tn5                     M - - I E Q D - - G L H A G S P A A W V E R L F G Y D W A Q Q T I G C S
4.  Tn903                   M S H I Q R E T S C S R P R L N S N M D A D L Y G Y K W A R D N V G Q S

                    40                                  60                                  80
1.  G A F V Y Q L T G G P E P Q P E L Y A K I A P R A P E N S A F D L S G E A D R L E W L H R H G I P V
2.  P A K V Y K L - G E N E N - - - L Y L K M T D S R Y K G T T Y D V E R E K D M M L W L - E G K L P V
3.  D A A V F R L S A Q G - - R P V L F V K - T D L S - - G A L N E L Q D E A A R L S W L A T T G V P C
4.  G A T I Y R L Y G K P D A - P E L F L K - H G K G - - S V A N D V T D E M V R L N W L - T E F M P L

                                     100                                120
1.  P R V V E R G A D D T A A W L V T E A V P G V A A A E E W P E H Q R F A - - V V E A M A E L A R A L
2.  P K V L H F E R H D G W S N L L M S E A D G V L C S E E Y E D E Q - S P E K I I E - L C G V H R A L
3.  A A V L D V V T E A G R D W L L L G E V P G Q - - - D L L S S H L A P A E K V S I - M A D A M R R L
4.  P T I K H F I R T P D D A W L L T T A I P G K T A F Q V L E E Y P D S G E N I V D A L A V F L R R L

        140                                     160                                    180
1.  H E - L P V E D C P S D R R L D A A V A E A R R N V A E G L - - V D L D D L Q E E R A G W T G D Q L
2.  S L H I D I S D C P Y T N S L D S R L A E L D Y L L N N D L A D V D C E N W E E D T P F K D P R E L
3.  H T - L D P A T C P F D H Q A K H R I E R A R T R M E A G L - - V D Q D D L D E E H Q G L A P A E L
4.  H S - I P V C N C P F N S D R V F R L A Q A Q S R M N N G L - - V D A S D F D D E R N G W P V E Q V

                                 200                               220
1.  L A E L D R T R P E K E D L V V C H G D L C P N N V L L D P G T C R V T G V I D V G R L G V A D R H
2.  Y D F L K T E K P E - E E L V F S H G D L G D S N I F V K H G - - K V S G F I D L G R S G R A D K W
3.  F A R L K A R M P D G E D L V V T H G D A C L P N I M V E N G - - R F S G F I D C G R L G V A D R Y
4.  W K E M H K L L P F S P D S V V T H G D F S L D N L I F D E G - - K L I G C I D V G R V G I A D R Y

        240                                     260                                    280
1.  A D I A L A A R E L E I D E D P W F G P A Y A E R F L E R Y G A H R V D K E K L A F Y Q L L D E F F
2.  Y Q I A F C V R S I - R E D I G E E - - Q Y V E L F F D L L G I - K P D W E K I K Y Y I L L D E L F
3.  Q D I A L A T R D I - A E E L G G E - - - W A D R F L V L Y G I A A P D S Q R I A F Y R L L D E F F
4.  Q D L A I L W N C L - - - - - G E F S P S L Q K R L F Q K Y G I D N P D M N K L Q F H L M L D E F F
```

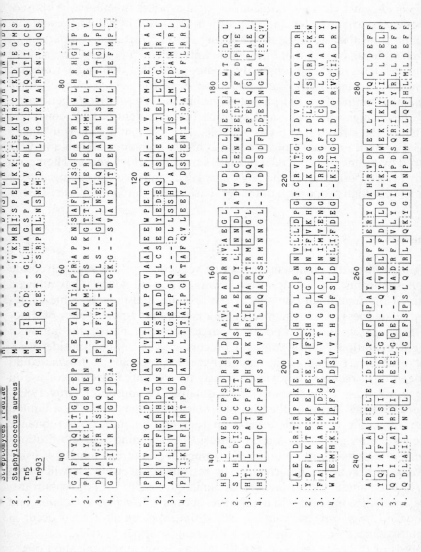

FIG. 3. Comparison of the amino acid sequences of aminoglycoside-3′-phosphotransferases of *S. fradiae*, *S. aureus* and the gram-negative transposons Tn*5* and Tn*903*. The amino acid sequences are taken from Thompson & Gray 1983. Sequence locations containing identical amino acid residues are enclosed within solid-line boxes and sites of favoured amino acid substitutions by dotted-line boxes. Gaps have been inserted (–) to improve sequence homology. The one-letter amino acid codes used are: A, alanine; C, cysteine; D, aspartic acid; E, glutamic acid; F, phenylalanine; G, glycine; H, histidine; I, isoleucine; K, lysine; L, leucine; M, methionine; N, asparagine; P, proline; Q, glutamine; R, arginine; S, serine; T, threonine; V, valine; W, tryptophane; and Y, tyrosine.

being the most homologous. A conserved cysteine residue (C) at position 145 is flanked by two regions containing predominantly acidic amino acids (aspartic and glutamic acid, D + E, at positions 115 + 175) which may function in the binding of the basic aminoglycoside antibiotics to the modifying enzymes. If these acidic amino acid residues function as proposed, the sequence variations located at position 115 and 175 would presumably determine the substrate range by changing the K_m. Detailed comparison of the nucleotide sequences that encode the four APH(3′) proteins reveals that all four proteins are evolutionarily related and that the gene from the neomycin-producing *S. fradiae* could have been the source of the widely disseminated aminoglycoside phosphotransferases. The APH(3′) gene of *S. fradiae* has a high guanosine and cytosine (GC) content (73%), consistent with that of the streptomycetes as a whole. The fact that a gene with such a high GC content encodes a protein that is highly homologous with a protein produced by a gene of 42% GC is perhaps surprising. However, as Thompson & Gray (1983) point out, in the *S. fradiae* gene sequence the coding triplets contain a very high proportion of the GC basis in the third (wobble) position.

Conclusions

Determination of the complete nucleotide sequences of the aminoglycoside 3′-*O*-phosphotransferase genes from several bacterial sources has allowed analysis of the functional domains of the proteins and their possible evolutionary relatedness. The sequence homologies found among genes of different gram-negative and gram-positive sources is consistent with the notion of divergent evolution from a common precursor. The aminoglycoside 3′-*O*-phosphotransferase gene of the aminoglycoside-producing *S. fradiae* could have been this precursor but was obviously not recruited directly for use in neomycin-resistant strains from clinical isolates (Thompson & Gray 1983). Different isozymic forms of the phosphotransferases may differ by a limited number of amino acid replacements in regions of homologous sequence between the enzymes of the varied sources. These amino acid substitutions presumably result in alteration in K_m values for different aminoglycoside substrates with concomitant changes in resistance spectrum. It will be of interest to examine the sequences of aminoglycoside 3′-*O*-phosphotransferases from the organisms that produce butirosin (*Bacillus circulans*) and lividomycin (*Streptomyces lividus*) to see if the expected alterations occur in the putative ancestral proteins for the different isozymes (Table 3).

Analysis of the gene sequences did not allow positive identification of the binding site of the aminoglycoside substrate on the enzymes; comparison with complete or partial nucleotide (protein) sequences of other aminoglycoside-modifying enzymes (acetyltransferases, adenylyltransferases) did not reveal any common sequences. The aminoglycoside-3′-*O*-phosphotransferases might be expected to show homology with other enzymes that use ATP as substrate—kinases, for example. However, our computer-assisted analysis has not yet revealed any such relationships. It should be noted that sequences determined for chloramphenicol acetyltransferases of different microbial sources failed to give clues as to acetyl-CoA binding sites (Horinouchi & Weisblum 1982). In fact, the chloramphenicol acetyltransferases do not form an obviously structurally related family. It will be of interest to examine the sequences of the aminoglycoside 3-*N*-acetyltransferase family (Table 3) as they become available; in this case the molecular size varies substantially for the different isozymes (S. Harford, personal communication 1980).

The antibiotic-resistance determinants of bacteria provide a rich source for the analysis of protein evolution and for the identification of enzymic functional domains by sequence comparison. Many of the genes for resistance to aminoglycosides, β-lactams, chloramphenicol, trimethoprim, and macrolides from a wide variety of microorganisms have now been cloned, and some have been sequenced. Since the transfer of these genes between different genera and species has been facile, and almost certainly due to strong selective pressures in the use of structurally related but not identical antibiotics, it is likely that numerous changes in primary sequence will have occurred as a result of adaptive requirements.

REFERENCES

Benveniste R, Davies J 1973 Aminoglycoside antibiotic-inactivating enzymes in actinomycetes similar to those present in clinical isolates of antibiotic-resistant bacteria. Proc Natl Acad Sci USA 70:2276-2280

Bryan LE, Van Den Elzen HM 1977 Effects of membrane-energy mutations and cations on streptomycin and gentamicin accumulation by bacteria: a model for entry of streptomycin and gentamicin in susceptible and resistant bacteria. Antimicrob Agents Chemother 12:163-177

Davies J, Kagan SA 1977 R-factors, their properties and possible control. Springer-Verlag, Vienna, p 207-219

Davies J, Smith DI 1978 Plasmid-determined resistance to antimicrobial agents. Annu Rev Microbiol 32:469-518

Davies J, Courvalin P, Berg D 1977 Thoughts on the origins of resistance plasmids. J Antimicrob Chemother 3(Suppl C):7-17

Dickerson RE 1972 The structure and history of an ancient protein. Sci Am 226(4):58-72

Horinouchi S, Weisblum B 1982 Nucleotide sequence and functional map of pC194, a plasmid that specifies inducible chloramphenicol resistance. J Bacteriol 150:815-825

Koch AL 1981 Evolution of antibiotic resistance gene function. Microbiol Rev 45:355-378

Komatsu K, Leboul J, Harford S, Davies J 1981 Studies of plasmids in neomycin-producing *Streptomyces fradiae*. Microbiology 1981, Am Soc Microbiol, Washington DC, p 348-387

Kopecko DJ, Brevet J, Cohen SN 1976 Involvement of multiple translocating DNA segments and recombinational hotspots in the structural evolution of bacterial plasmids. J Mol Biol 108:333-360

Okamoto S, Suzuki Y 1965 Chloramphenicol, dihydrostreptomycin-, and kanamycin-inactivating enzymes from multiple drug-resistant *Escherichia coli* carrying episome 'R'. Nature (Lond) 208:1301-1303

Shannon K, Phillips I 1982 Mechanisms of resistance to aminoglycosides in clinical isolates. J Antimicrob Chemother 9:91-102

Thompson C, Gray GS 1983 The nucleotide sequence of a streptomycete aminoglycoside phosphotransferase gene and its relationship to phosphotransferases encoded by resistance plasmids. Proc Natl Acad Sci USA, in press

Watanabe T 1971 The origin of R factors. Ann NY Acad Sci 182:126-140

DISCUSSION

Clarke: I was fascinated by the use of codons in *Streptomyces fradiae*, whose DNA has an extremely high guanosine and cytosine (GC) content, and leaves no room for synonymy in the genetic code. This reminded me of some early work by Sueoka (1965) on the amino acid compositions of organisms differing greatly in their GC content. In these organisms the change from one extreme of GC content to the other actually seems to have moved the amino acid compositions of the proteins. The correlations between GC content and the proportion of amino acids with high GC codons were very strong but not sufficient for the changes in amino acid composition to explain the changes in GC. Whatever is driving the DNA and pushing the GC content in one direction or another is so strong that it even moves the protein. So what is it?

Davies: In the examples I have given, we can compare in great detail enzymes with from 42% to 75% GC content. It is surprising how much similarity is retained at amino acid positions of various parts within the protein.

Clarke: If some driving force related to the DNA and some driving force related to the protein are opposed to each other, I wonder what is the driving force for the DNA.

Gressel: The sulphonamide drugs are synthetic and there is resistance to them. What is the origin of this resistance?

Datta: For resistance to either sulphonamides or trimethoprim a plasmid provides an extra or substitute enzyme. For sulphonamides it provides dihydropteroate synthase. The bacterial enzyme is inhibited by the sulphonamide, and the plasmid provides an alternative enzyme, from an unknown source.

Gressel: So it is quite different from sulphonamide drugs?

Davies: The mechanism is different, but the means by which the gene arrived in clinical isolates of bacteria could be very similar because it is known that some organisms possess sulphonamide-resistant dihydropteroate synthases.

Gressel: Sequencing of the gene would be fascinating because a probe could then be made and used to see if the gene possibly came from other such organisms.

Datta: Dr Ola Sköld in Sweden has done a lot of work on these substitute enzymes, but believes that a search for their source could be almost endless, through every living creature (personal communication).

Epstein: You mentioned, Dr Davies, one obvious mechanism for resistance, namely that the antibiotic cannot enter the organism, which is impermeable to it. Is this merely a possibility or are there actual examples?

Davies: There are such cases. Professor Datta mentioned tetracycline resistance, in which the antibiotic cannot stay in the microorganism because there is a positive mechanism of efflux. Surprisingly, there is no well documented case, other than by mutation, in which resistance occurs in clinical isolates as a result of a block in the uptake of the antibiotic into the cell. This is well illustrated by the aminoglycosides, which require active transport into the cells based on oxidative phosphorylation. If one specifically blocks oxidative phosphorylation, then the aminoglycosides do not enter the cells. This mechanism of resistance will determine cross-resistance to all the aminoglycosides. This is a clear example of a mutation that blocks antibiotic uptake; I know of no well characterized plasmid-determined mechanism that blocks uptake.

Datta: That mutational kind of blocking does occur in isolates from infections.

Kuć: We speak of resistance and sensitivity to, say, penicillin. Is the distinction sharp or do we have shades of resistance or shades of sensitivity?

Datta: When the resistance results from the production of penicillinase the distinction is sharp; but there are other examples of mutations to resistance e.g. in penicillin-binding proteins, where a gradation occurs. Most of the resistance seen in clinical isolates is of the penicillinase type.

Kuć: Would graded amounts of penicillin give sharp dose–response curves?

Datta: Yes.

Davies: The levels of resistance are almost always of the order of 100 times the minimum inhibitory concentration for the organism.

Wood: How do you distinguish a spontaneous point mutation from a phenotypic change caused by the insertion of a plasmid gene into the genome?

Davies: Insertion of DNA sequences into a functional gene sequence *is* a mutation, and like all mutations can be detected by a phenotype.

Wood: In that case, what is the rate of such a mutation? Is that kind of mutation more frequent than a base substitution, deletion or addition?

Davies: As far as I am aware, when a transposon is present in a bacterial cell, the frequency of mutation in that organism is increased above the basal spontaneous point mutation rate.

Wood: In pesticide resistance one is always interested in how frequently a 'mutation' occurs. We generally tend to assume that it is rare but, increasingly, one wonders whether it may be relatively common.

Sawicki: No-one has yet measured the *rate* of mutation of any gene for resistance in insects. Therefore any figure would fit, because there are no data.

Wood: There is some evidence. Kikkawa (1964) demonstrated mutation to parathion resistance in *Drosophila* after irradiation, and the rate of this mutation can be calculated to be 1.4×10^{-4} gametes. This suggests that natural mutation rates are probably one or two orders of magnitude less frequent than this.

Gressel: Has this not been tried in insect-tissue cultures, as in plant-cell cultures? In cell-suspension cultures it is much easier to do that sort of study.

Sawicki: No. As soon as one starts using tissue culture, one modifies very considerably the insect cell itself, which is no longer truly representative of the cell in the organism.

Gressel: But surely the genome will be identical?

Sawicki: Yes, up to a point.

Clarke: In *Drosophila*, a large proportion of what are called normal mutations are actually due to the insertion of transposable elements.

Hartl: It would be incorrect to assume that all insertions in or near eukaryotic genes are necessarily inactivations of the gene that was originally present. A transposable element inserting upstream or downstream from a eukaryotic gene can increase the overall activity of that gene or decrease the activity of the gene. Certain insertions in intervening sequences in genes in *Drosophila* are temperature-sensitive. So even the temperature-sensitivity of a phenotype no longer implies that the gene in question is a simple substitution.

Gressel: Are there any other known examples of the huge 'wobble-base' changes that you mentioned, throughout evolution, where there are similar amino acids yet totally different base sequences?

Davies: I'm sure that when more gene sequences from quite different organisms are determined there will be other examples. I doubt that the formation of the original transposon is a recent occurrence.

Gressel: Do you think these elements are moving around all the time?

Davies: That is the favoured hypothesis.

Clarke: The histones in sea urchins are a good example of constancy of amino acid sequence associated with quite dramatic changes in the nucleotides coding for those amino acids.

Datta: I can't remember what the guanosine and cytosine content of

Staphylococcus is, as opposed to gram-negative bacteria, but do the phosphotransferase genes come to resemble the DNA of the host bacteria?

Davies: No. In the case of *Streptomyces fradiae* the phosphotransferase gene is clearly a *Streptomyces* gene coming from an organism with a high GC content. In *Staphylococcus aureus*, the phosphotransferase has a different GC content from the average such content of the genome. In Tn*5*, the GC content is higher, and in Tn*903* it is lower than in the genomes of organisms with which these transposons are associated. The problem is how to define a wild-type organism: is it the first one that is found and studied by a microbiologist? Tn*5* was originally found in *Klebsiella*, and Tn*903* was originally found in *Shigella*. But we don't know if these transposons 'started' in these species.

Kuć: If a transposable element can move from one plasmid to another and then the plasmid itself can be transferred from one bacterium to another, can we be certain that we shall always see that character expressed in the second bacterium?

Datta: That is a difficult question. When it *is* expressed then we know that it is there. If it is not expressed then we don't know. One could identify it with a DNA probe, but I know of no example of a search for such an unexpressed gene.

Kuć: Would it be possible to transpose another element that would allow the expression of the original unexpressed element? Could this lead to the wrong conclusion? We may think we have transferred the element to *do* a particular event whereas in fact we have transferred an element that allows the full *expression* of the event.

Datta: That is conceivable, but in the examples studied, the plasmid contains the right gene for the right protein, and it is expressed in the host. It is straightforward and mechanistic.

Kuć: Are there no other influences on that expression?

Datta: Some of the penicillinase plasmids carry their own control mechanisms—enzyme synthesis is repressed except in the presence of the substrate.

Davies: If one looks at the same gene, or ostensibly the same biochemical function, as it is being passed around different organisms, one often finds changes in control functions as the gene passes through different species. In time, the gene may pick up a set of regulatory functions peculiar to the host organism at that time. But as to Professor Kuć's question of whether different hosts affect the expression of transposons, for penicillinase and other β-lactamases this doesn't seem to be the case. However, since the aminoglycoside-modifying enzymes are influenced by cyclic AMP levels within the cell, different hosts *can* affect the expression (level of resistance) to some extent, because of a general biochemical feature of the enzymes involved in resistance.

REFERENCES

Kikkawa H 1964 Genetic studies on the resistance to parathion in *Drosophila melanogaster*. II. Induction of a resistance gene from its susceptible allele. Botyu-Kagaku 29:37-42

Sueoka N 1965 On the evolution of informational macromolecules. In: Bryson V, Vogel H (eds) Evolving genes and proteins. Academic Press, New York, p 479-496

Accessory DNAs in the bacterial gene pool: playground for coevolution

DANIEL L. HARTL*, DANIEL E. DYKHUIZEN* and DOUGLAS E. BERG*†

**Department of Genetics and †Department of Microbiology and Immunology, Washington University School of Medicine, St Louis, Missouri 63110, USA*

Abstract. Chemostat studies of bacteria that harbour the prokaryotic transposable elements Tn*5* and Tn*10* and the temperate phages λ, Mu, P1 and P2 have shown that these accessory DNA elements confer a selective advantage on their hosts. We propose that similar selective effects provided the initial impetus for the evolution of nascent accessory DNA elements in primitive bacterial populations. In subsequent evolution the elements acquired or perfected the 'selfish' characteristics of over-replication and horizontal transmission. Such selfish traits led to the dissemination of accessory DNAs among commensal strains, species and genera, genetically interconnecting them to create a 'commonwealth' of species that potentially share a common gene pool. The involvement of accessory DNAs in genetic exchange provides selection at the population level for refinement and diversification of the elements and for regulation of their replication, transposition and transfer among cells. The diversity of intracellular environments encountered by the elements imposes constraints on their evolution while at the same time altering the selection pressures operating on conventional chromosomal genes. This process of coevolution of accessory DNAs with the genomes of their diverse hosts has led to a unique population structure and mechanism of genetic exchange among bacteria, which constitutes the most effective adaptive strategy yet devised by selection.

1984 Origins and development of adaptation. Pitman Books, London (Ciba Foundation symposium 102), p 233–245

The accessory DNAs of prokaryotes comprise a diverse collection of genetic elements that share two common properties. First, they are not normally essential for the reproduction of the bacterial cell that harbours them. Secondly, they are capable of either autonomous replication, separate from the host chromosome, or of over-replication resulting in an increase in their own copy number relative to other genomic sequences. Included among accessory DNAs are conjugative and non-conjugative plasmids, phages and transposable elements. Accessory DNAs can cause mutations and genome rearrangements, and many of them also mediate the exchange of genes between individuals in a population. Since each of these properties contributes to the adaptation of bacteria to temporally or spatially heterogeneous environments, accessory DNAs provide a fitting subject for this symposium on the origins of adaptations. Indeed, it has been argued by Reanney (1978) that accessory DNAs have provided bacteria with the most effective adaptive strategy yet devised by selection. Our purpose is to review recent data bearing on the origins of accessory DNA and to suggest how this remarkable adaptive strategy may have come about.

In addition to carrying determinants for transposition, replication or transmission between cells, many accessory DNA elements encode traits that dramatically enhance their hosts' survival or rate of growth and division in particular environments. Among these traits are: resistances to specific antibiotics, toxic metals, ultraviolet light, colicins and phages; enzymes used in certain steps in peripheral or major metabolic pathways; the specific restriction or modification of DNA; the production of enterotoxins or surface antigens; and the exclusion from the same cell of closely related accessory DNA elements. The only common feature among these traits is their dispensability in most environments.

In most cases analysed, the expression of accessory DNA functions is relatively independent of genotype in a broad range of bacterial hosts. For example, transposon- or plasmid-borne resistance to antimicrobial agents is often achieved by means of enzymes that modify or destroy the offending substance rather than by altering the site of action of the substance in the host. Such genes will function nearly equivalently in a broad range of hosts, and many accessory DNAs have evolved the capability of horizontal transfer—transfer not merely among clones or strains of a single bacterial species but also among different species and even genera. The importance of horizontal transfer is seen graphically in epidemiological studies demonstrating the dissemination of resistance-bearing plasmids among diverse bacterial pathogens in hospitals, and such transfer is also easily documented in laboratory experiments. Although retrovirus-mediated horizontal gene transfer in eukaryotes has been demonstrated (Jaenisch 1983), such transfer is exceedingly rare, and there is at present no reason to believe that it represents a major adaptive strategy in eukaryotes. It is thus the regular occurrence of interspecific gene transfer that decisively distinguishes population structure in prokaryotes from that in eukaryotes. This unique population structure implies that the concept of a prokaryotic species as a genetically isolated unit should be replaced with the concept of a genetic 'commonwealth' in which various strains, species and genera exchange genes on accessory DNA elements from a shared gene pool (Reanney 1978).

Evolution being notoriously opportunistic, the present functions of genes or structures need not necessarily reflect the evolutionary forces that fostered their original development. For example, the tiny halteres used today as balancers by dipterans in flight were once the second pair of wings that have become so prominent among the lepidopterans. Accessory DNA elements such as plasmids and transposons are genetically complex, highly evolved entities that must have originated ages ago and been retained in populations since that time (Campbell 1980, 1981). In the face of the long persistence of ancient accessory DNAs, it would not be surprising if prokaryotes had found many ways to use them. Consequently, some of the present functions encoded by accessory DNA, particularly the obvious ones such as antibiotic resistance, are almost certainly later acquisitions that have no significant bearing on the forces at work during the origin of the elements and, indeed, these later acquisitions would tend to obscure whatever functions the elements might originally have had.

A hypothesis concerning the evolution of certain classes of accessory DNA elements has been put forward recently and has generated considerable controversy (Doolittle & Sapienza 1980, Orgel & Crick 1980, Doolittle 1982). This is the 'selfish-DNA' hypothesis, and the fundamental idea is that genome evolution has a dynamic of its own that is relatively independent of phenotypic effects. According to Doolittle & Sapienza (1980), 'Natural selection operating within genomes will inevitably result in the appearance of DNAs with no phenotypic expression whose only "function" is survival within genomes. Prokaryotic transposable elements . . . can be seen as such DNAs, and thus no phenotypic or evolutionary function need be assigned to them . . . . [Transposability] itself ensures the survival of the transposed element, regardless of effects on organismal phenotype or evolutionary adaptability (unless these are sufficiently negative).' Actually, transposability itself ensures the survival of the transposed element only within a cell lineage because over-replication of the element will counterbalance loss through deletion. However, for an element to persist and spread throughout a population, horizontal transfer is also necessary. To be precise, if a transposable element originates in a cell in a population of size N individuals, and if the transposable element has no effect on the survival or reproduction of its host, then, in the absence of genetic exchange, the probability that the particular lineage carrying the transposable element eventually goes extinct is $1 - 1/N$, which is the same probability as for any other neutral gene, transposable or not. It is therefore not merely over-replication that is necessary for persistence, but also horizontal transfer between otherwise independent lineages. This consideration aside, it is not clear why the selfish-DNA hypothesis has been so controversial (see Doolittle 1982). Basically, the increase in frequency of selfish DNA elements in a population is formally similar to an increase in frequency of alleles exhibiting *meiotic drive*, a term applied to eukaryotes in which a heterozygote for two alleles produces unequal numbers of functional gametes carrying each allele, in contradiction to Mendel's first law (Sandler & Novitski 1957, Hartl 1977). For selfish DNA, a primary force impelling the element is over-replication; in meiotic drive it is over-transmission. What corresponds to horizontal transmission in the case of selfish DNA is the mating act itself in meiotic drive. Aside from these details, the two circumstances are remarkably analogous, yet the concept of meiotic drive has never created as much controversy.

An important alternative to the selfish-DNA hypothesis has been concisely expressed by Campbell (1981): 'Although an element with a negative effect on the host can be maintained by over-replication, long-term selection in competition with other hosts should ultimately eliminate it. Thus long-term survival may require that the element earn its keep, i.e. whatever cost is entailed in perpetuating the element should at least be balanced by some positive contribution to the organism's phenotype. Mere possession by the element of a gene conferring an advantage will not suffice. In that case, selection should favour retention of the gene and loss of the rest of the element. However, the elements themselves not only survive but, especially in the case of the larger phages and conjugative plasmids, have achieved a structural and regulatory complexity indicating a long and successful evolution. Hence their ability to replicate and disseminate themselves must itself be of some value to the host.' Campbell (1981) also points out that putative beneficial phenotypic effects of such elements may be sought at two levels: the level of individual selection, which involves direct effects of the element on the fitness of its host; and the level of the population, which involves the capacity of such elements to generate and disseminate novel gene combinations. In contrast, Doolittle & Sapienza (1980) argue that the selfish-DNA model is completely sufficient and that 'the search for other explanations may prove, if not intellectually sterile, ultimately futile'.

These models for accessory-DNA evolution are not mutually exclusive. On the one hand, the properties of over-replication and horizontal transmission confer a proliferative advantage on the element, which could counterbalance significant harmful fitness effects on the host, as is also the case with meiotic drive (Hartl 1977). On the other hand, elements that benefit their host will tend to spread more rapidly and widely than those that are neutral or detrimental. Thus, the ability to pick up and transmit genes that are favourable to the host would also be selectively advantageous to 'selfish' DNA. Consequently, the 'selfish' properties of over-replication and horizontal transmission become inextricably intertwined with the population effects of genetic exchange, which are themselves of benefit to the long-term survival of the population. An accessory DNA element that evolves *de novo* (or, more likely, from a pre-existing element) ought rapidly to become involved in the web of genetic interchange that characterizes the prokaryotic commonwealth of genomes. How much of the present prevalence of such elements should be attributed to their selfish nature, and how much to their ability to disseminate genes, then becomes a moot point, as they are both consequences of intrinsic properties of the elements. Of course, accessory DNAs and genetic exchange are not inevitably linked. Certain modes of genetic exchange, for example transformation, need not involve accessory DNA; and certain accessory DNAs, for example some of the virulent phages, are not intermediaries in genetic exchange. But for the kinds of elements under discussion here, selfishness and genetic exchange come together.

With regard to the importance of accessory DNA in mediating genetic exchange at the population level, arguments based on such well documented examples as the spread of infectious drug-resistance plasmids among hospital patients would seem to be unassailable (Mitsuhashi 1971). As a separate proposition, one can experimentally examine the additional possibility that these elements have direct effects on the individual fitness of their hosts. Earlier we mentioned that many accessory DNA elements carry genes such as antibiotic resistance that are essential to the survival of their hosts in particular environments. These conditionally essential genes are not the ones in question here, since, in most cases, they are likely to have been acquired secondarily in the course of evolution of the element. The acquisition of new genes by accessory DNA in prokaryotes usually occurs by the piecemeal addition and recombination of individually functional modules of genetic determinants, a process that Reanney (1978) has called 'coupled evolution' (for examples see Davey & Reanney 1980). We would like to know what effects accessory DNA elements may have other than those associated with these later acquisitions.

With conjugal plasmids or temperate phage, it is difficult to guess which genes presently carried might have been located on the element in its original state, as the great complexity of these elements suggests extensive evolution in a series of different hosts through simple nucleotide substitutions, repeated acquisition of blocks of genes, rearrangements, and probably through loss of genes as well. Nevertheless, some experiments relevant to basic evolutionary forces have been possible. For example, Zund & Lebek (1980) studied effects on the *Escherichia coli* growth rate of 101 R (resistance) factors in non-selective media. Only among the large (>80 kilobase) R factors was a significant increase in generation time ever found, but among about a third of these there was no detectable effect. The proposition that accessory DNA should be eliminated by selection because it represents an 'energetic burden' on the cell is facile but demonstrably untrue. Not all large plasmids decrease maximal growth rate, and among those that do, the effects could as well be due to physiological interactions between specific genes on the element and the particular host as to a generalized energy burden. For smaller plasmids, the effects on maximal growth rate are generally insignificant. Moreover, under the more stringent selective conditions of a chemostat (where a single nutrient is used to limit bacterial growth rate by being introduced into a bacterial culture at a continuous slow rate as a corresponding volume of exhausted medium and living cells is simultaneously removed) plasmid-free cells do not necessarily overgrow their plasmid-bearing counterparts (Melling et al 1977).

The effects of lysogeny (integration of prophage) for temperate phage are even more dramatic. Edlin et al (1975, 1977) and Lin et al (1977) have reported that lysogens of λ, Mu, P1 and P2 temperate phage are all *favoured* over their isogenic (but non-lysogenic) counterparts in energy-limited aerobic chemostats. The mechanism of this selective advantage is by no means clear, but for λ phage, it seems to be associated with the *rexA* function (Dykhuizen et al 1978, Dykhuizen & Hartl 1983). Whatever its mechanism, this effect flatly contradicts the naive expectation that lysogeny creates an energetic burden and that it would thus be at best selectively neutral. Rather, the effect may provide clues to the selective forces at work early in phage evolution and perhaps important under some conditions even today.

We postulate that effects on bacterial growth rate such as those conferred by prophages might be a general property of accessory DNAs and an important evolutionary force in the early stages of their evolution. It follows that a similar advantage might be conferred by the genetically simpler composite transposons and, if so, this possibility would be more amenable to direct analysis. Transposon Tn*5* was chosen for our tests because it is already being studied intensively, is widely used as a genetic tool, and because *E. coli* K12 normally lacks any sequences homologous to it. Tn*5* is a 5.7 kilobase (kb) composite transposable element, consisting of a central kanamycin/neomycin-resistance determinant bracketed by nearly identical 1534 base-pair (bp) insertion sequences, denoted IS*50L* and IS*50R* (Berg et al 1975, 1982). IS*50R* encodes two proteins, one a transposase and the other an inhibitor of transposition (Johnson et al 1982, Isberg et al 1982). IS*50L* is identical to IS*50R* except for an in-frame ochre terminator of polypeptide translation that abolishes the transposase and inhibitor activities while simultaneously creating the kanamycin/neomycin-resistance promoter. The Tn*5* element is therefore relatively simple, and our strategy of attack was first to create isogenic Tn*5* and non-Tn*5*-bearing strains of *E. coli* for competition in chemostats to ascertain whether an effect on growth rate could be detected in the absence of the relevant antibiotics.

Details of the experiments are described in Biel & Hartl (1983). A representative result is shown in Fig. 1, which involves competition for nutrient between isogenic Tn*5* and non-Tn*5*-bearing strains in chemostats limited for various carbon sources. For proline, glycerol or glucose as the limiting nutrient, the competing strains were SWB1100 (F^-*lac*::Tn5) *versus* SWB188 (F^-*lac tonA*); for lactose as limiting nutrient, they were SWB147 (F^-Tn5) *versus* SWB148 (F^-*tonA*). The left-hand axis is plotted as the natural logarithm of the ratio of the competing strains because, in theory, this quantity should change linearly with selection. As is evident in Fig. 1, Tn*5* has a substantial positive effect on bacterial growth rate in all carbon sources tested; the frequency of the Tn*5*-bearing strain increases from around 50% to over 90% in 40–50 h, which amounts to a conventional selection coefficient of

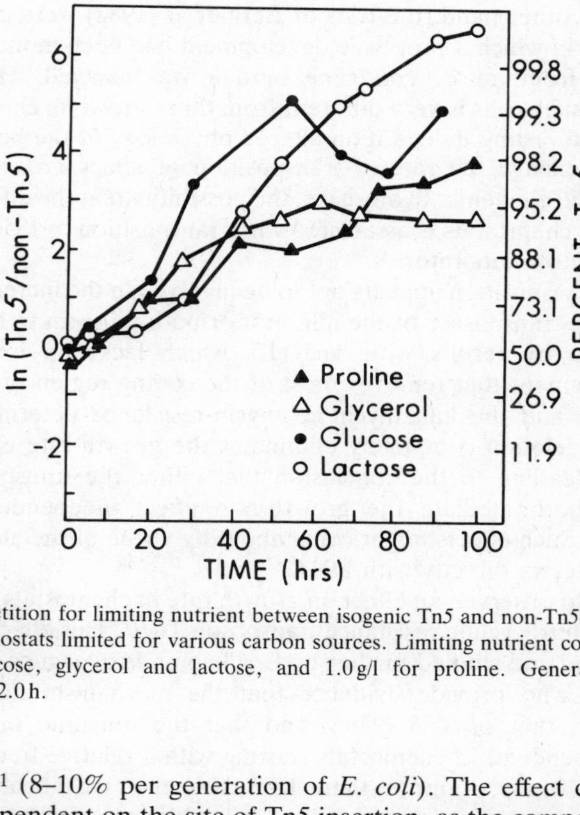

FIG. 1. Competition for limiting nutrient between isogenic Tn*5* and non-Tn*5*-bearing strains of *E. coli* in chemostats limited for various carbon sources. Limiting nutrient concentrations were 0.4 g/l for glucose, glycerol and lactose, and 1.0 g/l for proline. Generation times were approximately 2.0 h.

0.04–0.05 h^{-1} (8–10% per generation of *E. coli*). The effect does not seem markedly dependent on the site of Tn*5* insertion, as the competing strains in Fig. 1 include one with Tn*5* inserted in *lac* (SWB1100) as well as a strain with an insertion at a different (unknown) location (SWB147). Indeed, the same effect is observed in isogenic strains one of which carries Tn*5* inserted into the *lacP* site on an F′*lac* episome (Biel & Hartl 1983).

The increase in growth rate with F′*lacP*::Tn5 provides a convenient test of whether transposition is necessary for the occurrence of the growth-rate effect. In one experiment, F′*lacP*::Tn*5* clones were isolated from a chemostat population in which the Tn*5*-bearing strain had increased from 50% to over 90%. Curing (i.e. elimination) of the F′ in 35 such clones restored kanamycin sensitivity in all cases, indicating that the intact element had not been transposed to chromosomal sites in the course of selection. Although this experiment rules out transposition of the element as an intact unit, there remains the possibility that under chemostat conditions IS*50* itself might transpose. This seems rather unlikely because, where tested, IS*50* has been shown to transpose less frequently than the intact Tn*5* element (Berg et al 1982). On the other hand, the tests of Berg et al (1982) were carried out in broth cultures in which λ prophage development had been induced, and only transposition from an *E. coli* gene onto λ was assayed. However, the physiology of such cells is very different from those grown in chemostats, and it would be interesting indeed if an altered physiology of the host bacterium were able to change the rate of transposition of intact Tn*5* relative to its constituent IS*50* elements. In any case, the possibility that the effect of Tn*5* on growth rate in chemostats is associated with transposition of IS*50* is currently being tested in our laboratories.

Although transposition appears not to be involved in the increase in growth rate, either the transposase or the inhibitor or both do seem to be necessary. This is shown by results with Tn*5*-112, which lacks an internal 3.0 kb restriction fragment that removes most of the coding region of IS*50R* while leaving IS*50L* and the kanamycin/neomycin-resistance determinant intact. This internal deletion completely eliminates the growth-rate effect (Biel & Hartl 1983), leading to the conclusion that either the transposase or the inhibitor or both mediate the growth-rate effect independently of Tn*5* transposition. Such effects might come about by virtue of the ability of these proteins to interact directly with DNA.

We have also observed an effect on growth rate in chemostats conferred by the unrelated tetracycline resistance transposon Tn*10*. This effect of Tn*10* has been confirmed and studied further by L. Chao and colleagues (unpublished work, 1983), who provide evidence that the mechanism may be rather different from that of Tn*5*. They find that the outcome of selection is frequency dependent: in chemostats starting with a relative frequency of the Tn*10*-bearing strain of more than 1 in 10^3, the Tn*10*-bearing strain is favoured; at initial frequencies below 1 in 10^3, the Tn*10*-bearing strain decreases in frequency. This dependence of outcome on initial frequency is characteristic of strains bearing mutator alleles (Chao & Cox 1983), leading Chao et al to interpret the favourable selection for Tn*10* as due to a mutator effect of this transposon. Their interpretation seems to be supported by Southern blots, which indicate that the IS*10* component of Tn*10* transposes to new genomic locations in strains that had undergone selection in chemostats.

These examples involving Tn*5*, Tn*10*, λ, Mu, P1 and P2 seem to indicate a variety of direct selective effects of transposons of the sort that Campbell (1981) has postulated for accessory DNA elements that are 'earning their keep'. We believe that such effects may be quite general, although diverse in mechanism. Indeed, all five accessory DNA elements so far studied for effects on growth rate seem to have a beneficial selective effect under the appropriate conditions, and against this background it would be interesting to study other transposons and insertion sequences.

The molecular basis of these growth-rate effects warrants additional study. For λ, Mu, P1 and P2 the basis is unknown; for λ and Mu especially, analysis would be possible using the variety of phage and host mutations now available. Of course, whether the phage effects were associated with these elements at or near the time when they first evolved is unclear because of the present genetic complexity of the elements; effects we see today may also have evolved along the way, because even segments of selfish DNA will accumulate mutations that are favourable to the host. (The fact that most transposable elements regulate their own transposition may be taken as evidence of evolutionary feedback between the host and its accessory DNA.) Matters are simpler with insertion sequences (IS) and transposons because of their comparative simplicity; for example, a phenotypically favourable effect of a transposase could plausibly be argued to have been present from the earliest times of transposase evolution. Along these lines, we have speculated that the effect of Tn*5* may involve a secondary effect of transposase or inhibitor, which comes about from the ability of these proteins to interact with host DNA (Biel & Hartl 1983). The Tn*5* effect, it should be noted, is strain-specific; in lineages derived from *E. coli* strain CSH12, Tn*5* appears to be selectively neutral in chemostats. This opens up the possibility of genetic analysis of the host as well as the transposon, with particular regard to their interaction. If, for Tn*10*, the mutation-based interpretation of L. Chao et al (unpublished) is confirmed, it would provide an example of a different class of IS sequences having evolved as mobile mutators or, more generally, as *cis*-acting regulatory elements. Whether this mode of selection ought to be considered as selection at the individual level or at the population level is arguable, but it would imply a second way in which such transposable elements have been earning their keep from the very beginning of their existence.

All this leads to a model of accessory DNA evolution that is essentially coevolution, which Gilbert & Raven (1975) have defined as 'the dynamics of the evolutionary relationships that have led to a given situation and to the reciprocal modifications that have taken place in the participating organisms'. Here we are dealing not with independent organisms but with a commonwealth of organisms and a diverse group of potentially autonomous DNA elements, but the principles involved are quite similar. In our view, the subtle favourable effects of transposons observed in chemostats represent individual selection of the sort that favoured the establishment and proliferation of such DNA elements in the first instance. Once their over-replication and horizontal transmission became established, the elements took on a dynamic of their own, as elaborated in the theory of selfish DNA. However, at virtually the same time, the elements established and became part of the web of genetic exchange among members of the commonwealth, which provided, and continues to provide, an additional impetus of population-level selection, fostering further elaboration and refinement of the elements. Like many evolutionary hypotheses, this model of accessory DNA evolution may not be verifiable *in toto*, but it has the virtue of being specific, and various of its assumptions and implications are testable. As far as we are aware, the hypothesis is consistent with all pertinent evidence so far put forward.

Acknowledgements

This work was supported by NIH grants (GM24886, AI14267, and AI18980). Fig. 1 is based on data in: Biel SW 1981 The discovery and study of a selective advantage conferred by Tn*5* in carbon source limited chemostat cultures. PhD thesis, Purdue University, West Lafayette, Indiana.

REFERENCES

Berg DE, Davies J, Allet B, Rochaix JD 1975 Transposition of R factor genes to bacteriophage λ. Proc Natl Acad Sci USA 72:3628-3632

Berg DE, Johnsrud L, McDivitt L, Ramabhadran R, Hirschel BJ 1982 Inverted repeats of Tn*5* are transposable elements. Proc Natl Acad Sci USA 79:2632-2635

Biel SW, Hartl DL 1983 Evolution of transposons: natural selection for Tn*5* in *Escherichia coli* K12. Genetics 103:581-592

Campbell A 1980 Some general questions about moveable elements and their implications. Cold Spring Harbor Symp Quant Biol 45:1-9

Campbell A 1981 Evolutionary significance of accessory DNA elements in bacteria. Annu Rev Microbiol 35:55–83

Chao L, Cox EC 1983 Competition between high and low mutating strains of *Escherichia coli.* Evolution 37:125-134

Davey RB, Reanney DC 1980 Extrachromosomal genetic elements and the adaptive evolution of bacteria. Evol Biol 13:113-147

Doolittle WF 1982 Selfish DNA after fourteen months. In: Dover GA, Flavell RD (eds) Genome evolution. Academic Press, London, p 3-28

Doolittle FW, Sapienza C 1980 Selfish genes, the phenotype paradigm and genome evolution. Nature (Lond) 284:601-603

Dykhuizen DE, Hartl DL 1983 Selection in chemostats. Microbiol Rev 47:150-168
Dykhuizen D, Campbell JH, Rolfe BG 1978 The influence of a λ prophage on the growth rate of *Escherichia coli*. Microbios 23:99-113
Edlin G, Lin L, Kudrna R 1975 λ lysogens of *E. coli* reproduce more rapidly than non-lysogens. Nature (Lond) 255:735-737
Edlin G, Lin L, Bitner R 1977 Reproductive fitness of P1, P2 and Mu lysogens. J Virol 21:560–564
Gilbert LE, Raven PH 1975 General introduction. In: Gilbert LE, Raven PH (eds) Coevolution of animals and plants. University of Texas Press, Austin, p ix-xiii
Hartl DL 1977 Applications of meiotic drive in animal breeding and population control. In: Pollak E et al (eds) Proc Int Conf Quant Genetics, Iowa State Univ Press, Ames, p 63-88
Isberg RR, Lazaar AL, Syvanen M 1982 Regulation of Tn*5* by the right-repeat proteins: control at the level of the transposition reaction? Cell 30:883-892
Jaenisch R 1983 Endogenous retroviruses. Cell 32:5-6
Johnson RC, Yin JCP, Reznikoff WS 1982 Control of Tn*5* transposition in *Escherichia coli* is mediated by protein from the right repeat. Cell 30: 873-882
Lin L, Bitner R, Edlin G 1977 Increased reproductive fitness of *Escherichia coli* lambda lysogens. J Virol 21:554-559
Melling J, Ellwood DC, Robinson A 1977 Survival of R-factor carrying *Escherichia coli* in mixed cultures in the chemostat. FEMS (Fed Eur Microbiol Soc) Microbiol Lett 2:87-89
Mitsuhashi S 1971 Epidemiology of bacterial drug resistance. In: Mitsuhashi S (ed) Transferable drug resistance factor R. University Park Press, Baltimore, p 1-23
Orgel LE, Crick FHC 1980 Selfish DNA: the ultimate parasite. Nature (Lond) 284:604-607
Reanney DC 1978 Coupled evolution: adaptive interactions among the genomes of plasmids, viruses, and cells. Int Rev Cytol (suppl 8):1-68
Sandler L, Novitski E 1957 Meiotic drive as an evolutionary force. Am Nat 91:105-110
Zund P, Lebek G 1980 Generation time-prolonging R plasmids: correlation between increases in the generation time of *Escherichia coli* caused by R plasmids and their molecular size. Plasmid 3:65-69

DISCUSSION

Davies: Was the effect on the λ lysogen the same order of magnitude as you see with Tn*5*?

Hartl: Yes.

Davies: So would you argue that you may actually select for lysogens in that way?

Hartl: That is the argument of Edlin et al (1975). The experiments with phages λ and Mu were done before the full panoply of λ and Mu mutations were available and it was never analysed genetically. Nowadays enough mutations are available to allow the genetic basis of the effect to be sorted out. The few experiments that have been done suggest that the effect in λ requires the *rexA* cistron.

Clarke: It would be nice to know if there was an advantage of having two genetic elements. Do we have the dose–response curve, so to speak?

Hartl: We've done experiments of that sort and at least for the small number of copies that we have looked at—one, two and three—the effect of IS*50* is roughly additive. That is, having two copies doubles the rate of selection; having three copies triples it. I'm sure that there would eventually be a plateau, and if enough copies of the same element are accumulated they would probably become detrimental.

Gressel: If you used a copy of a different element, e.g. Tn*10* or Tn*5*, what would happen?

Hartl: We have not done this.

Datta: You lumped P1 together with the other elements but that does not insert into the chromosome, does it?

Hartl: There are two problems with that: P1 is maintained as a plasmid rather than inserted into the chromosome; and the particular derivative of P1 that was used carries Tn*9*, so a transposon effect is involved as well.

Datta: If the selective advantage was provided by the plasmid, that might account for plasmids in the Murray collection of old bacteria (Hughes & Datta 1983) having survived there. On the other hand they are likely to contain transposases; we haven't looked at that.

Southwood: Your paper reminds me of Lotka-Volterra competition equations between co-existing animal species. There is one condition where the outcome of competition depends very much on the starting densities. If you start a bit later on, i.e. with different densities, do you get a different outcome? Have you looked at this at all against the Lotka-Volterra framework?

Hartl: We have looked at the effect of initial frequency and have not found any with Tn*5*. For Tn*10* there does seem to be an effect of initial frequency that grossly resembles what one finds with mutator genes. With mutator genes in chemostats, it turns out that genes that increase the mutation rate confer an advantage because the mutator-bearing strain of *Escherichia coli* is more likely to sustain a mutation that is favourable for growth in the chemostat. The most likely mutations to be selected are those conferring an ability to stick to chemostat walls, or to use some of the citrate in the buffer, or for the cells to clump together and, hence, not to wash out of the chemostat. If one initiates a chemostat with the frequency of the mutator strain that is below a certain threshold, then the normal strain will be more likely to sustain that mutation spontaneously because of its vast numerical preponderance. In those cases the mutator gene will lose, in competition. At an initial frequency above that threshold, the mutator gene will begin to increase in frequency by 'hitch-hiking' along with the favourable mutations. With Tn*10* if the initial frequency of the Tn*10*-bearing strain is less than 1 in 10^3, the Tn*10*-bearing strain loses. If it is greater than 1 in 10^3, the strain increases in frequency (L.

Chao et al, unpublished work, 1983). We have done similar experiments with Tn*5*, relative to initial frequencies, but have not found anything similar.

Graham-Bryce: You say that the antibiotic resistance-bearing element in Tn*5* is essentially fortuitous, but if one were to follow Dr Davies's line of thought, that element might have arisen originally from some producer species and somehow been carried along. If that is correct, is it likely that the function of the middle sections of Tn*5*, for which you have not yet identified any purpose, might be revealed if you were to challenge with the right kind of material?

Hartl: That is conceivable.

Southwood: Or you could challenge with some other organism, again bearing in mind Lotka-Volterra types of experiment. Here you have something close to two beetles of the same genus; if you were to bring in another unrelated beetle, this might suddenly reverse the relative performances of the two. Professor Datta has mentioned that although there is not always a great disadvantage attached to one of these antibiotic-resistance genes, once it is left in a person's gut without the antibiotic challenge then the non-resistant form tends to become dominant again.

Hartl: What I meant is that it is *fortuitous* that it happened to be IS*50* that flanks the kanamycin-resistance gene. The selection for kanamycin resistance is extremely important. The selective forces impinging on genes *now* need not necessarily reflect the kinds of forces that impinged on them originally. In my paper I was trying to suggest that the things that impinge on the insertion sequences by themselves are quite distinct from the selective forces that impinge on them once they mobilize an antibiotic-resistance gene, or a heavy-metal resistance gene, or something else that is of phenotypic advantage to the organism.

Datta: So the conclusion is that the genes are not entirely selfish because they pay 'rent'?

Hartl: Yes, as I said, they earn their keep!

REFERENCES

Edlin G, Lin L, Kudrna R 1975 λ lysogens of *E. coli* reproduce more rapidly than non-lysogens. Nature (Lond) 255:735-737

Hughes VM, Datta N 1983 Conjugative plasmids in bacteria of the 'pre-antibiotic' era. Nature (Lond) 302:725-726

Final general discussion

Clarke: We have seen how studies in one area can illuminate studies in another. People working on insecticides can perhaps gain new insights from those working on herbicides. Studies on microorganisms, which can be analysed in greater experimental detail, may contribute much to our understanding of evolution in higher organisms. In passing, there has been mention of heavy-metal resistance genes in bacteria, carried by plasmids. Can any parallels be drawn between heavy-metal resistance in bacteria and in plants?

Datta: The mechanism for mercury resistance in bacteria is very clearly worked out: it is a complex detoxification, with several enzymic steps (Summers & Silver 1978). Perhaps similar mechanisms operate in plants, but I don't know.

Bradshaw: I am ignorant of resistance to mercury in higher plants so my reaction is to examine the published work on microorganisms in order to understand the plant possibilities more.

Davies: In yeast, levels of heavy-metal resistance can be amplified specifically by tandem duplications. This can happen also in bacteria for some resistance mechanisms but has not been reported for metal resistance. The biochemical basis to the mechanism of mercury resistance in yeast is not known. The tandem duplication was demonstrated by specific hybridization experiments (Fogel & Welch 1982).

Hartl: Has anyone looked to see whether metal mine sites have been colonized by soil bacteria?

Bradshaw: I am not sure about metal-resistant bacteria, but several people have looked at fungi and found them to have specific resistances.

Davies: Lead and copper mine sites are full of interesting bacteria, in fact, and are valuable sources of resistant organisms. Silver-resistant bacteria, in addition, can be obtained from the effluent of photographic film manufacturing and developing factories. The resistant organisms convert silver salts to the metal that precipitates around colonies on agar plates.

Bradshaw: Nevertheless, we must realize that there are limitations to

---

1984 Origins and development of adaptation. Pitman Books, London (Ciba Foundation symposium 102) p 246-252

adaptation. The copper-resistant lawns that grow around the copper refinery that I mentioned in my paper are growing on an enormous thickness of springy, undecayed organic matter. The vegetation that was growing there in 1955 can be found perfectly preserved underneath. So although there may be evolution of resistance by the microorganisms, this evolution is limited. Of course we exploit this by using copper as a wood preservative.

Gressel: Most degradation of lignin is done by fungi. Professor Georgopoulos has pointed out that resistance to copper sulphate (Bordeaux mixture) has not occurred in fungi that have been sensitive to copper.

Georgopoulos: I would expect that the fungi found in copper mines are resistant *species*, and not resistant *mutants* of sensitive species. This is probably true also of fungi growing in the polluted parts of Canada that Professor Hutchinson described.

Hutchinson: The ponds at the Smoking Hills that I mentioned (see Hutchinson, this volume) contain a chemical soup at a pH of about 1.8 (Havas & Hutchinson 1983). We can isolate the alga *Euglena mutabilis* from there, but it always contains a fungal (yeast) contaminant. When we finally were able to get rid of this contaminant and produce axenic *Euglena*, the growth of the *Euglena* dropped dramatically. It turns out that the contaminant is always a particular *Cryptococcus*. Collections of *E. mutabilis* from six other very acidic sites in North America always had the same yeast growing with it. When the two organisms are grown together in the laboratory there is a remarkable mutualistic effect, and they will grow even at pH 1.0. The heavy-metal tolerance of each is also remarkably enhanced by the other. We can grow the two species in solution at pH 2 in the presence of 1000 p.p.m. of aluminium, lead, cobalt or zinc etc. An obvious question is how the yeast enhances the metal tolerance of *Euglena*, and whether the reported occurrence of intracellular bacteria in *E. mutabilis* also functions in this. There could even, speculatively, be a transfer of genetic material between all these. Furthermore, we have isolated an alga *Chlorella saccharophila* from the roots of the grass *Deschampsia cespitosa*, growing at Sudbury. The alga has, like the grass, a multiple metal tolerance, including co-tolerances that we could not have predicted from the substrate on which they occur in the field. So here may be another similar example for the study of genetic transfer. In our studies of *E. mutabilis*, we needed a control and so we used a stock culture of *E. mutabilis* from the Cambridge algal collection. This grew miserably until we put it together with the yeast from the Smoking Hills and then it grew extremely well. Researchers in Germany have studied intracellular bacteria in *E. mutabilis*, and found that the same stock of Cambridge *Euglena*, through 30 years of culturing, had lost the intracellular bacteria as well as the *Cryptococcus*. It is very likely that in nature, *E. mutabilis* depends on a combination of organisms for survival in these harsh environments which mutually provide the chemical environment and nutrients for survival. Studies of single organisms may often overlook such natural selective mixtures.

Foy: Many stress-tolerance differences in plants may be related to the microbial populations on their roots. For example, rice can oxidize manganese and protect itself from manganese toxicity by excluding the element from the interior of roots. Soya bean cannot do this. If soya bean and rice are grown together, the rice roots attract, oxidize and detoxify the excess manganese and therefore protect the soya bean roots (Doi 1952). This oxidation of manganese may well be microbial in origin. However, in varieties of beans, cotton and wheat that we have studied, high internal tolerance to manganese appears more important than exclusion as a protective mechanism.

Wheat and barley varieties that tolerate aluminium raise the pH of their root zones and cause detoxification of excess aluminium by precipitation. Hence, an exclusion mechanism seems to be involved. Aluminium-sensitive varieties of the same species tend to lower the pH, which increases aluminium solubility and toxicity. If aluminium-tolerant Dayton barley and aluminium-sensitive Kearney barley are grown in the same pot of vigorously aerated nutrient solution, the final solution pH is about midway between the pH values for the two pure cultures of Dayton and Kearney; but the Dayton still grows better than the Kearney when the shared solution contains aluminium. This suggests either that pH change is not important in aluminium tolerance or that each variety maintains its own and different root-zone pH which is not destroyed even when the solution is swirled around the plant roots. I support the latter view. A pH increase of only 0.1 unit in a thin layer of solution surrounding the root can decrease the aluminium solubility by 50%, and hence reduce toxicity.

Aluminium is known to interfere with DNA replication, presumably by binding the esteric phosphate component of the molecule. But before aluminium can reach the DNA it must cross two membranes, the plasmalemma and the nuclear membrane, or membranes of other DNA-containing organelles such as mitochondria. In certain barley genotypes aluminium tolerance is controlled by one major, dominant gene (Reid 1971). Ideally, we need to characterize the membranes of aluminium-tolerant and sensitive barley genotypes structurally, electrochemically and enzymically to understand aluminium tolerance mechanisms.

Graham-Bryce: I have been struck by the range of resistance mechanisms suggested and demonstrated in bacteria. One tends to think of bacteria as simple compared to higher organisms but they have a diverse range of resistance mechanisms, some of which presumably are possible only because of that simplicity and might not operate so easily with greater degrees of organization.

Foy: In selecting plant genotypes for tolerance to various stress factors, perhaps we should look specifically at microbial associations on their roots, and particularly at different species and strains of mycorrhizae which can markedly increase the uptake of phosphate and water (Nelson & Safir 1982).

Bell: Recently, Chiariello et al (1982) described the transfer of phosphorus from one plant to another through common mycorrhizal filaments.

Foy: Mycorrhizae are found on the roots of most crop plants growing in soils. Perhaps we have been ignoring something important here because we seldom study plants under sterile conditions. The tolerance of plants to stress could be greatly influenced by specific mycorrhizal populations and their interactions with specific plant genotypes, as well as by the inherent properties of the host plant itself (Davis et al 1983).

Clarke: I guess that the outcome of associations between organisms can appear to be genetic characteristics of the host, with simple inheritance. We might be looking in the wrong places.

Southwood: One of the parallels throughout this symposium has been between the adaptation of insects to different host plants and the similar phenomenon of the development of resistance in pests to herbicides or insecticides, and to drugs on the part of bacteria. Let us not forget that many of these plant-feeding insects have inside them internal symbionts, which may be very important. We may be looking only at an outside package, when it is the inside parts that are changing, using some of these mechanisms that we have heard about for microorganisms.

Clarke: I believe that the insect *Mayetiola destructor* has a gene-for-gene relationship with maize (its host), very similar to the gene-for-gene relationships between fungi and their plant hosts (Hatchett & Gallun 1970).

Foy: In relation to induced responses such as production of phytoalexins in response to stress, is it possible that we are talking about one mechanism of toxicity control for metals? This could be a chelation mechanism by which the plant uses natural organic acids or other binding compounds to metabolically detoxify a metal that is already in a position to do some damage. Would metal stress condition the plant to produce more of these compounds, which would then protect it? We have preliminary evidence, from high-pressure liquid chromatography, that aluminium-tolerant Dayton barley responded to the metal by producing more total organic acids, while the aluminium-sensitive Kearney barley did not (E.H. Lee & C.D. Foy, unpublished data). Is the organic acid content of a plant fixed as an inherent property, or is it inducible? Isoenzymes would fall into the same category.

Clarke: One knows that some plant enzymes are inducible, such as alcohol dehydrogenase in roots under anaerobic conditions.

Foy: Proline accumulates under drought stress.

Hartl: Our meeting has revealed the remarkable diversity of adaptations that are possible. I wonder whether any general principles of adaptation have emerged over the last few days. It struck me in Dr Davies's paper that bacteria seem to use the same set of enzymes over and over again, and they steal enzymes from some other species or some other genus when needed. It is conceivable, surely, that a synthetic antibiotic could be produced against which no enzyme is available, or which would require too many simultaneous steps to evolve.

Bell: This is how a general protein toxin behaves; it denatures all proteins within a living cell. This is why heavy-metal tolerance is limited; all proteins in the cell will eventually be denatured by certain ions. This is in contrast to those molecules that affect one metabolic step only.

Graham-Bryce: But decreased permeability of the organism could act as a general mechanism of resistance which could affect compounds of almost any class.

Clarke: We ought to remember here the archaebacteria being dug out of hot springs under the Atlantic. Their maximum growth rate is at a temperature much higher than the boiling point of water (Baross & Deming 1983).

Bell: Well presumably some of the relevant protein has not been denatured at that temperature!

Bradshaw: Darwin, in chapter 15 of his 'Origin of species', said he could see no limits to this process of adaptation. But despite the extraordinary adaptations that we have heard about, I still think we must accept that limits exist.

Hutchinson: I personally feel that there probably are no limits to adaptation. The two constraints on the biochemical or genetic possibilities are: opportunity and time. If we take these constraints into account, then I would suggest there are no limits.

Wolfe: We have to beware of trivialization in this argument. Older plant pathology textbooks commonly advise that if diseased material is present it should be uprooted and burnt. It is difficult to imagine many pathogens adapting to that constraint.

Epstein: Practically all the examples of adaptation discussed at this symposium are in response to stresses imposed on organisms. We have said little about responses to opportunities such as the eutrophication of a lake, which gives it a high concentration of nutrient salts and, hence, a high productivity. In an oligotrophic lake there may be low levels of nutrients and little growth, but human development can lead to the eutrophication of the lake. The organisms in the lake then adapt, and use the new opportunity to expand. In this kind of adaptation, new niches are formed. We have said very little about the exploitation of such niches. Perhaps the imposition of a stress is a more clear-cut and quantifiable type of study for experimentalists! But, in nature, and through human activity as well, new niches *are* created and organisms *do* adapt.

Bowers: So we could ask if there is such a thing as evolution towards a more adaptable or more fit condition which is entirely dissociated from stress.

Epstein: In nature it is probably always some sort of a mixture.

Kuć: I wonder how productive it is to probe deeply into early evolution. I feel that if we probe too deeply into the beginnings of things we come up against spontaneous generation! Rather, we should emphasize problems of today. Nature tends to bury toxic materials far below the surface, as the human race now tries to do.

Sawicki: Although adaptation is beneficial to a species because it enables it to survive in a changing environment, the survival of such a species can be detrimental to other species of its biotype. The development of insecticide resistance in insect pests is one such example. Resistance has made pest control more difficult and expensive. The need to overcome this form of adaptation was stressed in the seventh report of the Royal Commission on Environmental Pollution (1979). As scientists we should find ways of overcoming this most unwelcome form of adaptation.

Clarke: If all life were wiped off the earth tomorrow, I believe that in a few million years' time, even the Sudbury smelter area would be populated with organisms.

Southwood: This is where the influence of time, which Professor Hutchinson mentioned, becomes apparent.

Bradshaw: Most coastal mudflats have no plants on them. Yet they have been available to angiosperms for as long as they have existed. There have been one or two spectacular exceptions such as mangroves, but that is all. In Britain the startling evolutionary event of the last 50 years has been the appearance of *Spartina anglica* on these mudflats. We know that this has happened by hybridization and polyploidy (Marchant 1968). But it remains extraordinary that the habitat has defied evolution so long.

Hutchinson: One of the problems to be faced in Britain is that we are dealing with a depauperate flora, in which the number of species available to perform these tricks of adaptation is remarkably small, compared with those in most other parts of the world. There are geological and historical reasons for that. If the exposed mine sites in the UK were in Africa, I would suggest that they would be vegetated fairly quickly.

Bradshaw: Perhaps this is true. One of the interesting strengths of bacteria, therefore, is that they have a gigantic gene pool at their disposal, unlike the plant life in the UK, for example!

REFERENCES

Baross JA, Deming JW 1983 Growth of 'black smoker' bacteria at temperatures of at least 250 °C. Nature (Lond) 303:423-426

Chiariello N, Hickman JC, Mooney HA 1982 Endomycorrhizal role for interspecific transfer of phosphorus in a community of annual plants. Science (Wash DC) 217:941-943

Davis EA, Young JL, Lindermann RG 1983 Soil lime level (pH) and mycorrhizal effects on growth responses of sweetgum seedlings. Soil Sci Soc Am J 47:251-256

Doi Y 1952 Studies on the oxidizing power of roots of crop plants. 1. The differences of crop plants and wild grasses. 2. Interrelation between paddy rice and soybean. Proc Crop Sci Soc Jpn 21:12-13, 14-15

Fogel S, Welch JW 1982 Tandem gene amplification mediates copper resistance in yeast. Proc Natl Acad Sci USA 79:5342-5436

Hatchett JH, Gallun RL 1970 Genetics of the ability of the hessian fly, *Mayetiola destructor*, to survive on wheats having different genes for resistance. Ann Entomol Soc Am 63:185-194

Havas M, Hutchinson TC 1983 The Smoking Hills: natural acidification of an aquatic ecosystem. Nature (Lond) 301:23-27

Marchant CJ 1968 Evolution in *Spartina (Gramineae)*. II. Chromosomes, basic relationships and the problem of *S.X. townsendii* agg. J Linn Soc Lond Bot 60:381-409

Nelson CE, Safir GR 1982 Increased drought tolerance of mycorrhizal onion plants caused by improved phosphorus nutrition. Planta (Berl) 154:407-413

Reid DA 1971 Genetic control of reaction to aluminum in winter barley. In: Nilan RA (ed) Barley genetics II. Washington State University Press, Pullman (Proc Int Barley Genet Symp 1969) p 409-413

Royal Commission on Environmental Pollution 1979 (Chairman: Sir Hans Kornberg), Seventh Report. Agriculture and pollution, 7644. HMSO, London

Summers AO, Silver S 1978 Microbial transformations of metals. Annu Rev Microbiol 32:637-672

Chairman's closing remarks

B. C. CLARKE

Department of Genetics, School of Biological Sciences, University of Nottingham, University Park, Nottingham NG7 2RD, UK

Professor Arthur Cain, as I have mentioned already (p 14), wrote a very illuminating paper called 'The perfection of animals' (Cain 1964), and I would like to start these remarks with a plea for its wider circulation. Cain's argument is that organisms can respond much more easily to the evolutionary demands of their environments than most people suppose. It is an important point to make, because there has been much counter-argument in favour of random genetic drift and neutrality. An illustration of the virtue, and the difficulty, of this point is given by our attempts to understand the evolution of proteins, discussed here by Julian Davies.

It has always seemed to me a beautiful and amazing thing that when the amino-acid sequences of similar proteins from different species are lined up with each other, we can count the numbers of differences and can make a 'tree' of resemblance that roughly corresponds to the evolutionary tree derived from comparative anatomy and palaeontology. Contained within a single gene is a potted history of the organism that carries it. There are two alternative explanations for this extraordinary circumstance. One is that protein evolution is, essentially, a stochastic process, producing an evolutionary clock that 'ticks' every time an amino-acid substitution occurs: the substitutions take place at random, but can nevertheless measure time in the long-term in a manner similar to the measuring of geological time by the random process of radioactive decay. A second explanation is that each protein in each organism is highly constrained by the genetic, biochemical and ecological environment in which it finds itself, that there is consequently a 'viscosity' in its evolution, and that where a molecule can go depends on where it has come from. Thus even a molecule entirely subject to natural selection could contain information about its own history. Studies of haemoglobin by Max Perutz and others have shown that many parts of the molecule have important functions: the molecule is not just carrying oxygen; it is

1984 Origins and development of adaptation. Pitman Books, London (Ciba Foundation symposium 102), p 253-257

binding diphosphoglycerate and chloride ions, binding other haemoglobin molecules, influencing the structure of the erythrocyte, affecting susceptibility to malarial parasites, and so on. Yet this molecule still contains a history of the organism.

These observations stress the importance of pre-adaptation. Indeed, a first lesson from this symposium is that pre-adaptation is more important than we had supposed. Tony Bradshaw has discussed genostasis, and has noted that many plants have apparently reached a limit to their adaptation. Others, however, have progressed further because of a predetermined ability to do so; an ability that, although it may be a consequence of natural selection, is fortuitous with respect to the new challenge. The case for pre-adaptation is strengthened by Tom Hutchinson's remarks about the super-resistant *Deschampsia* that appears to have a ready-made tolerance to several environmental threats. The same case has also been reinforced by Spyros Georgopoulos, who described the 'low-risk' fungicides, to which no resistance has yet evolved, and by Jonathan Gressel's report on the relative inability of insects to become resistant to the phenoxy herbicides compared to the triazines (although there are, as he has pointed out, many complexities in explaining these differences).

In our last discussion Tony Bradshaw mentioned the grass *Spartina anglica* which now thrives in a habitat that has been unoccupied by grasses for several million years. The appearance of *S. anglica* seems to have been a chance concatenation of two genomes that enabled it to invade a new habitat. This, again, is a sort of pre-adaptation.

I can illustrate the perils of pre-adaptation by considering the glazing of my mother's windows. She lives in a small village, with very narrow streets and with a large density of transient lorries. She has installed double-glazed windows to keep out the noise, but these windows also help to save heating costs and to deter burglars. If the price of oil should rise to new heights, the effect would probably be: (1) to decrease the incidence of lorries; (2) to increase the savings on fuel; and (3) probably, to increase the incidence of burglars consequent upon the ensuing economic depression. Thus an observer in the future might with reason interpret her double-glazing purely as a fuel-saving or a burglar-preventing device, yet this would not be a true reflection of its original purpose and history. Double-glazed houses are pre-adapted to saving fuel, although in Britain this role is not yet cost-effective. Where I live we have no lorries; therefore we have not double-glazed our windows. If and when the price of fuel goes up we might well go 'extinct' (bankrupt) because our house is not pre-adapted to fuel-saving or preventing burglary.

In using this example I am simply illustrating that something selected for one purpose can subsequently turn out to be useful for another. This is a common phenomenon in the economics of organisms, but one that is often forgotten.

For me, the second lesson of this meeting is its support for the view that the coevolution of predators and prey, of hosts and parasites, and of plasmids and bacteria is a very potent evolutionary force, as illustrated by Arthur Bell's elegant examples of allelopathy, Dick Southwood's fascinating discussion of *Heliconius* butterflies and passion-flower vines, and Dan Hartl's experiments on the exotic adjustments between accessory DNAs and their bacterial hosts. Coevolution may be a major factor in generating the pre-adaptations mentioned above. An insect that has been subject, say, to juvenile hormone analogues produced by plants may well be less susceptible to Bill Bowers's evil brews than an insect that has not been so exercised.

A third lesson is about the precision of co-evolution and its consequences. Joseph Kuć's paper on phytoalexins seems to me to approach for the first time a physiological and biochemical explanation of gene-for-gene relationships between plant parasites and hosts, and it suggests that these relationships are based, as in animals, on the distinction between 'self' and 'non-self'. There is, perhaps, a general rule that if the 'intelligent' parasite wants, so to speak, to persuade the host that it is not foreign, it must mimic the host, either through mechanisms such as those described by Joseph Kuć, or in vertebrate animals through immunological mechanisms, or in other organisms through more conventional biochemical processes. The immediate effect on a genetically uniform host of a parasite that specializes in mimicking that host, or at any rate in adapting to its biochemical environment, is to generate a selective pressure on the host which favours genetic diversity. An antigenic or biochemical change may make the parasite recognizable, or less well adapted, and therefore more easily overcome. There would then be, for the host, an advantage in being different from the norm. This point was first made by J. B. S. Haldane in 1949 and thereafter has been much neglected. It draws attention to what may be a very potent force in generating diversity. An example of a polymorphism probably maintained by this force is the major histocompatibility complex, a highly polymorphic group of genes that occurs in most, perhaps all, vertebrates. The histocompatibility polymorphism is known to be associated with disease resistance, and appears to be particularly concerned with viral infections. Recently it has been found that there are special molecular mechanisms for increasing the amount of histocompatibility variation by the process of gene conversion.

If it pays to be different, parasite–host interactions can generate polymorphisms in parasites as well as in hosts, and can also cause divergence between species. Host organisms in such conditions need to be different from their neighbours; otherwise they will catch diseases from them. Naomi Datta's incompatibility groups of plasmids probably exemplify a force causing divergence between parasites. Mutant forms that differ in the control of replication may be favoured because they escape from repression by related plasmids. Mechanisms of this sort will cause organisms that do not share a gene pool to diverge from each other, and may explain a good deal of evolutionary diversification.

Geneticists have recently been much interested, for practical reasons, in the restriction enzymes of bacteria, which are widely used in genetic engineering. However, they also present a fascinating evolutionary problem. What forces have driven them to recognize different sequences of DNA? We have already mentioned, in the discussion of Julian Davies's paper, the forces that drive the GC compositions (the percentages of guanosine and cytosine) in bacteria. One might ask whether plasmids have driven the GC contents of their hosts by requiring the hosts to diverge from each other, the selective force being the consequences of incorporating foreign DNA into the host genome. This, in turn, could influence the composition of plasmid DNA.

The fourth lesson is less easy to state simply. The mechanisms that generate adaptive characteristics are much more diverse and ingenious than we had supposed: Roman Sawicki described a response to selection by apparent duplication of genes; Charlie Foy spelled out the complexities of adaptations to manganese, aluminium and iron; Ernie Hodgson discussed delicate adjustments among the cytochromes; and Julian Davies and Naomi Datta both reported responses to selection by transfers of information between species. It is a sign of the times, and of the progress made in the last 25 years, that we are now seriously discussing the transfer of genes, even between widely divergent groups. We know that transposable elements can be incorporated into the prokaryote and eukaryote genome, and that there may be movement between mitochondrial DNA, chloroplast DNA and nuclear DNA. There are 'uncomfortable' resemblances between plant haemoglobins and insect haemoglobins that may suggest transfers of genes even between the kingdoms. If transfers between different eukaryotic groups do indeed turn out to be common, they may help to explain many hitherto puzzling phenomena. Perhaps the evolution of resistance to antibiotics in bacteria is a model, for example, of the evolution of melanism, more or less simultaneously, in two hundred species of moths (Lees 1981). Perhaps there is a role for transposable elements in the origin and evolution of supergenes. The ability of whole ecosystems to respond to particular environmental changes may be aided by rare but important interchanges of DNA. The possibility of interspecific gene transfer as a mechanism that generates some of the building blocks of adaptation is perhaps the most interesting and disturbing thought that I shall take away from this symposium.

REFERENCES

Cain AJ 1964 The perfection of animals. Viewpoints Biol 3:36-63
Haldane JBS 1949 Disease and evolution. Ric Sci Suppl 19:68-76
Lees DR 1981 Industrial melanism: genetic adaptation of animals to air pollution. In: Bishop JA, Cook LM (eds) Genetic consequences of man-made change. Academic Press, London, p 129-176

Index to contributors

Entries in **bold** *type indicate papers; other entries refer to discussion contributions*

Bailey, J. A. 49, 114, 116, 117
Bell, E. A. 38, **40,** 48, 50, 51, 96, 115, 116, 133, 134, 135, 148, 149, 150, 215, 249, 250
*Berg, D. E. 233
Bowers, W. S. 48, 49, 51, 97, 98, **119,** 132, 133, 134, 135, 136, 164, 251
Bradshaw, A. D. **4,** 15, 16, 17, 18, 32, 35, 36, 67, 68, 69, 71, 92, 95, 97, 99, 134, 136, 150, 151, 186, 187, 200, 201, 202, 214, 246, 250, 251

Clarke, B. C. **1,** 14, 32, 34, 37, 38, 48, 51, 67, 87, 89, 90, 91, 92, 93, 97, 98, 113, 115, 132, 134, 135, 136, 137, 147, 148, 150, 162, 163, 164, 165, 184, 186, 187, 188, 202, 203, 214, 217, 228, 230, 243, 246, 249, 250, 251, **253**

Datta, N. 92, 93, 114, 115, 151, 201, 202, **204,** 214, 215, 216, 217, 218, 228, 229, 230, 231, 244, 245, 246
Davies, J. E. 15, 16, 69, 87, 95, 97, 114, 117, 200, 215, 216, 217, **219,** 228, 229, 230, 231, 243, 246
*Denholm, I. **152**
Dittrich, V. 202
*Dykhuizen, D. E. **233**

Elliott, M. 17, 95, 134, 135, 136, 137
Epstein, E. 33, 36, 99, 229, 250, 251

Fowden, L. 32, 69, 71, 147, 149
Foy, C. D. **20,** 32, 33, 34, 35, 36, 37, 48, 67, 70, 71, 91, 248, 249

Georghiou, G. P. 88, 95, 116, 162, 184, 185, 187, 203, 217
Georgopoulos, S. G. 17, 48, 88, 89, 93, 97, 185, **190,** 200, 201, 202, 203, 247
Graham-Bryce, I. J. 16, 71, 90, 92, 117, 136, 150, 163, 201, 202, 215, 216, 245, 248, 250
*Gray, G. **219**
Gressel, J. 15, 33, 35, 38, 69, **73,** 87, 88, 89, 90, 94, 96, 98, 99, 116, 117, 133, 135, 137, 163, 164, 165, 185, 200, 203, 228, 229, 230, 244, 247

Harborne, J. B. 50, 137
Hartl, D. L. 15, 35, 91, 96, 97, 135, 150, 184, 185, 202, 203, 214, 216, 217, 230, **233,** 243, 244, 245, 246, 250
Hodgson, E. 95, 114, 163, 165, **167,** 185
Hutchinson, T. C. 17, 34, 35, 37, **52,** 67, 68, 69, 70, 71, 72, 96, 97, 201, 202, 216, 247, 250, 251

Kuć, J. 18, 48, 50, 51, 69, 98, **100,** 113, 114, 115, 116, 117, 135, 148, 149, 151, 165, 203, 216, 217, 229, 231, 251

*Motoyama, N. **167**

Sawicki, R. M. 37, 50, 51, 90, 96, 116, 134, 135, **152,** 163, 164, 165, 166, 184, 185, 186, 187, 214, 217, 230, 251
Southwood, T. R. E. 49, 132, 134, 137, **138,** 148, 149, 150, 151, 244, 245, 249, 251

Wolfe, M. S. 33, 36, 250
Wood, R. J. 38, 87, 88, 94, 95, 149, 183, 184, 186, 201, 229, 230

*Non-participating co-author
Indexes compiled by John Rivers

Subject index

Abies balsamea 125, 135
Acer pennsylvanicum, dust deposition and 55
Acer rubrum, dust deposition and 55
Acer saccharum
  dust deposition and 55
  reproduction, acid rain and 64
Acetylcholinesterase 153, 154, 165, 166
  altered 168–170, 179
  inhibition 169, 170, 176
  insensitive 158, 159, 168
Achillea borealis, adaptation to altitude 12
Acid rain 54, 56
  foliar buffering capacity and 64
  pollination and 64
  SO_2 and 63, 71
  tolerance to 64, 65
Acinetobacter spp. 207, 210
Acylalanines 195
Adaptation
  evolutionary 1–3, 4, 9, 10, 11, 12, 91
    responses to opportunities 250, 251
    See also Fitness
  lethal environment and 1, 2, 11, 12
  limits to 9, 12, 18, 250
  reciprocal 141
  See also Animal, Insect, Plant
Adaptationists 2, 3
Adenostoma fasciculatum 44
Ageratum, anti-juvenile hormonal compounds in 127–129, 133
Agrobacterium tumefasciens 216
Agrostis spp., metal-binding proteins in 70
Agrostis stolonifera, copper-tolerant populations 7
Agrostis tenuis
  heavy metal-tolerant populations 5, 6, 17, 18, 92
  selection against tolerance 9
  variability for tolerance 10, 36
Air pollution
  plant adaptation to 52–72
    mechanisms of 64, 65
  pollutant interactions 63
  responses of communities and taxa 54–58
  responses of species and populations 58–63
  species tolerant to 53
Alfalfa *See Medicago sativa L.*
Algae, metal tolerant 62, 63, 69, 70
Aliesters 173
Alkaloids, plant 49, 50, 124, 135, 149
Allelopathy 43–45, 136
  non-protein amino acids in 45
Alopecurus myosuroides, triazine-resistant 82
Alopecurus utriculatus, triazine-resistant 82
Alternaria solani 42
Aluminium tolerance
  accumulators 70
  cultivars 23, 24, 26, 27
    short-strawed 23, 24
    yield and 23, 24, 33
  exclusion and 70
  genetic control 25
  mechanisms 28, 248, 249
  nutrient efficiency and 24
  selection for 26, 27, 35
Aluminium toxicity 21, 22, 24, 62
Amaranthus hybridus, atrazine-resistant 87
Amaranthus retroflexus 79, 80
  triazine-resistant 82
Amikacin 221, 224
  resistance 207
Amino acids, non-protein, allelopathy, in 45
Aminoglycosides 207–212, 221
  enzymes modifying 221–227
Aminoglycoside acetyltransferases 223, 227
Aminoglycoside phosphotransferases 223, 224, 226
  amino acid sequences 224, 225, 228, 230
  evolution 224
  gene, nucleotide and protein sequences 224, 226, 227, 230, 231
    guanosine–cytosine (GC) content 226, 256
  nucleic acid sequence homologies 224, 226, 227
  protein sequence homologies 224, 226, 227
  resistance factors 224
Aminoglycoside resistance 221
  biochemical mechanisms 222, 223
  genes 227
  mutational blocking in 229

α-amino-β-methylaminopropionic acid 46
α-amino-β-oxalylaminopropionic acid 46
α- and β-aminoproprionitrile 46
Ampicillin resistance 209, 210, 211
Anabasine 124
Anguria and *Passiflora* vines, *Heliconius* butterflies and 141, 142
Animal adaptation, local environment, to 37, 38
Anthoxanthum odoratum
  copper tolerance in 7
  heavy-metal-tolerant populations 5, 8
  rapid and local evolution 12
Anthracnose, *C. lindemuthianum*, immunization against 109
Antibiotic-producing organisms 220, 224, 226
Antibiotic resistance 204–218
  cross-resistance in 215, 216
  determinants 220, 227
  evolution 219–232, 235
  mechanisms 219, 220, 221, 224
  mutations and 205, 206
  plasmids and 206–209, 212, 217
  resistance genes and 210–212
  transposons and 209, 210
Anti-ecdysone 132
Anti-juvenile hormones in plants 121, 123, 127–129
Apramycin 221
Arachidonic acid, phytoalexins and 117
Arcostaphylos glandulosa 44
Arsenic toxicity to plants 62
Artemesia californica 43, 44
Artemesia tilesii, tolerance to acid rain and SO_2 64, 96, 97
Asclepias syriaca 124
Aspergillus nidulans 197, 199
Aspergillus terreus 129
Astragalus spp. 45
Atmospheric pollution *See Air pollution*
Atrazine resistance 79, 83, 87
Atropa bella-donna 124
Azetidine-2-carboxylic acid 44, 45, 46

Bacillus spp. 207, 208
Bacillus circulans 226
Bacteria
  antibiotic-resistant 204–218
  genetic stability 216, 217
  growth rate, accessory DNAs and 237–242
  heavy metal-resistant 246
  resistance mechanisms 248
  restriction enzymes of 256
Bacterial conjugation 205, 206, 208
Bacterioides spp. 207
Bacteriophage, evolution 238, 241, 243, 244
Baikiain 46
Balsam fir 125, 135
Barley
  aluminium tolerance in 25, 34, 35, 36, 249
  saline tolerance in 33
  variability in 36
  See also Hordeum
Bellis perennis, limits to herbicide resistance in 12
ben-A gene 197
Benomyl 193, 195, 197
Benzimidazoles 193, 194, 195
  resistance gene for 197
Benzofurans 102
Benzyl penicillin resistance 210
Betula alleghaniensis, dust deposition and 55
Betula papyrifera, reproduction, acid rain and 64
Bipyridillium resistance 82
Boophilus microplus 153
Brachypodium distachyon, triazine-resistant 82, 84
Brassica campestris 80
Buddleia, insects on, normal host-plant relationships of 144
Bussea spp. 45
Butirosin 224, 226
Butterflies
  cabbage white 143
  Helioconius, Passiflora and *Anguria* vines and 141, 142
  swallow-tail, coumarins in plants and 141, 144
  trap lining by 141

Cabbage looper, resistance to 51
Cadmium, interactions with other pollutants 63
Cadmium-excluding plants 22
Caesalpina tinctoria 45
Caffeic acid 48
  derivatives 102
Calcium efficiency in plants 34
  genetic control 26
Callosobruchus maculatus 148, 149
Caloplaca cerina, air pollution and 56
Caloplaca phlogina, air pollution and 56
Camphene 44
Camphor 44
Campylobacter spp. 207
Canavanine 45, 46, 149
Capsidiol 103
Carbamates 153
  N-methyl 170
  N-propyl 170
  resistance 154, 156, 163, 170

Carbamate inhibitors 169
Carbaryl 176
Carbendazim 197
Carbenicillin resistance 210
Carboxamides, resistance to 197, 198
Carboxin resistance 197, 198
Carboxylesterase E4 154–156, 163, 164, 165, 166, 173
  gene amplification in 155, 163
3-Carboxytyrosine 46
Carya spp., dust deposition and 55
Caryedes brasiliensis 149
Catalytic centre activity 153, 154, 155
Catechol 41
Cement works, effect on forestation 53
Cephalosporin resistance
  mutation to 205
  plasmid-determined 207, 208
Ceratitis capitata, cross-resistance in 183, 184
Cercis canadensis, dust deposition and 55
Cercospora beticola
  benzimidazoles and 193, 199
  dithiocarbamates and 196
  fentins and 195, 196, 199
Chagas' disease vector 128
Chemostat 238, 240, 241, 242, 244
Chenopodium album 80
  triazine-resistant 82, 83
Chickens, limits to adaptation 12
Chloramphenicol acetyltransferases 227
Chloramphenicol resistance 205
  genes 227
  plasmid-determined 207, 208, 210, 211
Chloramphenicol-treated fungi 198, 200
Chlorella saccharophila 247
  heavy-metal-tolerant 63
1-chloro-2, 4-dinitrobenzene (CDNB) 172, 173
Chlorogenic acid 42
Chlorosis, iron-deficiency 21, 26, 27
Cholinesterase, altered 168–170, 179
Chromenes 127
Chromones 102
Chromosomal breakage and recombination, stress and 96, 97, 98
Chrysanthemum, source of natural insecticides 122, 123, 134, 135, 136, 137
Chrysopa carnea, tolerance for pyrethroids 89
1,8-Cineole 44
Cladosporium cucumerinum 105, 106
Clostridium spp. 207
Cobalt toxicity 62
Coevolution, host–parasite 141, 144, 148, 149, 150, 151, 255
  diffuse 142, 148, 149
Coleophoridae 139
Coleoptera 139, 140
Colletotrichum circinans 41
Colletotrichum lagenarium 105, 106, 110
Colletotrichum lindemuthianum 104, 108, 109
Colonization
  result of evolution 5, 6, 12
  speed of 6–8, 16
Compactin 129
Compositae, phytoalexins in 101, 102
Conjugation, bacterial 205, 206, 208
Convallaria majalis 45
Convolvulaceae, phytoalexins in 101, 102
Copper tolerance 7, 9, 15, 16, 17
  accumulation in cell 70
  binding proteins and 70
  copper requirement in 93
  cotolerance for other metals 18
  grass species, in 9, 10, 11, 247
  supertolerance 9, 18
Copper toxicity to plants 5, 15, 16, 54, 62
Corpus allatum 127, 128
Corynebacterium spp. 207
Cotton, manganese tolerance in 28
Coumarins in plants, swallow-tail butterflies and 141, 144
Cruciferae 144
Cryptococcus 247
Cucumber mosaic virus 106, 116
Cucurbits, immunization against disease 106, 109, 110, 114, 116
Culex pipiens 166
Culex quinquefasciatus
  esterase resistance in, regression of 95
  phosphate-resistant, gene amplification in 162
Cyanogenic glucoside/enzyme systems in legumes 136
Cyanogenic glycosides 42
Cytochrome *P*-450 177, 181
  -dependent monooxygenase system 168, 176, 177
  genetic variants 179, 180
  isozymes of 176, 180, 181
Cytotoxins 128, 129

Dactylis glomerata, rapid and local evolution 12
Darwinian evolutionary theory 3, 4, 6, 13
DDT resistance 157, 179, 186
DDT-resistant mosquitoes, fitness and 94, 95
DDT dehydrochlorinase 156, 157
Dectes texanus 143, 144
Defensive chemicals 121

Deguelin 122
Deh gene 186
Dehydro-juvabione 125
Delphinium, limits to colour variability 12
Demeton-*S*-methyl 163
2′-Deoxyluteone 41
Deschampsia cespitosa
metal binding proteins 70
populations 69
tolerance
air pollution, to 53, 69
heavy metals, to 62, 69
multi-metal 62, 63, 68, 202, 247, 254
SO_2, to 62, 63, 202
Deschampsia flexuosa, heavy metal tolerance in 68
α, γ-diaminobutyric acid 46
α, β-diaminopropionic acid 46
Diazinon 157, 158, 172, 173
resistance 157, 158, 176
Diazoxon 176
1,2-Dichloro-4-nitrobenzene (DCNB) 172, 173
Dictyoneurida 139
Dieldrin resistance 183, 184
Dieldrin-resistant mosquitoes, fitness and 94, 95
Dienochlor 202
Digitaria sanguinalis 79
Dihydrostreptomycin 221
3,4-dihydroxyphenylalanine 46
Dimethoate 157, 158
resistance 158
Dimethoxon 158, 159
Dioclea megacarpa 149
o-Diphenol 41
Diptera 140
Dithiocarbamates 196
DNA, accessory elements, evolution 233–245
bacterial growth rate and 237–242
gene transfer, horizontal, in 234, 235, 242
genetic exchange and 236, 237, 242
over-replication in 236, 242
selfish DNA 235, 236, 241, 242
traits acquired by 237
traits carried by 234, 237
transposable elements 233, 235, 238–242
See also Bacteriophage, Plasmids, Transposons
Drosophila
cross-resistance in 183
hybrid dysgenesis in 217
mutation to resistance 230
resistance
gene location and 184
multifactorial 184
Drought tolerance 22, 26, 28, 34
Drug resistance, multiple 204, 205
Dry bean, iron efficiency in 26
Dust deposition, effects on trees and shrubs 54, 55

Echinochloa crus-galli 79
triazine-resistant 83
Endomeiosis 165
Endothia parasitica 41
Enduracididine 46
Enterobacteriae, resistance in 212
Environment
animal adaptation to 37, 38
plant adaptation to 10, 11, 12, 37, 38, 49, 91, 92
Environmental variables 2
Enzymes
detoxication 180, 181
isozymes of 181
non-oxidative 176, 179
oxidative 132, 133, 176
xenobiotic-metabolizing 168*ff*.
Equisetum arvensa 124
Eragrostis curvula (Schrad.) Nees, 22, 23
iron-efficient 27, 28, 34
Erwinia tracheiphila 106
Erysiphe gramins f. sp. *hordei* 196
Erythromycin resistance 207, 208
cross-resistance 215
Escherichia spp. 207
Escherichia coli 205, 209, 210
chromosomal resistance to streptomycin 96
plasmids and 237
transposon Tn*5* and 238–241
transposon Tn*10* and 244
Esterases 162, 163, 164, 165, 168, 173–176, 184, 187, 188
functions in resistance 165, 176
resistance 95
See also Acetylcholinesterase, Carboxylesterase
Ethirimol 196
O-Ethyl *O*-*p*-nitrophenyl phenylphosphonothioate (EPN) 171
Ethylene, exposure to 72
Euglenia mutabilis, heavy-metal-tolerant 63, 247
Evolution
colonization as result of 5, 6, 12
limits to 11

Farnesal 125
β-Farnesene 147

Farnesol 125
Fentins 195
Fenvalerate 174, 176
Festuca arundinacea, aluminium tolerance in 27
Festuca ovina, evolutionary cost 12
Fish poisons 122
Fitness 8, 13, 76, 77, 78
  acquisition 90, 91
  definition 92
  environment and 91
  relative, stability of resistance and 94–99
  resistance and 92, 93, 164, 165, 193, 194
Flavonones 102
Fluoride pollution, tolerant varieties 58
Food shortages 29
Formamide hydrolase 42
Fungal infection, recognition of self and non-self 108
Fungal pathogens, interactions with pollutants 63
Fungi
  adaptation to fungicides 190–203
  copper tolerance and 17, 247
  resistance to 48, 49, 50, 51
  toxicity 49
  See also under named fungi
Fungicidal compounds in plants 48, 49
Fungicides 190
  acylalanines 195
  adaptation of fungi to 190–203
    'all-or-none' adaptation 191
    relative adaptation 194, 195
  benzimidazoles 193, 194, 195, 197
  carboxamides 197
  dithiocarbamates 196
  effectiveness, loss of 191, 193, 194, 195, 196
  'high-risk' 191
  'low-risk' 196, 200, 201, 203, 254
  mechanism of action 190, 191
  'moderate-risk' 194, 195, 203
  multifunctional 203
  organotonins 195
  persistence 201
  phthalimides 196
Fungicide resistance
  biochemical mechanisms 196–199
  cross-resistance 200
  degrees of 191, 195
  genetic control 196–199
Furanoacetylenes 102
Furanocoumarins 141, 144
Fusarium oxysporum f. sp. *cucumerinum* 106
Fusidic acid resistance 207

Gall-forming insects 139
Gene amplification 153, 155, 156, 162, 163, 164
Gene transfer, horizontal 149, 150, 234, 247, 256
Gene variability *See Variability*
Genostasis 13, 254
Gentamycin 221
  resistance 207, 208, 210, 211, 212
Geranium carolinianum, genetics of SO_2 response 59
Germination, spaced 75, 76
Germplasm, stress-tolerant, development and release 26, 27
Gilia capitata, adaptation to serpentine soils 12
Glucan elicitors 106, 107, 109
Glucosinolate 42
Glutamic acid 46
α-Glutamyl-β-aminopropionic acid 46
Glutathione, conjugation 170–173
Glutathione *S*-transferase 168, 171, 176, 181, 184
  isozymes of 173
Glyceollins 102, 103
Glycine max. L. 21, 143
Gossypol, antifungal factor, as 42
Gossypium hirsutum L.
  aluminium tolerance in 23
  manganese tolerance in 28
Grass(es), heavy-metal-tolerant 62, 69, 70
Grass tetany 22, 24
Grasshoppers, tolerance for DDT 89
Green bean, immunization against anthracnose 109

Haemophilus spp. 207, 210
Haemophilus influenzae 211, 216
Hamamalis virginiana, dust deposition and 55
Heliconius butterflies 141, 142
Helminthosporium sacchari, resistance to 51
Hemarthria altissima (Poir.) Stapf & C. E. Hubb
  aluminium and cold-tolerant 23, 37
  local environment and 37
Hemiptera 140
Herbicides
  degradation 75, 84, 88
  fitness 76, 77, 78
  persistence 75, 76, 88
  phenoxy- 73, 74, 75
  resistance
    enrichment for 75–80, 85, 88, 90
    evolution 73–93
    factors affecting 75, 76

Herbicides (*contd.*)
resistance (*contd.*)
incidence and conditions for 80–82
mode of action 82
types 74, 88, 89
selection pressure 75, 76, 77, 78, 80, 88, 89
selectivity 74, 75
spaced germination and 76
tolerance 79, 80, 88, 89
See also Triazine
High input: maximum output agriculture 21
Homoarginine 46
Hordeum spp., limits to disease resistance 12
Hordeum vulgarum L. 23
See also Barley
4-Hydroxyarginine 46
Hydroxymethylglutaryl-CoA reductase, inhibitors of 129
3-(3-Hydroxymethylphenyl)alanine 45
5-Hydroxynorleucine 46
4-Hydroxyproline 46
Hylemya antiqua 140

Immunization in plants 106, 109, 110, 111, 112, 114
specificity 115
transmission 116
Inhibitins 41, 42
Insect(s)
endocrine system
plant chemistry and 121
plant secondary compounds and 124–129
gall-formers 139
hormonal mimics in plants 123, 124–127, 132
–insecticide adaptation 152–166
juvenile hormones 121, 123, 125, 164
analogues 124–127, 133, 134
biosynthesis 128, 129
interference in production, transport and action 127
leaf-miners 139
moulting hormones (ecdysones) 125
phytophagous *See Phytophages*
–plant adaptation 138–151, 249
coevolution 141, 142, 144, 148, 149, 150, 151
gene transfer in 149, 150
host-plant shift, 143
plant defences 141, 143
pre-adaptive insect characteristics 145
removal of plants to new areas 142, 143, 150
species–area relationship 143
–plant interactions 119–137
competition and cooperation 120
resistance mechanisms in 168 *See also Insecticide resistance*
stenophagous 143
Insecticide(s) 152
carbamate 153, 154, 156
multi-functional 203
natural 121–124
organophosphorus 153, 154, 156, 157, 158, 163, 164, 168–176
plants, in 48, 49
Insecticide resistance 152–166, 251
biochemical mechanisms 167–189
gene linkage in 181, 183, 184
regulator genes 179, 180
cross-resistance 156–159, 174, 176, 178, 183
definition 159, 160, 162
detection 160, 161
gene amplification in 153, 155, 156, 162, 163, 164
gene selection for 187
kdr gene in 157
knockdown resistance 157, 179
multifactorial 184, 185
multiple 156, 157, 159
multiplicate 156, 157, 159
origins 153
polyploidy in 164
R (resistance) and S (susceptible) enzymes 153, 156, 158, 159
catalytic centre activity 153, 154, 155
instability of production 155, 156, 164
qualitative differences 153, 154
quantitative changes 154–156
recommended dose and 159, 160
Insertion sequences 209, 210, 220
bacterial growth rate and 238–241, 244, 245
Ipomeamarone 103
Iron deficiency 21, 22, 24, 33, 48
Iron efficiency 26, 27, 34, 48
mechanisms of 28
Iron toxicity 21
Isoflavonoids 102
Isothiocyanate 52, 51
Isowillardine [β-(uracil-3-yl)alanine] 46
Isoxazolin-5-ones 46

Juglans nigra 43, 44
Juglone 43
Juvabione 125, 126, 132, 133, 135, 136
Juvadecene 125, 126
Juvenile hormone *See Insect juvenile hormone*
Juvocimenes 125

Kanamycin 221, 224
  resistance 207, 208, 210, 211
  resistance gene 238, 240, 245
Kasugamycin 194, 195
kdr gene 157, 186, 187
Kievitone 102, 103
Klebsiella spp. 207, 231
Kleidocerys resedae 144
Kochia scoparia, triazine-resistant 82

β-Lactams, resistance genes 227
β-Lactamases 208, 211, 221, 231
Lactuca sativa 45 *See also Lettuce*
Lathyrus aphaca 46
Lathyrus odoratus 46
Lead
  -tolerant plant species 59
  toxicity to plants 5, 16
Leaf-miner insects 139
Leaf-rust resistance 34
Leaf structure
  natural selection and 147
  resistance to insect predators and 147
Lecanora conizaeoides, SO_2-tolerant 57
Lecanora dispersa, pollution-tolerant 57
Leguminosae
  coumarins in 141
  phytoalexins in 101, 102, 107
Lepidoptera 139, 140, 141
Leptinotarsa decemlineata 143
Lettuce
  disease resistance in 48, 49, 50
  manganese tolerance 26
  toxicity 48, 49, 50
Leucaena leucophylla 46
Lichens, air pollution and 53, 55, 56
Limestone quarrying, effect on forestation 53, 54, 55
Lincomycin resistance 207
  cross-resistance in 215
Liriodendron tulipifera, dust deposition and 55
Lividomycin 221, 224, 226
Lolium spp., adaptivity 15
Lolium perenne 5
  copper tolerance 16
  herbicides, tolerance to 15
  SO_2 tolerance and 59, 60
Lolium rigidum
  herbicides, tolerance to 15
  triazine-resistant 82
Lophochloa phleoides, triazine-resistant 82
Lotus corniculatus 42
Lubimin 103
Lupinus spp., pollution and 59
Lupinus albus 41
Luteone 41
Lycopersicon esculentum Mill., iron deficiency in 21
Lycopodium clavatum 124
Lysopersicum spp. 42

Macrolides, resistance genes 227
Macropiper excelsum 126
Magnesium efficiency in plants 22, 34
Maize, aluminium tolerance, genetic control 25
Major histocompatibility complex 255
Malathion 174, 176
  resistance 184
Maneb 196, 201, 203
Manganese tolerance 26
  mechanisms of 28, 248
Manganese toxicity 24
Marchantia polymorpha, lead tolerance in 59
Mayetiola destructor–maize relationship 249
Medicago sativa L. 22, 23
  aluminium tolerance in 35
  manganese tolerance in 26
  selection for tolerance 27
Meiotic drive 235
Melandrium rubrum, limits to adaptation 12
Melanism, industrial, in moths 13
Mephospholan 164, 202
Mercury resistance 211, 214
  bacteria in 246
Metalaxyl 192
Metals, heavy
  tolerance in bacteria 246
  tolerance in plants
    cost of 8, 9, 32, 33
    cotolerances 17, 18, 62
    definition 18
    development 5, 6–8, 13, 246, 249
    fitness and 97
    generalized 18
    level of 16
    limits to 9, 10, 18
    mechanisms 17
    multiple 62, 70
    plant root and 17, 28
    selection against 9
    specific 17, 18
  toxicity, plant adaptation to 4–19, 22, 202
  See also under specific named metals
Metallothionein 70
6-Methoxymellein 103
Methylbutyrate 173
Methyl chavicol 126
Methyl cinnamate 126
Mevinolin 129
Mexican bean beatle, DDT-tolerant 89

Microorganisms, resistance in 95, 96
Microsomes, housefly, oxygenase activity in 176, 177, 178, 179
Mimosine 46
Mimulus guttatus, tolerant to air pollution 53
Mine sites, adverse characteristics of 202
Mineral stress
  plant adaptation to 20–39
  research on 24
  tolerance to 23, 24
    genetic control 25
    mechanisms of 27–29
    yield and 33
Minimum input agriculture 21, 23, 24, 33
Mites, adaptation in 202
Mitochondria, chloramphenicol-treated 198, 200
Mitochondrial electron transport 197, 198
  cyanide-insensitive 199
Molybdenum deficiency 21
Monooxygenase reactions 176
Mosquitoes, relative fitness and resistance 94–99
Mosses, air pollution and 55, 56, 57
Mudflats, plant life on 251
Musca domestica 154, 157, 158, 169, 171, 176, 186, 187
  resistant strains
    gene location in 179, 184
    monooxygenase activity 177–180
  sex determination in 188
Mustard oils 144
Mutant aliesterase hypothesis 173
Mutation to drug resistance 205, 206, 229, 230
Mycobacteria spp., resistance in 206
Mycorrhizae, plant stress tolerance and 249
Mycosphaerella melonis 106
Mycosphaerella pinodes 109
Myzus persicae
  phosphate resistance, gene amplification in 154–156, 186
  progeny of, variation in 217

Nalidixic acid resistance, mutation to 205
Naphthalene, exposure to 72
Naphthylacetate 154, 156, 165, 173
Neisseria spp. 207
Neisseria gonorrhoeae 205, 211, 212
Neomycin 221, 224
  resistance 207, 226
  resistance gene 238, 240, 245
Neonotonia (Glycine) wightii 46
Nephotettix cincticeps 169, 173
Netilmycin 221
Nickel toxicity to plants 54, 62, 68
Nicotine 49, 50, 124, 135
Ninhydrin reacting compounds 46
Nitrogen efficiency in plants 22
  genetic control 26
Nornicotine 124

Oats, iron efficiency in 26
Obtusaquinone 130
Obtusastyrene 130
Ocimum basilicum 126, 132, 133
Omethoate 163
Opl gene 159
Optical difference spectra 177
  Type I 177–180
Organophosphate inhibitors 169
Organophosphorus insecticides 153, 154, 156, 157, 158, 163, 164, 168–176
  resistance to 154–156, 157, 162
Organotonins 195
Oryza spp., limits to disease resistance 12
Oryza sativa L. 21
Oscinella frit 140
Ostrya virginiana, dust deposition and 55
Oxidant pollution 57, 58
Oxidant tolerant varieties 58
Oxidases, mixed function 123, 124, 128, 129, 132, 133, 176, 178, 184
Oxydendrum arboreum, dust deposition and 55
Oxyria digyna, evolutionary cost 12
Ozone
  exposure 59
  –SO_2 interactions 63
  tolerant varieties 58, 59

Papilio machaon complex 141, 144
Paraoxon 154, 155, 159, 174
Paraquat
  persistence lacking in 82
  resistance 82
Parasitism, evolution and 2
Parathion 157, 158, 172
  resistance 157, 158, 230
Paromomycin 221
Parthenium argentatum 44
Passiflora and *Anguria* vines, *Heliconius* butterflies and 141, 142
Pasteurella spp. 207
Penicillin resistance 214, 211, 229
  cross-resistance in 215
  fitness and 93
  mutation to 205
  plasmid-determined 207, 208, 219
Penicillinase 205, 215, 221, 229, 231
  TEM 209, 210, 211
Penicillium brevicompactin 129

Pentac 202
Isopentenylphenols 130
Permethrin 159
  resistance 95
Permian period, plant feeding by insects in 138, 139
Pesticide resistance 88, 89, 90
  competitive ability and 90, 92
  insects, in 13
Phages 233–245
Phalaris paradoxa, triazine-resistant 82
Pharbitis, arachidonic acid and 117
Phaseollin 102, 103
Phaseolus aureus 45
Phaseolus lunatus 41
Phaseolus vulgaris L.
  aluminium tolerance in 23, 36, 37
  manganese tolerance in 28
Phenolic compounds 41, 44, 51
Phenylpropanoid phenols 102
Pheromone alarm, aphids, of 147, 148
Phorodon humuli 163
Phosphate resistance *See Organophosphorus resistance*
Phosphorus
  efficiency in plants 22, 34
    genetic control 26
  fixation in soils 22
Phosphotriesterase 173, 174
Phthalimides 196
Phytoalexins 41, 48, 100–118, 249
  accumulation 101, 104, 105, 107, 114
    direct stimulation and 106
    necrotization of plant tissues in 107, 111
    plant–fungus interactions, in 106, 107, 111
    suppression of 109
  anti-microbial activity 104, 106
  detoxification by fungi 109
  elicitation 105, 107, 111, 117
    suppression of 117
  induced susceptibility and 108, 109
  response to mechanical damage 115, 116
  selective toxicity 105
  sensitization to 106, 111, 113, 114, 115
  stress metabolites, as 107, 111
  structural considerations 101–106
    specificity and 102, 103, 110, 114, 115
  synthesis 104, 105
    cell death and 117
  virus diseases unaffected by 111
Phytophages
  attachment and 139, 141
  desiccation and 139, 140
  food quality and 140, 141
  plants affected by 140
Phytophthera infestans 41, 42, 106, 108, 109, 149
Phytophthera megasperma f. sp. *glycinea* 108
Phytophthera phaseoli 41
Phytotoxins 106, 107, 114
Phytotuberin 103
Pieris brassicae 143
Pieris rapae 143
α- and β-Pinene 44
Pinus banksiano, reproduction, acid rain and 64
Pinus ponderosa
  oxidant pollution and 57
    tolerant species 54
Pinus strobus
  dust deposition and 55
  oxidant-sensitive 57
  reproduction, acid rain and 64
  SO_2 and 54
Pisatin 102
  mutant pea plants producing 114
  suppression of production 109
Pisum sativum 46
Plant adaptation
  air pollution, to 52–72
  heavy-metal toxicity, to 4–19, 246
  insects, to *See Plant–insect adaptation*
  local environment, to 37, 38
  mineral stress, to 20–39
  stress, to 100, 101
    microbial factors 248, 249
    selection and propagation and 22, 35
Plant alkaloids 49, 50, 124, 135, 149
Plant breeding
  fertilizer effectiveness and 22
  specific traits, for 22
Plant cell regeneration, selection in 98, 99
Plant compounds, fungicidal and insecticidal 48, 49
Plant defensive chemicals 120–124, 130
Plant, disease resistance in 100, 101, 104, 107, 108, 110
  genetic potential for 101, 111, 114
  hypersensitive response in 111
  immunization and 106, 109, 110, 111, 112
  multicomponent aspects 101, 110
  specificity and 108, 110, 111
Plant–insect adaptations 138–151, 249
  coevolution 141, 142, 144, 148, 149, 150, 151, 255
  gene-for-gene relationships 255
  gene transfer in 149, 150
  host–plant shift 143
  plant defences 141, 143
    protease inhibitors and 148
  pre-adaptive insect characteristics 145

Plant–insect adaptations (*contd.*)
removal of plants to new areas 142, 143
species–area relationship 143
Plant, insect feeding and 140
defences against 140, 141
Plant–insect interactions 119–137
competition and cooperation 120
Plant pathogens, susceptibility to *See Susceptibility*
Plant–plant interactions 40–51
disease resistance in 40, 41, 48, 49, 50, 51
ecological situation and 48
higher plant–higher plant 43–46
phytotoxins and 44–46
higher plant–lower plant 40–43
beneficial associations 43, 49
inhibitins 41, 42
parasitism and 48, 49
post-inhibitins 41, 42
prohibitins 41
response to attack 48
Plant populations, selection in, inbreeding and 96
Plant secondary compounds 120–124, 141, 144
ecological role 134, 135, 149
insect endocrine processes and 124–129, 130
metabolic dumps, as 134, 136, 137, 141
Plantago lanceolata
heavy-metal-tolerant populations 5
lead tolerance in 59
Plasmids 205, 206–209, 233–245
classification 208, 209
conjugative pili 206, 209, 214
-determined antibiotic resistance 206–212, 214–217, 219
dissemination 234, 237
evolution 219
F factors 212, 215, 216
gene evolution in 237, 238
genes, transposable *See Transposable elements, Transposons*
incompatible 208, 209, 211, 214, 255
mechanisms of resistance 207, 208, 215, 216, 217
pre-antibiotic bacteria, in 212, 214, 244
promiscuous 208
R factors 205, 211, 212
resistance genes 206, 207, 209, 210–212, 214, 224
surface exclusion 208
Plausibility 2, 3
Poa annua
evolutionary cost 12
tolerance to herbicides 15
Poa pratensis L. 43
variation of tolerance to stresses 22
Pohlia nutans, acid tolerance in 54
Pollination, acid rain and 64
Pollutant–fungal pathogen interactions 63
Pollutant–pollutant interactions 63
Pollutant–virus interactions 63
Polyacetylenes 102
Polygonatum multiflorum 45
Polygonum cilinode, acid tolerance in 54
Polypogon monspeliensis, triazine-resistant 82
Post-inhibitins 41
Preadaptation 253, 254, 255
Precocenes 127, 128, 129
oxidative activation 130, 133
Proallatotoxins 127, 128
synthetic 129, 130
Prohibitins 41
Proline, salt stress, in 48
Prolyl-tRNA synthetases 45
Propoxur (oxidase) resistance 95
Prostaglandins, phytoalexin elicitation and 117
Protease inhibitors, plant resistance to insects and 148, 149
Protocatechuic acid 41
Proteus spp. 207
Providencia spp. 207
Pseudomonas spp. 207, 209
Pseudomonas aeruginosa 210, 211
Pseudomonas lachrymans 106
Pseudoperonospora cubensis 106, 192
Pyrethrins 122, 123, 136
plant protection and 134, 135, 137
resistance to 95
Pyrethroids 154, 157, 159, 160
resistance 157, 184, 186, 187
Pyricularia oryzae 194
Pyrrhocoris apterus 125

Quercus spp.
insect adaptation to 145
tannin production in 148
Quercus prinus, dust deposition and 54, 55
Quercus rubra, dust deposition and 54, 55

Ranunculus acris, limits to climatic adaptation 12
Reproduction, plant, pollutants and 64
Resistance
acquired 201, 202
definition 88–90, 167, 186
energetic cost 95, 96
fitness and 92, 93

genes 201
evolution 206, 207, 209, 210, 214, 219–221, 224
world-wide spread 210–212
gene frequency and 185, 186
knockdown 157, 179
Mendelian 184, 185, 186
multifactorial 184, 185
multiple 204, 205
polyfactorial 185
polygenic 185
stability, relative fitness and 94–99
Resorcinol-type compound 41
Rhodnius prolixus 128
Rhododendron maxima, dust deposition and 55
Ribosomal protein mutants 96, 97
Ribostamycin 221
Rice, stress tolerances in 34
See also Oryza
Rifampicin resistance, mutation in 205
Rishitin 103
Root-zone pH
aluminium tolerance and 34, 35, 248
iron efficiency and 48
Rotenone 122
Rumex acetosella, metal-tolerant populations 5
Rumex obtusifolia, SO_2 tolerance in 59
Rye, aluminium tolerance in, genetic control 25

Safynol 102
Salinity 22
tolerance 33, 99
Salix caprea 144
Salmonella spp. 207, 210, 211, 214
Salvia leucophylla 43, 44, 48
Sassefras albidum, dust deposition and 55
Scopolin 42
Scytalidium, copper tolerance and 17
Secondary plant compounds 120–124
ecological role 134, 135
insect endocrine processes and 124–129
Selection
against tolerance 8, 9
for tolerance 7
habitat developing 11, 12, 13
indirect 135
natural 2, 3, 10, 13
Selection pressure 75, 76, 77, 78, 80, 87, 88, 89, 90
Selenium
accumulator plants 45, 70
toxicity 62
Selfish DNA hypothesis 235, 236, 241, 242
Senecio vulgaris 75, 80
triazine-resistant 83, 84
Sensitization *See Immunization*
Serratia spp. 207
Sesamex 176
Sesamin 123, 125, 136
Sesamolin 123, 125
Sesquiterpenoid phytoalexins 102, 109
Sheep, environmental adaptation 38
Shigella spp. 207, 214, 231
Shigella flexneri 205, 211
Silene vulgaris, tolerant populations 5, 16
Silver, toxicity to plants 62
Silver-resistant bacteria 246
Simazine tolerance 80
Sisomycin 221
Smog, effect on plant life 52, 53
Snails, environmental adaptation 37
Soil
acidity 21, 22, 23, 26, 36, 54, 62, 63
alkalinity 22
conditions
crop development and 20, 21
plant adaptation and 22, 23, 24
variations in 32
dry 22, 26
hard 22
heavy-metal pollution 22, 59, 62, 63
saline 22, 33
wet 22
See also Subsoil
Solanaceae
alkaloids 41
phytoalexins in 101, 102
Solanum spp. 42, 143, 147
limits to disease resistance 12
Solanum berthaultii, hairy leaf structure 147
Solanum nigrum, triazine-resistant 83
Sorghum, stress-tolerant populations 26, 35
Soybean
aluminium tolerance and 25, 35
iron efficiency in 23, 26, 27, 29
manganese tolerance and 26, 28
See also Glycine max. L.
Spartina anglica 251, 254
Spectinomycin 210, 221
Sphagnum spp., acidic pollution and 57
Stachys 144
Staphylococcus spp. 207, 208, 209
Staphylococcus aureus 204, 215, 224, 231
Stellaria media, limits to herbicide resistance 12
Stemphylium loti 42, 48
Steroid glycoalkaloids, potato, in 149
Stilbenes 102
Streptococcus spp. 207, 209

Streptococcus fradiae 224, 226, 228, 231
Streptococcus lividus 226
Streptogramin B resistance 207
  cross-resistance 215
Streptomycetes, antibiotic-producing 224, 226
Streptomycin 221
  resistance 204, 205
    chromosomal 96, 97
    mutation in 205
    plasmid-determined 207, 208, 210, 211
Stress tolerance 23, 24
  mechanisms of 27–29, 33, 34
Stylosanthes spp.
  aluminium tolerance in 35, 37
  plant adaptation and 37
Subsoils, problems of, plant adaptation in 21, 22
Succinate dehydrogenase 197, 198, 199
Sudbury smelter
  effect on humans 67
  effect on plants 53, 54, 61, 67
    adaptation and 67, 68
Suicide substrates 128, 129, 130
Sulphonamide resistance 215, 228, 229
  plasmid-determined 207, 208, 211
Sulphur secretion, wild horseradish, in 49, 50
Sulphur dioxide
  interactions with other pollutants 63
  pollution 54, 56, 61
  selective factor, as 54, 58
  sensitivity, acute and chronic 71
  tolerant varieties 58, 59, 60, 65
Superoxide dismutase, SO_2 tolerance and 65
Super-tolerance 9, 18
Susceptibility of plants to pathogens
  induced 108, 109
  induced resistance and 109, 111
  phytoalexin metabolism and 109
  recognition of self and non-self 108, 111, 115, 255
  specificity and 115
  stress mechanisms and 107, 108, 111
  suppression of resistance and 109, 164
  toxin production and 109
Sweet basil 126, 132, 133
Synergists 123

Tanacetum cinerariifolium (chrysanthemum) 122, 123, 134, 135, 136, 137
Tannins
  antifungal agents, as 41
  British oaks, in 148
Temephos resistance 95
Terpenes, pines and conifers, in 136
Terpenoids 102, 129
Tetrachlorvinphos 169
Tetracycline resistance 205
  plasmid-determined 207, 208, 210, 211, 219
Tetrahydrolathyrine 46
Tetranychus urticae 154, 159
Thiabendazole 197
Thuja plicata 126
Thujic acid 125, 126
Tobacco, insecticide, as 124
Tobacco horn worm 49
Tobacco necrosis virus 106, 110
Tobacco plant
  disease susceptibility 48, 49, 50
  immunization against blue mould 109, 114
  ozone tolerance in 58, 59
  susceptibility to insect attack 135
Tobramycin 221
  resistance 207
Tolerance
  cost 8, 9, 12
  definition 88–90, 167, 186
  evolution 6–8
  limits to 9, 10
  mechanisms 17
  selection against 8, 9
  variability and 10, 12, 14, 15
  See also Resistance
Tomato
  aluminium tolerance in 25, 36
  iron efficiency and 26, 29
Toxins, microbial, plant disease, in 109
Trans-cinnamic acid 44
Transposable elements 233, 235 *See also Transposons*
Transposases 238, 240, 241, 244
Transposons 209, 210, 211, 212, 214, 215, 216, 224, 230, 231, 233–245, 256
  expression of 231
  Tn*1* and Tn*3* 210
  Tn*5* 224, 231
    bacterial growth rate and 238–241, 245
  Tn*7* 211
  Tn*10*, bacterial growth rate and 241, 244
  Tn*903* 224, 231
Trap lining 141
Tree bark, lichen substrate, as 56, 57
Triazine
  binding protein 83
  resistance 74, 75, 79, 81, 82, 85, 88, 89
    biochemistry and genetics of 82–85, 87
    chloroplast membrane protein and 87
    gene flow in 87
    maternal inheritance 83, 88
  selection pressures 80, 87, 88

tolerance 80, 83, 88, 89
biochemistry and genetics of 82–85
See also Herbicides
Trichomes 141, 144, 147
Trifolium repens
adaptation to low-fertility soil 12
rapid and local evolution 12
Trimethoprim resistance 205, 206
plasmid-determined 207, 208, 210, 211
resistance genes 227
Triphenyltin acetate 195
Triphenyltin hydroxide 195
Triticale, aluminium tolerance, genetic control 25
Triticum spp., limits to disease resistance 12
Triticum aestivum L. 21, 45
Tropaeolum majus 143
Trypsin inhibitor in cowpea resistance to bruchid beetle 148
Tsuga canadensis
dust deposition and 55
reproduction, acid rain and 64
β-Tuberculin 197
Tyrosyl-tRNA synthetase 45

Umbelliferae, coumarins in 141
Urtica 144
Ustilago maydis 197, 200
ants mutants 197, 198, 199
oxr-1 mutants 198, 199

Variability for tolerance
appropriate and inappropriate 12, 13
limits 10, 12, 13
mutation and 15
occurrence 14, 15, 36
persistence 15
Verbascum 144
Veronica peregrina, rapid and local evolution 12
Verticladiella procera 58
Vibrio spp. 207
Viburnum acerifolium 55
Vicia benghalensis 46
Vigna unguiculata, bruchid beetle and 148
Vines, *Heliconius* butterfly and, coevolution 141, 142
Virus, plant, immunization against 110, 116
Virus–pollutant interactions 63

Warfarin-resistant rats 68
Wheat
aluminium tolerance in 23, 33
genetic control 25, 34, 35, 36
selection for 26, 27
manganese tolerance in 28

Xanthium 143
Xenobiotic-metabolizing enzymes 168*ff*.

Yeast
copper tolerance and 17
heavy-metal tolerance enhanced by 247
mercury tolerance and 246
Yersinia spp. 207

Zea mays L., iron deficiency in 21
Zinc
tolerance in *A. odoratum* 8
toxicity to plants 5, 16